Urban Climatic Suitability Design and Risk Management

Urban Climatic Suitability Design and Risk Management

Guest Editors

Lin Liu
Genyu Xu
Jing Du
Xiaoyu Tian

MDPI

Basel • Beijing • Wuhan • Barcelona • Belgrade • Novi Sad • Cluj • Manchester

Guest Editors

Lin Liu
School of Civil and Transportation Engineering
Guangdong University of Technology
Guangzhou
China

Genyu Xu
School of Architecture and Planning
Yunnan University
Kunming
China

Jing Du
College of Biochemical Engineering
Beijing Union University
Beijing
China

Xiaoyu Tian
Guangzhou Institute of Energy Conversion
Chinese Academy of Sciences
Guangzhou
China

Editorial Office
MDPI AG
Grosspeteranlage 5
4052 Basel, Switzerland

This is a reprint of the Special Issue, published open access by the journal *Buildings* (ISSN 2075-5309), freely accessible at: https://www.mdpi.com/journal/buildings/special_issues/4IULLJC2EQ.

For citation purposes, cite each article independently as indicated on the article page online and as indicated below:

Lastname, A.A.; Lastname, B.B. Article Title. *Journal Name* **Year**, *Volume Number*, Page Range.

ISBN 978-3-7258-7084-4 (Hbk)
ISBN 978-3-7258-7085-1 (PDF)
https://doi.org/10.3390/books978-3-7258-7085-1

Contents

About the Editors

Lin Liu

Lin Liu is an Associate Professor and a master's supervisor at the School of Civil and Transportation Engineering, Guangdong University of Technology, China. She obtained her B.S., M.S., and Ph.D. degrees from Harbin Institute of Technology. She has been dedicated to urban climate research since 2013, conducting in-depth investigations into fine-scale numerical simulation of urban climate multi-physics, formation mechanisms of spatiotemporal distributions of thermal environments and pollutants, and coupling effects of urban climate on energy consumption and carbon emissions. Her work provides a scientific basis and technical support for the multi-objective synergistic governance of urban cooling, energy conservation, and carbon reduction. She is the project leader of more than 10 research projects and has published more than 70 academic papers in international journals such as *Building and Environment*, *Urban Climate*, *Applied Energy*, etc.

Genyu Xu

Genyu Xu is a lecturer and master's supervisor at the School of Architecture and Planning at Yunnan University. He is dedicated to comprehending the complex interactions within urban landscape ecology and human settlements, with a particular focus on the thermal environments of buildings and urban spaces. His primary interests also encompass green building design, the development of ecological cities, and the physiological and psychological mechanisms of human thermal comfort and adaptation. He has expertise in a multitude of advanced computational and analytical techniques, including multi-objective parametric optimization, Urban Weather Generator (UWG) simulations, and the integration of remote sensing with deep learning for scalable urban energy modeling.

Jing Du

Jing Du is a lecturer at the College of Biochemical Engineering of Beijing Union University in Beijing, China. She graduated from Harbin Institute of Technology, majoring in Heating, Ventilating, and Air Conditioning. Her main research interests include urban micrometeorology theory and modeling, and human thermal comfort improvement theory and practice.

Xiaoyu Tian

Xiaoyu Tian, Ph.D., works at the Guangzhou Institute of Energy, Chinese Academy of Sciences. Her main research interests include distributed microgrid technology and terminal energy efficiency improvement technology. She has published over 30 articles and has been awarded two Hunan Provincial Natural Science Excellent Paper Awards and the Hunan Province Outstanding Doctoral Dissertation.

buildings

MDPI

Article

Assessing the Integrated Impacts of Outdoor Thermal Environment and Air Quality on Thermal Comfort in Residential Areas: A Case Study of Wuhan

Wei Liao [1], Dixin Wu [2], Bo Pan [3], Congyue Qi [4], Yingtian Xie [1], Xinling Zhan [1] and Shen Xu [5,6,7,*]

[1] School of Architecture and Urban Planning, Guangdong University of Technology, Guangzhou 510090, China; liaow@gdut.edu.cn (W.L.); 13031506x@gmail.com (Y.X.); zhanxinling1229@163.com (X.Z.)
[2] China Nuclear Industry 22nd Construction Co., Ltd., Wuhan 443101, China; dysindar@163.com
[3] Wuhan Lingyun Building Decoration Engineering Co., Ltd., Wuhan 430040, China; thepb@163.com
[4] China Construction Third Engineering Bureau Group Co., Ltd., Wuhan 430070, China; qicongyue@126.com
[5] School of Architecture and Urban Planning, Huazhong University of Science and Technology, Wuhan 430074, China
[6] Hubei Engineering and Technology Research Center of Urbanization, Wuhan 430074, China
[7] The Key Laboratory of Urban Simulation for Ministry of Natural Resources, Wuhan 430074, China
[*] Correspondence: xushen@hust.edu.cn

Abstract

Climate change and rapid urbanization are intensifying global environmental challenges, particularly the nexus of urban heat stress and air pollution, which collectively impact human health and outdoor livability. This study investigates the spatiotemporal distribution of the outdoor thermal environment and PM2.5 concentrations in residential areas of Wuhan, a city with a hot-summer–cold-winter climate, and evaluates their combined effects on outdoor thermal comfort. Field measurements of microclimatic parameters and PM2.5 levels were conducted in two typical residential communities during winter and summer, supplemented by 582 valid questionnaires to assess residents' subjective thermal responses. Key findings include (1) residents' satisfaction decreases by approximately 1 unit for every 90-unit increase in AQI, and PM2.5 concentrations can effectively substitute for AQI in characterizing the impact of air quality on satisfaction. (2) The explanatory power of the UTCI-MTSV model (R^2: 0.3152–0.7763) is pollutant-dependent; winter thermal comfort adheres to the linear law of UTCI, while dispersed summer votes indicate a non-linear effect of PM2.5 on comfort. (3) Wuhan residents show high thermal tolerance across AQIs. The lower limits of the Thermal Acceptability Range (TAR) are 7.1 °C (AQI-I) and 8.8 °C (AQI-II), both below neutral ranges, while the summer TAR-UTCI is 30.9 °C (above the neutral range). Better air quality improves the reliability of the thermal acceptability–UTCI fit. (4) TCV peaks at approximately 16 °C, increasing then decreasing with UTCI; at identical UTCI levels, better air quality enhances comfort, particularly within the 0–10 °C range. This study provides empirical evidence to inform urban design strategies for mitigating heat stress and pollution in hot-summer–cold-winter regions.

Keywords: hot-summer–cold-winter region; residential areas; outdoor thermal comfort; outdoor air quality; subjective comfort perception; thermal acceptability

1. Introduction

Climate change and urbanization have precipitated significant environmental challenges, including global warming, exacerbated air pollution (particularly PM2.5), and increased frequency of extreme weather events. Data from the IPCC Sixth Assessment Report indicate that the global mean surface temperature rose by 1.2 °C between 1901 and 2021, with heat-wave-related deaths reaching 26,800 in 2019 [1]. In China, PM2.5 pollution caused an average of 0.971 million premature deaths annually up to 2017, incurring economic losses of 2.75 trillion CNY [2]. Urban residential areas, as critical components of the living environment, face dual pressures of declining thermal comfort and deteriorating air quality, directly affecting residents' health and activity patterns [3]. Research aimed at ameliorating these issues can enhance the urban environment and positively influence public well-being. Among environmental concerns, diminished comfort and poor air quality are particularly pressing. Improving air quality reduces residents' exposure to pollutants, while enhancing outdoor thermal comfort encourages physical activity and informs urban design and renewal.

Field measurements based on meteorological stations are frequently employed to assess outdoor environmental performance. Van Hove et al. [3], using observations from Rotterdam, demonstrated a pronounced link between construction intensity and the urban heat island effect. Perini et al. [4] confirmed through parametric building experiments that density metrics are positively correlated with air temperature, whereas increasing building height can amplify the cooling capacity provided by vegetation and green spaces. Thermal comfort is typically evaluated via on-site measurements combined with questionnaire surveys, grounded in thermal environment characteristics and human physiological response.

Key physical indicators influencing thermal comfort include air temperature (T_a), the core determinant of human heat balance, while relative humidity (RH), wind speed (Va), and other variables regulate the local environment, with seasonal variations in their effects [5–7]. Subjective evaluation indices such as the Thermal Sensation Vote (TSV), Thermal Comfort Vote (TCV), and Thermal Acceptability Vote (TAV) are also considered, with corresponding metrics provided by ISO 10551 [8]. Evaluation frameworks often employ indices like the Physiologically Equivalent Temperature (PET) and the Universal Thermal Climate Index (UTCI). For example, Wang Yi [9] found that UTCI exhibits a strong correlation with summer thermal sensation votes in Shanghai.

In assessing outdoor pollutants, researchers often rely on indicators such as mean PM2.5 concentration and the Air Quality Index (AQI) [10–12]. Particulate matter (PM2.5) is the dominant pollutant affecting air quality in Wuhan's urban areas [13]; thus, mean PM2.5 is frequently employed as a neighborhood-scale indicator of outdoor air quality. Parameters like the AQI (defined by the U.S. EPA) integrate concentrations of common pollutants to provide an intuitive snapshot of pollution levels [14]. The AQI also estimates health risks from short-term pollutant exposure, making it a widely adopted field metric. This clear delineation offers a reliable basis for evaluating pollution risks in residential outdoor environments, facilitating investigation of their impact on living conditions and residents' health.

Numerous studies have demonstrated mutual influence between thermal comfort and air quality, as illustrated in Figure 1. Strong solar radiation and high-temperature conditions intensify photochemical reactions, resulting in elevated concentrations of photochemical pollutants such as ozone [15–17]. Nevertheless, existing research has paid limited attention to the spatiotemporal heterogeneity of thermal comfort and pollutants. Most prior studies on residential environments employed a single-perspective approach, focusing on how individual microclimatic parameters affect human thermal perception outdoors [18–20]. Approximately 60% of these studies concentrate on outdoor thermal comfort indices (e.g., UTCI, PET, SET), while 40% address air-quality parameters (e.g., PM2.5, pollutant concentrations), primarily using on-site measurements. However, outdoor thermal comfort is governed by the combined effects of multiple factors. Historically, thermal comfort research has considered key microclimatic elements—air temperature, wind speed, humidity, and thermal radiation—as primary determinants, examining their individual influences on human sensation. The impact of air quality on thermal comfort perception has often been overlooked. Variations in outdoor environmental factors are complex and occur over wide spatial and temporal scales, with interdependent parameters; consequently, a single microclimatic variable cannot fully account for changes in thermal comfort. Wind-tunnel experiments on street-canyon prototypes by researchers such as Xie [21], Lin [22], and Mei [23] have revealed the dual mechanism by which building aspect ratios and layout density simultaneously affect buoyancy-driven ventilation efficiency and pollutant dispersion capacity. Their findings indicate that as independent variables change, air temperature rises while pollutant residence times lengthen, as detailed in Table 1.

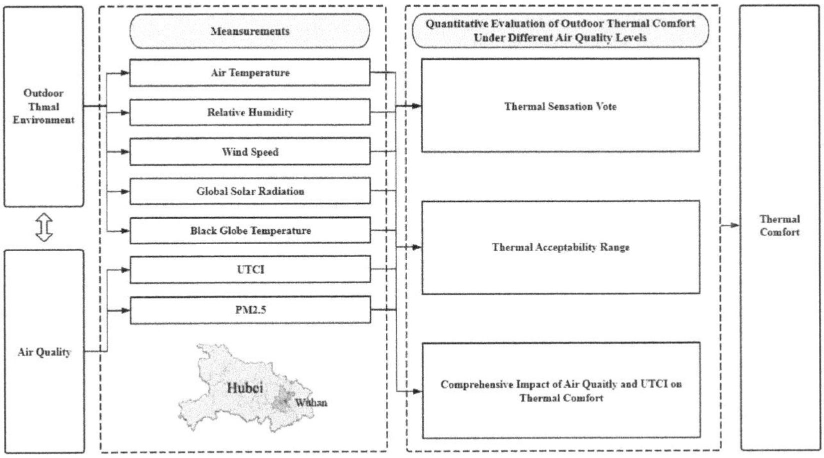

Figure 1. Mechanism diagram of the influences on thermal comfort and pollutants.

Moreover, the literature reveals a paucity of integrated investigations into outdoor thermal environment and air quality both domestically and internationally. The team led by Chunping Miao [24] demonstrated a linear negative relationship between summer street-tree canopy density and PET values in Shenyang, while the group headed by Liyan Rui [25] confirmed via modeling in Nanjing that green-space configuration exerts simultaneous effects on both thermal conditions and air quality, highlighting the potential for synergistic regulation through residential design elements. Empirical evidence shows that these design attributes play dual roles in thermal comfort and pollutant dispersion. In Tianjin, Bian et al. [26] quantified a 1-grade decline in air satisfaction for every 230-unit increase in AQI and identified a neutral PET threshold of 21.68 °C, significantly lower than that reported for hot-humid regions. Employing spatial overlay analysis, Sevgi Yilmaz et al. [27] uncovered a conflicting pattern: building frontage ratios of 60–75% mitigate heat stress yet simultaneously reduce PM2.5 dispersion rates. At the mesoscale, Wowo Ding's team [28] used large-scale gridded simulations to quantify the trade-off between street-canyon aspect ratios and improvements in thermal comfort versus prolonged PM2.5 residence times, further developing a coupled UTCI-AQI model. Collectively, these findings indicate that current research has yet to establish a multi-parameter framework for assessing outdoor thermal comfort that incorporates the impact of varying air pollution levels. Reconciling thermal environment and air-quality conflicts is therefore required to guide sustainable environmental design in high-density urban residential districts.

To address this empirical gap, the present study focuses on residential communities in Wuhan, China, as the study area and employs field measurements to investigate the spatiotemporal patterns of outdoor thermal comfort and PM2.5, thereby exploring their relationship. The study compares associations between different air pollutant concentrations and thermal comfort, and develops regression models relating UTCI to TCV under varying air pollutant concentration classes. The main goal is to gain an in-depth understanding of the distributional characteristics of outdoor thermal comfort and air quality in Wuhan, provide scientific evidence for optimizing the development of healthy residential districts, and offer a reference design system.

Table 1. Literature review on outdoor thermal comfort and outdoor air pollutants.

Author	Core Focus	Research Content	Study Morphology	Evaluation Parameters	Indicators and Methods
Miao C, He X, Gao Z, et al. [29]	Thermal Comfort and Air Quality	Vertical distribution of thermal comfort in relation to PM2.5 and O_3 concentrations	street canyon	PET, PM2.5, O_3, NO_2	Temperature, humidity, pollutant concentration (field measurement)
Xie X, Liu C H, Leung D Y C [21]	Air Quality	Influence of street-canyon morphology on air quality	prototype/street canyon	pollutant concentration	Air temperature, wind speed (wind-tunnel experiment)
Martilli A, Nazarian N, Krayenhoff E S, et al. [30]	Thermal Comfort	Bulk Richardson number, ambient wind parameters, and surface-heat distribution	prototype/building layout	air temperature	Air temperature, wind speed (wind-tunnel experiment)
Lin L, Hang J, Wang X, et al. [22]	Air Quality	Street-canyon morphology, bulk Richardson number, ambient wind parameters, and surface-heat distribution	prototype/street canyon	pollutant concentration	Wind speed, turbulent kinetic energy (wind-tunnel experi-ment)/pollutant concentration (wind-tunnel experiment)
Mei S J, Liu C W, Liu D, et al. [23]	Air Quality	Effects of street-canyon morphology on wind environment and air quality	prototype/street canyon	air temperature/air exchange rate/pollutant residence time/pollutant concentration	Wind speed, air temperature (empirical formula)
Bian G, Gao X, Zou Q, et al. [26]	Thermal Environment and Air Quality	Combined impacts of urban-park thermal environment and PM2.5 on thermal comfort	urban park	PET, AQI, TSV, TCV	Temperature, wind speed, PM2.5 (field survey)
Yilmaz S, Sezen I, Irmak M A, et al. [27]	Thermal Comfort and Air Quality	Differences in thermal comfort and SO_2/PM10 levels between urban centers and suburbs	urban center/suburb	PET, SO_2, PM10, wind speed	Temperature, elevation, building density (GIS analysis)
Rui L, Buccolieri R, Gao Z, et al. [31]	Microclimate and air quality	Regulation of thermal environment and PM_{10} by different residential vegetation layouts	residential area	PM10, temperature, wind speed	Green-space coverage ratio, isolation index (ENVI-met simulation 5.6)

2. Methodology

2.1. Cases of the Urban Areas

Wuhan is located at the confluence of the Yangtze and Han Rivers (geographical coordinates 29°58′–31°22′ N, 113°41′–115°05′ E) and serves as the capital of Hubei Province. In recent years, Wuhan's urban development has advanced steadily. According to the 2022 Wuhan Statistical Yearbook, urban construction continues alongside population growth, with about 80.49% of Wuhan's population residing in urban areas by the end of 2022. Consequently, issues such as pollution and thermal environment problems have increased, posing threats to residents' physical and mental health. Meteorological data analysis results are as follows: The Wuhan Statistical Yearbook 2022 [32] data are shown in Figure 2, showing that August is the hottest month in summer, with temperatures reaching up to 40 °C, while January is the coldest month in winter, with average temperatures around 3 °C and lows below zero. Previous studies on the annual distribution of PM2.5 pollutants in Wuhan, based on measurement times in the Wuhan Environmental Air Quality Monthly Report (Wuhan Ecological Environment Monitoring Center, 2023; Wuhan Ecological Environment Monitoring Center, 2024), indicate that the average PM2.5 concentration in July 2024 was 13 μg/m^3, while in December 2023, it was 60 μg/m^3.

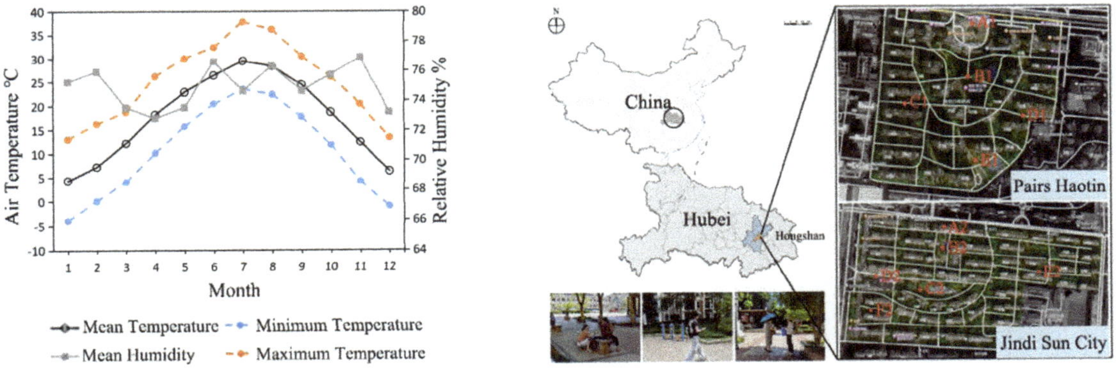

(a) (b)

Figure 2. Statistical chart of monthly meteorological conditions in Wuhan over the years (**a**). Schematic diagram of the residential communities location and distribution of measurement points (**b**).

This study selected two typical residential communities in Wuhan as measurement sites: Paris Haoting Community and Jindi Sun City Community. Paris Haoting Community is located at No. 2 Luoyu East Road, Hongshan District, Wuhan, China; Jindi Sun City Community is located at No. 369 Guanshan Avenue, Guanshan, Hongshan District, Wuhan, China. As shown in Figure 2, both communities are the typical post-2000 residential areas in urban Wuhan, which have similar spatial and morphological structures, representing prevalent medium-to-high-density residential forms developed during China's rapid urbanization. More broadly, Wuhan is situated within China's vast hot-summer–cold-winter region, a strategic climatic zone that encompasses many of the country's major metropolitan areas and serves as a critical hub for population and economic activity. Consequently, both communities, surrounded by traffic arteries in the central urban area, are subject to characteristic urban heat island effects and anthropogenic pollution, making them relevant and strategic cases for investigating the research problem.

2.2. Research Methodology

This study employs a combination of microclimate measurements and thermal comfort surveys to investigate the microclimate and thermal comfort in residential areas of Wuhan.

2.3. Microclimate Measurement

The study selected representative meteorological days in winter (20–23 January 2024) and summer (27–30 July 2024) to conduct research in two representative communities. To eliminate interference from extreme weather, measurement dates avoided periods of strong convective weather such as rain or snow, and were deliberately chosen on sunny or cloudy days. Considering the potential impact of traffic volume differences between weekdays and weekends on pollutants, measurement times covered two weekdays and weekends in both the coldest months (December–January) and the hottest months (June–August) to enhance data representativeness. Furthermore, considering human activity patterns, the measurement period was set from 9:00 to 17:00, during which outdoor thermal comfort and pollutant distribution were investigated.

To capture the spatiotemporal heterogeneity of microclimate and air quality, 12 fixed measurement points were systematically established across the two sites. Points were selected to cover a spectrum of common outdoor space types, stratified by key morphological variables known to influence the physical environment. Selection accounted for variations in ground cover (pavement, lawn) and the presence or absence of tree shading; actual locations are shown in Figure 2. Additionally, a monitoring point was set up on the rooftop of the residential area to record background data. At each measurement point, a PC-8/8A mini weather station and a TM-188D handheld comprehensive heat index meter were installed to obtain meteorological parameters at a height of 1.1 m above ground, including ambient temperature (T_a, °C), relative humidity (RH, %), globe temperature (T_g, °C), wind speed (V, m/s), and wind direction. Data were automatically recorded at 1 min intervals. The layout of measurement points and instruments used are shown in Appendix A.

2.3.1. Questionnaire Survey

The questionnaire survey consisted of three main parts. Detailed information is presented in Appendix B.

(1) Personal Information

The first part collected basic personal information from respondents, including fundamental physical characteristics, clothing thermal resistance, and activity status. According to ASHRAE data [33], half of the respondents had a seated metabolic rate of 1.0 met, while the other half were in a standing slow-walking state with a metabolic rate of 1.6 met.

(2) Thermal Environment Evaluation and Air Quality Assessment

Subjective thermal responses were obtained through the questionnaire, reflected by thermal sensation votes. Thermal sensation was assessed using the traditional ASHRAE 7-point scale (−3: very cold; −2: cold; −1: slightly cool; 0: neutral; 1: slightly warm; 2: warm; 3: very warm). Thermal comfort was evaluated using a 5-point scale (−2: very uncomfortable; −1: uncomfortable; 0: neutral; 1: comfortable; 2: very comfortable). Thermal acceptability was represented on a 5-point scale (−2: completely unacceptable; −1: unacceptable; 0: neutral; 1: acceptable; 2: completely acceptable). Preferences for meteorological conditions such as air temperature, wind speed, relative humidity, and solar radiation were recorded using a 5-point scale. Perceptions of pollutants were expressed on a 5-point scale (−2: very poor; −1: poor; 0: neutral; 1: good; 2: very good).

(3) Additional Information

Beyond basic physiological information, researchers sought a more accurate understanding of individuals' outdoor conditions and comfort perceptions. This involved statistical analysis of residents' outdoor thermal comfort and sensitivity to pollutants, as well as recording the criteria by which residents assess air quality. Furthermore, other activity information of the subjects was documented, with detailed inquiries into typical residential activity specifics.

2.3.2. Thermal Comfort and Air Quality Indicators

(1) Outdoor Thermal Environment Indicators

Previous studies have demonstrated the high applicability of the UTCI in hot-summer–cold-winter regions. Therefore, this study employs UTCI to comprehensively assess the outdoor thermal environment, with specific adjustments for Wuhan. The study utilizes RayManPRO3.1 to calculate UTCI, inputting air temperature (T_a), relative humidity (RH), wind speed (V), mean radiant temperature (T_{mrt}), and individual human parameters. Among meteorological parameters, T_a and RH are directly measured, while wind speed V at 10 m is calculated.

T_{mrt}, as a crucial factor in human energy balance and thermal comfort, is commonly calculated using the black globe temperature (T_g). According to ISO7726 standards [34], the formula for calculating T_{mrt} is given as Equation (1), and the formula for calculating wind speed at 10 m is as follows Equation (2).

$$T_{mrt} = \left[(T_g + 273.15)^4 + \frac{1.10\times10^8 V^{0.6}}{\varepsilon D^{0.4}}(T_g - T_a)\right]^{0.25} - 273.15 \quad (1)$$

where T_g is the black globe temperature (°C), V is the wind speed (m/s), T_a is the air temperature (°C), D is the sphere diameter (0.05 m), and ε is the black globe emissivity (0.95).

$$v_x = v_m \times \frac{log\left(\frac{z}{z_0}\right)}{log\left(\frac{m}{z_0}\right)} \quad (2)$$

where v_x represents the wind speed at height z, m/s, v_m is the wind speed at the measured height, m/s, and z_0 is the roughness parameter, with a value of 0.1.

In addition to UTCI, the Thermal Acceptability Range (TAR) is another key metric for quantifying outdoor thermal comfort, illustrating the dynamic adaptability of residents to thermal conditions in their environment. ASHRAE Standard 55 [33] specifies that acceptable thermal conditions are those in which at least 80% of respondents find the conditions acceptable.

(2) Air quality indicators

The Air Quality Index (AQI) is a quantitative measure used to specifically assess air quality levels. According to the NAAQS definition [35], air quality can be categorized into six classes, as shown in Table 2.

Table 2. Air pollutant concentrations and corresponding air quality AQI classification evaluation standard.

PM2.5 Concentration (μg/m³)	PM10 Concentration (μg/m³)	AQI	Air Quality Level	Air Quality Grade
0.0–12.0 (24 h)	0–54 (24 h)	AQI ≤ 50	Excellent	I
12.1–35.4 (24 h)	55–154 (24 h)	50 < AQI ≤ 100	Good	II
35.5–55.4 (24 h)	155–254 (24 h)	100 < AQI ≤ 150	Mild Pollution	III
55.5–150.4 (24 h)	255–354 (24 h)	150 < AQI ≤ 200	Moderate Pollution	IV
150.5–250.4 (24 h)	355–424 (24 h)	200 < AQI ≤ 300	Heavy Pollution	V
250.5–350.4 (24 h)	425–504 (24 h)	300 < AQI ≤ 400	Severe Pollution	VI
350.5–500.4 (24 h)	505–604 (24 h)	400 < AQI ≤ 500	Severe Pollution	VI

Specifically, based on the concentration of PM2.5, the corresponding sub-item relative index is derived according to a specified calculation formula and is designated as the Air Quality Index (AQI) value for this pollutant. Meanwhile, sub-item indices for other pollutants, such as PM10 and CO, are incorporated into the reference framework. If the sub-item relative indices of these pollutants are greater than or equal to the sub-index corresponding to PM2.5, the larger value is selected as the final AQI. This approach ensures that the AQI accurately reflects the comprehensive impact of pollutants on air quality. The corresponding relationship between the AQI limit values of PM2.5 and its mass concentration limits is presented in Table 2. Accordingly, in this study, meteorological and pollution data obtained from field measurements were converted into the corresponding AQI values using the prescribed calculation formula. The computational method for pollutant values adopts an exponential interpolation calculation, as expressed in Equation (3):

$$AQI = \frac{Ihigh - Ilow}{\rho high - \rho low}(\rho - \rho low) + Ilow \tag{3}$$

where I is the Air Quality Index (AQI), which is the output value, ρ is the daily average mass concentration of PM2.5, which is the input value, I_{low} is the index limit corresponding to ρ_{low}, a constant, I_{high} is the index limit corresponding to ρ_{high}, a constant, ρ_{low} is the mass concentration limit less than or equal to ρ, a constant, and ρ_{high} is the mass concentration limit greater than or equal to ρ, a constant.

Among them, the constants of the index limits and mass concentration limits are shown in Table 3.

Table 3. Air quality index limits and mass concentration limits for PM2.5.

Air Quality Grade	AQI	Mass Concentration Limit ρ/(μg/m³)
I	0	0
II	50	35
III	100	75
IV	150	115
I	200	150
VI	300	250
VI	400	350
VI	500	500

(3) Overall Comfort Vote

The Overall Comfort Vote (OCV) is an index employed to comprehensively evaluate human overall comfort in a given environment, encompassing multiple perceptual dimensions such as thermal comfort, air quality comfort, visual comfort, and acoustic comfort [36]. Unlike indicators focusing on a single environmental factor, OCV emphasizes the combined effects of multiple environmental factors and their collective influence on human perception. In studies of the built environment, OCV is typically obtained through questionnaires or subjective evaluation scales, whereby participants are asked to rate their overall comfort according to their personal perceptions. Quantification is generally conducted using a Likert scale, such as a 7-point or 5-point scale [37].

3. Results

3.1. Measured Physical Environmental Characteristics of Outdoor Spaces

On-site measurements of the outdoor environment were conducted over 8 days. See Table 4. The definitions for the environmental parameters in this table are as follows: T_a (Air Temperature, °C); RH (Relative Humidity, %); Va (Wind Speed, m/s); G (Global Solar Radiation, W/m^2); T_g (Black Globe Temperature, °C); UTCI (Universal Thermal Climate Index, °C); PM2.5 (μg/m^3).

Table 4. Winter physical environment measurements.

		A1	B1	C1	D1	E1	F1	A2	B2	C2	D2	E2	F2
	Max	2.4	1.5	2.3	10.1	2.7	3.2	8.5	8.6	8.5	9.3	7.7	7.4
T_a	avg.	1.0	−0.1	0.8	1.6	−0.1	0.7	5.0	4.8	5.4	5.3	4.4	4.6
	Min	−2.6	−3.6	0.6	−3.6	−3.2	−3.0	1.3	0.3	0.6	0.7	0.5	1.1
	Max	47.8	47.6	47.6	51.5	46.5	45.3	53.5	56.5	52.4	54.3	57.9	55.3
RH	avg.	38.4	38.4	38.8	38.8	37.9	36.4	37.0	35.7	35.1	34.2	35.5	34.8
	Min	27.3	26.5	27.3	27.0	27.7	24.9	28.8	27.9	28.7	26.5	28.7	28.2
	Max	1.9	2.2	1.6	5.9	1.6	1.5	2.3	1.2	1.3	1.5	1.0	0.8
Va	avg.	0.7	1.3	0.5	1.3	0.6	0.6	0.4	0.5	0.5	0.7	0.5	0.4
	Min	0.0	0.1	0.0	0.0	0.0	0.0	0.0	0.0	0.1	0.0	0.0	0.0
	Max	685.9	788.1	4441.8	147.7	149.3	618.0	95.7	693.2	72.8	615.2	400.8	465.5
G	avg.	138.1	139.9	550.4	15.3	67.4	211.7	51.2	269.6	41.2	271.9	94.9	55.2
	Min	6.3	3.3	0.0	0.0	7.4	0.0	18.7	24.6	4.4	0.0	20.0	0.0
	Max	4.9	3.1	19.5	18.4	2.8	20.2	17.9	17.0	23.4	23.6	17.9	17.9
T_g	avg.	1.4	1.5	2.8	1.9	1.5	9.9	6.4	8.1	12.4	13.9	6.4	6.4
	Min	0.2	0.0	0.0	0.0	0.0	1.4	1.7	1.3	2.3	4.0	1.7	1.7
	Max	9.1	6.4	23.8	19.1	3.7	24.3	23.8	20.8	27.6	26.4	21.5	19.6
UTCI	avg.	3.2	0.1	2.8	−5.5	1.3	5.6	6.3	8.9	14.2	16.1	6.6	6.7
	Min	−8.4	−11.3	−7.2	−23.2	−6.1	−6.2	−1.5	0.5	1.4	1.6	1.8	2.1
	Max	31.0	34.0	35.0	49.0	35.0	33.0	58.7	64.7	64.7	64.1	66.1	62.5
PM2.5	avg.	18.6	17.3	18.3	14.5	19.0	19.8	47.7	49.1	49.8	49.3	51.8	51.4
	Min	1.0	4.0	0.0	1.0	1.0	8.0	38.4	41.1	41.1	40.6	43.4	43.0

Note. Definitions for the environmental parameters in this table: T_a (Air Temperature, °C); RH (Relative Humidity, %); Va (Wind Speed, m/s); G (Global Solar Radiation, W/m^2); T_g (Black Globe Temperature, °C); UTCI (Universal Thermal Climate Index, °C); PM2.5 (μg/m^3).

Table 5 presents the average, maximum, and minimum values of air temperature, relative humidity, wind speed, global solar radiation, black globe temperature, UTCI, and PM2.5 concentration at each measurement point in (1) Paris Haoting and (2) Jindi Sun City, respectively.

Regarding air temperature, the mean value in winter was 2.83 °C (maximum: 10.1 °C, minimum: −3.63 °C). The average temperature in Paris Haoting (1) was 3.32 °C lower than in Jindi Sun City (2), indicating a significant temperature difference in Wuhan between cold spell periods and non-cold spell periods in winter. The summer average temperature was 35.27 °C (maximum: 38.6 °C, minimum: 29.9 °C). The average temperature in Paris Haoting was 1.47 °C higher than in Jindi Sun City, reflecting noticeable inter-community temperature differences during summer.

In terms of relative humidity, the winter average was 38.16%, with a maximum of 51.5% and a minimum of 24.9%, showing relatively small variation. The summer average was 53.75%, with a maximum of 73.1% and a minimum of 37.3%, showing larger variation, indicating that humidity has a more significant impact on thermal comfort in summer. The winter average black globe temperature was 1.43 °C, with a maximum of 19.5 °C and a minimum of 0 °C; the summer average was 41.75 °C, with a maximum of 54.4 °C and a minimum of 30.0 °C, reflecting the significant influence of solar radiation on the thermal environment. The winter average wind speed was 0.68 m/s, with a maximum of 5.9 m/s and a minimum of 0 m/s; the summer average was 0.69 m/s, with a maximum of 2.1 m/s and a minimum of 0 m/s. Variations in wind speed significantly impact thermal comfort, especially in summer, where higher wind speeds help alleviate thermal discomfort.

Table 5. Summer physical environment measurements.

		A1	B1	C1	D1	E1	F1	A2	B2	C2	D2	E2	F2
	Max	38.6	35.8	36.1	36.7	37.0	36.2	35.3	38.5	35.2	36.6	36.0	36.3
T_a	avg.	36.8	35.2	35.0	35.3	35.6	35.4	34.5	35.9	34.5	35.8	33.8	35.3
	Min	33.5	34.1	33.2	33.6	32.3	34.0	32.2	31.7	32.5	33.8	29.9	32.9
	Max	63.9	61.8	65.3	60.6	63.3	55.8	67.7	71.2	66.3	67.0	73.1	57.3
RH	avg.	53.1	56.1	56.9	54.5	53.4	50.7	57.4	47.2	57.4	58.4	47.8	50.2
	Min	48.2	54.3	52.7	51.1	49.1	47.4	53.2	38.5	53.8	55.9	37.3	46.4
	Max	1.0	1.1	0.8	2.1	0.7	0.9	0.1	0.8	0.8	0.6	0.5	1.2
Va	avg.	0.7	0.7	0.6	1.3	0.6	0.7	0.1	0.6	0.7	0.4	0.3	0.9
	Min	0.4	0.5	0.4	0.9	0.3	0.6	0.0	0.4	0.5	0.3	0.2	0.7
	Max	582.7	671.9	96.1	557.1	609.7	561.1	33.4	761.5	361.5	411.0	320.6	508.8
G	avg.	373.3	299.6	25.3	357.2	477.0	311.5	22.1	524.9	137.9	168.1	139.9	71.3
	Min	46.9	32.2	8.2	79.6	40.3	0.0	6.9	93.2	0.0	11.4	46.5	0.0
	Max	49.8	44.0	54.4	48.3	42.8	41.3	35.5	40.0	36.8	37.8	37.2	38.0
T_g	avg.	41.5	38.9	42.0	43.1	36.3	38.9	34.7	37.5	35.5	36.0	34.0	36.5
	Min	33.6	35.5	35.6	37.2	31.8	37.1	32.3	32.0	32.5	33.9	30.0	32.9
	Max	49.3	44.3	50.0	47.7	42.1	41.5	41.2	45.5	41.0	43.0	42.8	43.5
UTCI	avg.	42.8	40.2	41.9	43.6	38.2	39.7	39.8	42.3	39.7	41.5	39.0	41.2
	Min	36.4	37.7	38.0	38.9	34.2	38.5	37.0	35.5	36.8	38.5	34.0	37.0
	Max	36.4	37.7	38.0	38.9	34.2	38.5	38.4	8.1	8.3	7.3	34.1	24.0
PM2.5	avg.	36.4	37.7	38.0	38.9	34.2	38.5	26.9	6.9	6.8	6.5	29.6	18.9
	Min	36.4	37.7	38.0	38.9	34.2	38.5	18.7	5.9	5.2	5.4	25.2	9.9

Note. Definitions for the environmental parameters in this table: Ta (Air Temperature, °C); RH (Relative Humidity, %); Va (Wind Speed, m/s); G (Global Solar Radiation, W/m^2); T_g (Black Globe Temperature, °C); UTCI (Universal Thermal Climate Index, °C); PM2.5 (μg/m^3).

For relative humidity, the winter average was 38.16% (maximum: 51.5%, minimum: 24.9%), showing relatively limited variation. In contrast, the summer average was 53.75% (maximum: 73.1%, minimum: 37.3%), exhibiting greater fluctuation, which suggests a more significant impact of humidity on thermal comfort in summer. The average black globe temperature was 1.43 °C in winter (maximum: 19.5 °C, minimum: 0 °C) and 41.75 °C in summer (maximum: 54.4 °C, minimum: 30.0 °C), underscoring the substantial influence of solar radiation on the thermal environment. The average wind speed was 0.68 m/s in winter (maximum: 5.9 m/s, minimum: 0 m/s) and 0.69 m/s in summer (maximum: 2.1 m/s, minimum: 0 m/s). Variations in wind speed significantly impact thermal comfort, with higher speeds providing greater relief from thermal discomfort in summer.

The average UTCI was 3.21 °C in winter (maximum: 23.8 °C, minimum: −23.2 °C), indicating generally poor thermal comfort conditions during the humid-cold winter. The summer average was 41.25 °C (maximum: 50.0 °C, minimum: 34.0 °C), reflecting the strong stress of high temperatures on thermal comfort. The average PM2.5 concentration was 18.43 μg/m^3 in winter (maximum: 68.7 μg/m^3, minimum: 0 μg/m^3) and 26.6 μg/m^3 in summer (maximum: 38.9 μg/m^3, minimum: 5.2 μg/m^3). Elevated PM2.5 concentrations in winter may be associated with heating activities and traffic flow, whereas lower summer concentrations are likely related to increased rainfall and vegetation coverage.

3.1.1. Analysis of Measured Thermal Environment Results

Based on the measurement results (see Figure 3), winter temperatures are characterized by lower values and greater fluctuations, while summer features higher temperatures with significant humidity variations, both seasons markedly affecting thermal comfort. Distinct seasonal differences in black globe temperature and wind speed further influence the thermal environment. UTCI results indicate poor thermal comfort in winter and significant thermal discomfort in summer. Seasonal variation in PM2.5 concentration reflects changes in air quality, with high winter concentrations linked to heating and traffic emissions, and lower summer levels correlated with rainfall and vegetation cover. These measurements provide a crucial scientific foundation for understanding seasonal variations in thermal comfort and pollutant status within typical Wuhan residential areas, facilitating the evaluation of outdoor environmental quality.

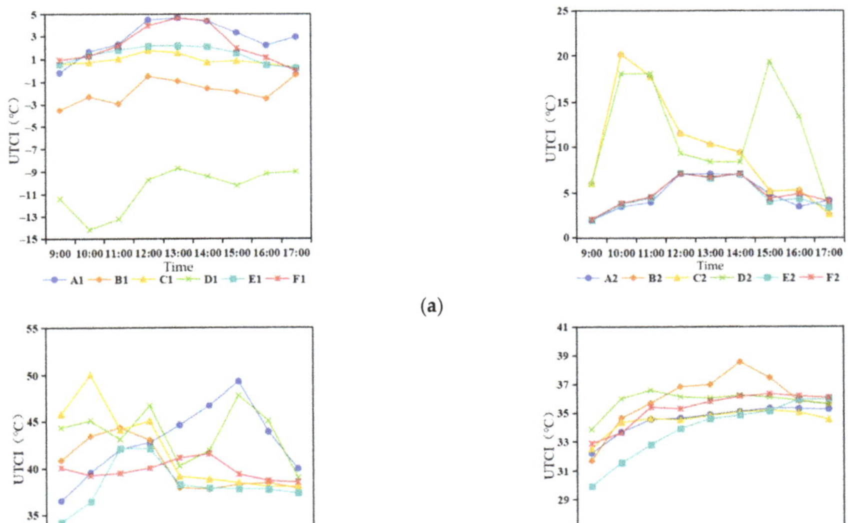

(**a**)

(**b**)

Figure 3. Temporal and spatial variations in UTCI at the measurement points of the two communities: (**a**) winter; (**b**) summer.

Further analysis of Figure 3 elucidates key spatiotemporal patterns driving thermal discomfort. Winter UTCI (Figure 3a) shows high sensitivity to solar access and urban morphology, with exposed sites (e.g., A1, F1) experiencing brief daytime warming while shaded areas remain cold. In stark contrast, summer (Figure 3b) is characterized by prolonged periods of severe heat stress (UTCI > 40 °C), with minor relief provided only by dense shading. This persistent thermal stress directly shapes the subjective thermal perceptions reported by residents.

3.1.2. Analysis of Measured Pollutant Results

Further analysis of PM2.5 concentrations across different measurement points within the same residential communities (see Figure 4) revealed significant spatial heterogeneity. Specifically, point F2 exhibited the highest average PM2.5 concentration (36.5 $\mu g/m^3$), whereas B1 and D2 had relatively lower averages (13.8 $\mu g/m^3$ and 22 $\mu g/m^3$, respectively). Diurnal variations were also observed, with concentrations peaking around 9:00 and reaching a trough at 15:00, potentially related to patterns in human activities and traffic flow.

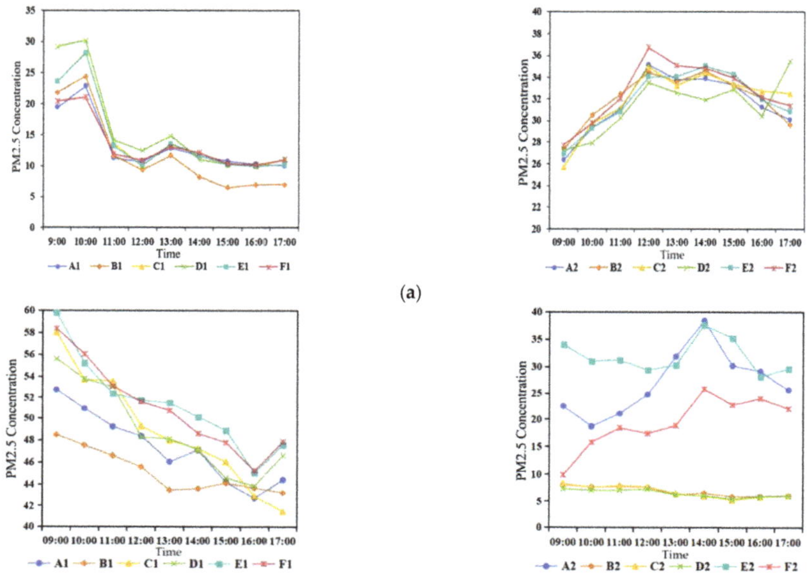

(**b**)

Figure 4. Temporal and spatial variations in PM2.5 concentration at measurement points in the two communities: (**a**) winter; (**b**) summer.

Considering combined spatiotemporal variations, PM2.5 concentrations at both sites exhibited periodic characteristics. Human activities were identified as a significant influencing factor. Overall, PM2.5 concentrations were higher in Jindi Sun City than in Paris Haoting, potentially attributable to differences in population density, traffic flow intensity, and surrounding anthropogenic emission sources. Additionally, the diurnal fluctuation range of PM2.5 concentration was greater in Jindi Sun City, a phenomenon possibly related to specific climatic conditions and traffic pollution levels during the measurement period. Analysis of other meteorological data over time showed a weak correlation between air temperature and pollutant distribution within the two communities.

The spatiotemporal patterns in Figure 4 reveal the direct impact of local sources and sinks on PM2.5. The consistently high concentrations at point F2 are attributable to its proximity to major traffic arteries and limited green space, inhibiting dispersion. In contrast, points with better ventilation and vegetation (e.g., B1, D2) maintained lower levels. The characteristic diurnal cycle—peak during morning rush hour and afternoon minimum—closely follows traffic emissions and atmospheric mixing conditions. Notably, winter (Figure 4a) exhibits both higher concentrations and greater variability than summer (Figure 4b), consistent with seasonal emission patterns and meteorological stagnation.

These findings provide the physical context for the residents' air quality satisfaction votes and its interplay with thermal sensation.

Distribution statistics for all PM2.5 concentration data from both communities (see Figure 5) indicate that on the measurement day (20 January 2024), the overall median concentration was 30 μg/m^3 (Paris Haoting: 24 μg/m^3; Jindi Sun City: 32 μg/m^3). The two studied communities share similar orientation and layout forms and are surrounded by roads with relatively consistent traffic flow, enhancing the representativeness of the distribution results.

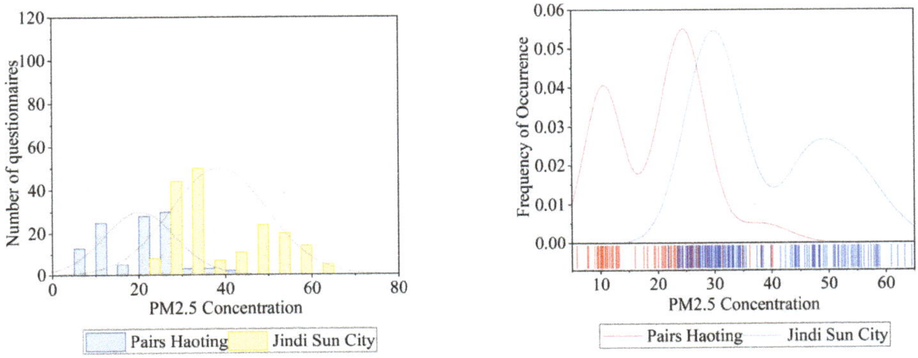

Figure 5. PM2.5 concentration distribution range and box–whisker plot analysis for the two residential community.

On the other hand, the statistical summary in Figure 5 quantifies the significant inter-community difference in PM2.5 exposure. Jindi Sun City's higher median, larger interquartile range, and elevated maximum concentration demonstrate a more polluted and variable environment compared to Paris Haoting. This validated contrast between the two sites is fundamental to our subsequent investigation, as it creates a natural experiment for assessing how different AQI levels modulate the relationship between thermal environment and comfort.

3.2. *Outdoor Voting Characteristics of Wuhan Residents*

3.2.1. Basic Data Analysis of Wuhan Resident Respondents

(1) Respondent Age and Gender Distribution

A total of 597 questionnaires were distributed across winter and summer, with 582 valid responses retained after screening. The survey controlled for gender and age factors, as shown in Table 6. Males accounted for 51% of respondents and females 49%, indicating a balanced gender ratio. Minors were excluded from the analysis as their environmental perceptions differ significantly from adults and they constituted a small proportion of the sample.

Table 6. Gender ratio of respondents in residential areas and proportion of age of respondents in residential areas of Wuhan.

Gender	Percentage	Age	Percentage
		<18	10%
male	51%	18–35	30%
		35–50	16%
female	49%	50–60	13%
		>60	10%

(2) Characteristics of Residents' Activity Time Distribution

The primary outdoor activity times for residents in each season are shown in Figure 6. In summer, peak activity concentrated in the evening (18:00–20:00), with the proportion of votes particularly high at 19:00 (13%). In winter, the peak period occurred in the afternoon (15:00–17:00), with the highest vote proportion at 16:00 (13%). This pattern is likely related to sunlight availability, temperature variations, and resident lifestyles. Summer evenings are relatively comfortable, encouraging outdoor activities, while winter afternoons offer ample sunshine and warmer temperatures, making them more favorable for going out.

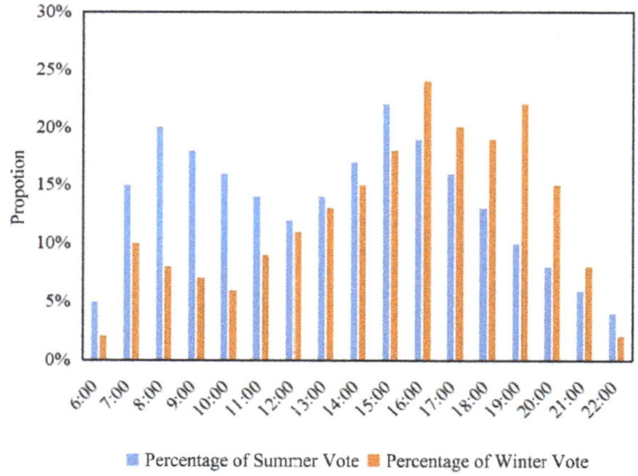

Figure 6. Statistics on the distribution preferences of residents' outdoor activity time in different seasons.

The characteristics of residents' daily outdoor activity duration are shown in Table 7. Analysis of vote distribution across different time intervals (min/day) revealed differences in the concentration of activity times between seasons. In summer, the highest proportion of votes (34.92%) was for the 30–60 min/day interval; in winter, the highest proportion was for the >60 min/day interval. This suggests that summer activity times are more concentrated, typically within half an hour to an hour, while winter activity times are more dispersed, with a larger proportion exceeding one hour.

Table 7. Statistics on residents' daily average activity duration preferences in different seasons.

Average Activity Duration (Min/Day)	Percentage of Summer Votes	Percentage of Winter Votes
0–14	6.61%	6.22%
15–29	14.02%	10.27%
30–60	34.92%	22.16%
>60	44.44%	61.35%

Based on these findings and considering the impact of seasonal differences, 9:00 and 15:00 were selected as typical analysis periods for both winter and summer.

3.2.2. Relationship Between Thermal Sensation Vote (TSV) and Thermal Comfort Vote (TCV)

Regression analysis was employed to fit curves and examine the quantitative relationship between the Thermal Sensation Vote (TSV) and Thermal Comfort Vote (TCV) for residents in Wuhan residential areas (see Figure 7). The results reveal a non-linear relationship. As the TSV value increases, the TCV value first increases and then decreases, exhibiting a clear threshold. When TSV is between -1.35 and 1.88, TCV ≥ 0, with TCV reaching its maximum at TSV = 0.26. This indicates that Wuhan residents experience the highest comfort within the "slightly cool" to "slightly warm" thermal sensation range, peaking under nearly neutral thermal conditions.

Further comparative analysis incorporating different pollutant concentrations and seasons (see Figure 8) found that while the pollutant concentration gradient did not drastically alter the overall shape of the TSV-TCV curve, differences in data distribution between winter and summer suggest an indirect influence on the comfort threshold. For instance, the peak TCV value corresponding to TSV = 0.26 in summer was slightly lower than in winter, implying that hot and humid conditions may attenuate the comfort experienced in neutral thermal states. The model's overall explanatory power (R^2 = 0.6393) is moderate, indicating that besides thermal sensation, other environmental variables such as air pollution and wind speed exert superimposed effects on thermal comfort perception, warranting further quantification.

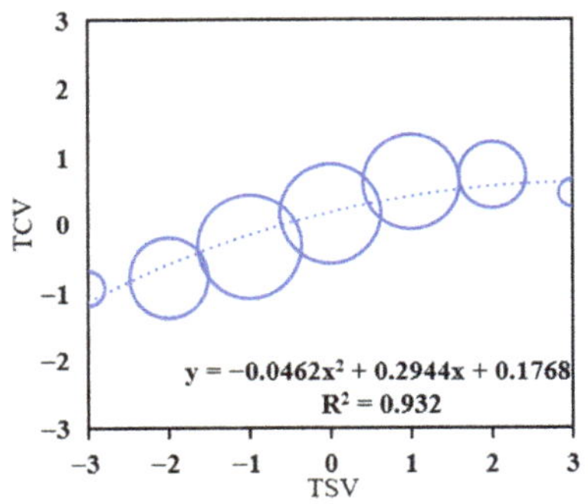

Figure 7. The relationship between TCV and TSV of outdoor space residents in residential areas of Wuhan.

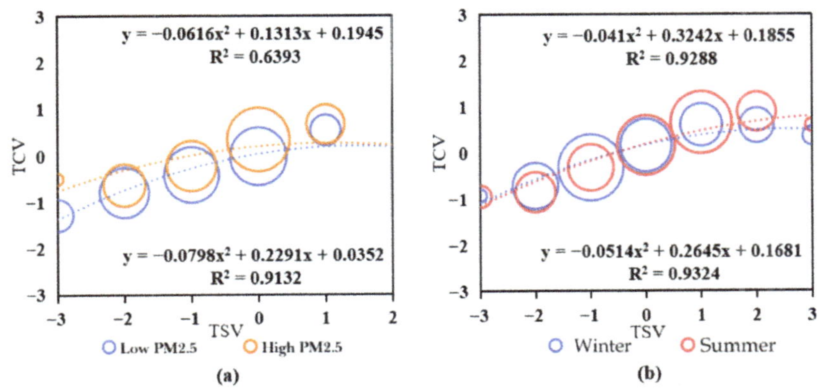

Figure 8. Relationship between TCV and TSV of residents in outdoor spaces of residential areas in Wuhan: (**a**) different pollutant concentrations; (**b**) winter and summer.

3.2.3. Relationship Between Air Quality Index (AQI) and Air Satisfaction Vote (ASV)

First, according to the AQI calculation method introduced earlier in Section 2.3.2, the measured PM2.5 concentration data were converted into the corresponding air quality categories shown for the pollutant indicator section and averaged. Then, by fitting AQI and ASV data, Equation (4a) for the correlation between ASV and PM2.5 concentration and Equation (4b) for the correlation between ASV and AQI were ultimately obtained through statistical methods, as shown below:

$$ASV = 1.10061PM2.5^2 + 1.6087PM2.5 + 29.494(R^2 = 0.9227) - 273.15 \quad (4a)$$

$$ASV = -0.0061AQI + 0.8723(R^2 = 0.9791) \quad (4b)$$

Within a certain range, there is a good linear relationship between Air Satisfaction Vote (ASV) and the Air Quality Index (AQI). Simultaneously, the difference in the correlation coefficient (R^2) compared to that between PM2.5 concentration and ASV is small, proving that the effect of these two indicators on ASV is extremely similar. PM2.5 concentration can largely substitute for AQI's role in characterizing air quality. From Figure 9, it can be seen that for every increase of approximately 90 in AQI, satisfaction decreases by one unit, indicating that Wuhan residents perceive air quality accurately. Better air quality leads to higher satisfaction. When ASV = 0, the predicted neutral AQI is calculated as 131, meaning no tendency towards satisfaction or dissatisfaction with the air environment. Even when AQI is 0, satisfaction with air quality does not reach a very high level, implying that for Wuhan residents, acceptance of air quality within their residential communities is relatively high, and also that factors other than PM2.5 may influence residents' judgments.

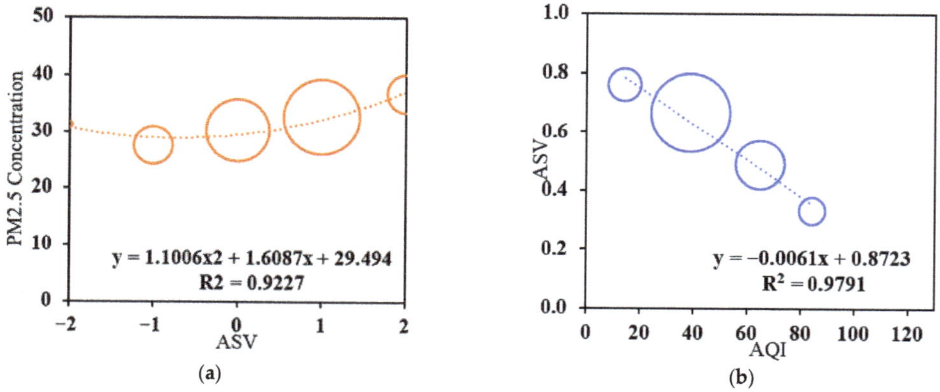

Figure 9. Relationship between air quality and air satisfaction of all respondents (**a**); PM2.5 concentration and ASV (**b**); the correlation between ASV and AQI.

4. Quantitative Evaluation of Outdoor Thermal Comfort Under Different Air Quality Levels

4.1. Thermal Sensation Vote (TSV)

By analyzing the thermal acceptability range under different AQI levels, the Mean Thermal Sensation Vote (MTSV) was plotted for each 1 °C interval of the Universal Thermal Climate Index (UTCI) and compared with a quadratic curve. The influence of air quality levels corresponding to different PM2.5 concentrations on thermal comfort evaluation and the associated patterns were analyzed. Based on the AQI classification method described in Table 2, thermal sensation regression models were developed for winter, summer, and annual conditions under varying pollutant levels (see Figures 10 and 11).

The results indicate seasonal differences in the linear relationship between UTCI and MTSV (see Table 8), with model explanatory power varying with pollutant levels. Under winter conditions, the regression model MTSV = 0.0486UTCI − 1.1822 for the AQI-II level exhibits a high goodness-of-fit (R^2 = 0.7763), suggesting a strong correlation between thermal sensation and UTCI in low-temperature environments, where PM2.5 concentration has a stable impact on thermal comfort evaluation. Under summer conditions, the model

MTSV = 0.0982UTCI − 3.5146 (Equation (5c)) corresponding to the AQI-I level has the largest slope, but the R^2 value decreases to 0.5058. This indicates that in high-temperature and high-humidity environments, the interfering effect of pollutant concentration on thermal sensation increases, reducing the model's explanatory power. In the comprehensive annual analysis, the R^2 of MTSV = 0.0542UTCI − 1.9597 (Equation (5d)) under the AQI-II level is only 0.3152, further confirming that seasonal factors significantly modulate thermal comfort evaluation.

Under the same pollutant level, the R^2 values of winter models (e.g., Equations (5e) and (5f)) are generally higher than those of summer models (e.g., Equations (5c) and (5d)). This implies that thermal comfort evaluation in winter is more likely to follow the linear relationship with UTCI, while in summer, due to the influence of combined environmental factors, thermal sensation votes exhibit greater dispersion. Furthermore, the fluctuation of R^2 values (0.3152–0.7763) across different AQI levels reveals the non-linear influence mechanism of PM2.5 concentration on thermal comfort evaluation. Deteriorating air quality may weaken the statistical correlation between UTCI and MTSV by altering human thermal adaptation behaviors.

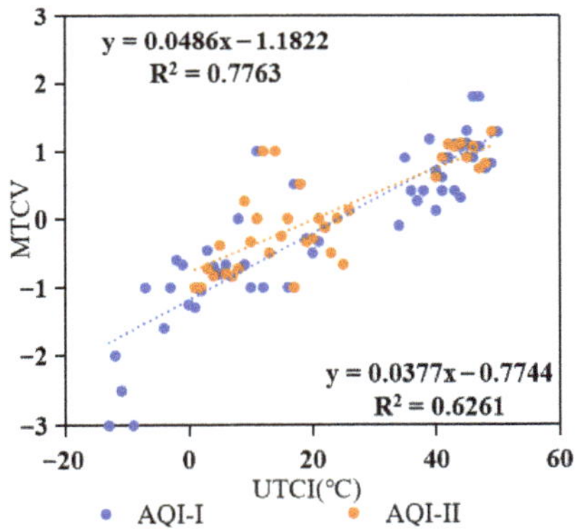

Figure 10. Regression analysis between MTSV and UTCI.

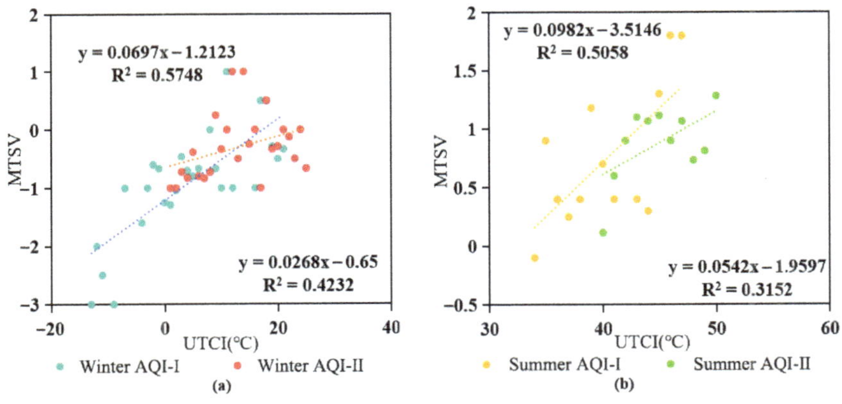

Figure 11. Regression analysis between MTSV and UTCI (**a**) Different pollutant levels in winter; (**b**) Different pollutant levels in summer.

Table 8. Regression equations of UTCI and mean thermal sensation (MTSV) under different seasons and AQI levels.

Thermal Comfort Index	Classification		Regression Model of Universal Thermal Climate Index (UTCI) and Mean Thermal Sensation	
UTCI	AQI-I		$MTSV = 0.0486UTCI - 1.1822(R^2 = 0.7763)$	(5a)
	AQI-II		$MTSV = 0.0377UTCI - 0.7744(R^2 = 0.6261)$	(5b)
	Summer	AQI-I	$MTSV = 0.0982UTCI - 3.5146(R^2 = 0.5058)$	(5c)
		AQI-II	$MTSV = 0.0542UTCI - 1.9597(R^2 = 0.3152)$	(5d)
	Winter	AQI-I	$MTSV = 0.0697UTCI - 1.2123(R^2 = 0.5748)$	(5e)
		AQI-II	$MTSV = 0.0268UTCI - 0.65(R^2 = 0.4232)$	(5f)

The obtained equations are as follows:

4.2. Thermal Acceptability Range (TAR)

To investigate human thermal comfort within specific temperature ranges, the thermal acceptability range is commonly used for evaluation. Following the Thermal Acceptability Range (TAR) defined by ASHRAE Standard 55 [33], this study adopted the UTCI temperature range acceptable to 80% of respondents. The thermal acceptability range under different AQI levels was analyzed, and the percentage of thermal unacceptability for each 1 °C interval of UTCI was plotted and compared with a quadratic curve.

Overall, residents in Wuhan exhibit relatively high acceptance of the winter thermal environment (see Table 9). The quadratic curve fitted to the data in Figure 12 shows an R^2 of 0.5414 between thermal acceptability and UTCI. Based on the intersection points between the curve and the 20% unacceptability threshold, the overall 80% thermal acceptability range is determined to be 13.3–27.1 °C. Figure 13a,b plot the intersection points between two quadratic fitting curves and the 20% unacceptability threshold under different AQI levels (AQI-I and AQI-II) in different seasons. The R^2 values of the two fitted quadratic curves are 0.563 and 0.258, respectively, indicating that under the AQI-I air quality level, the relationship between thermal acceptability and UTCI is more reliable. The lower limit of TAR for AQI-I is 7.1 °C, and that for AQI-II is 8.8 °C, both lower than the lower limits of their respective overall neutral ranges. Similarly, the summer UTCI (30.9 °C) is higher than the upper limits of their respective neutral ranges. This indicates that residents in Wuhan exhibit considerable tolerance to changes in the outdoor thermal environment.

Table 9. Acceptable UTCI ranges for residents in Wuhan in different seasons.

Season	Acceptable UTCI Range for Residents in Wuhan (°C)
Summer	19.17–33.98
Winter	7.12–20.15

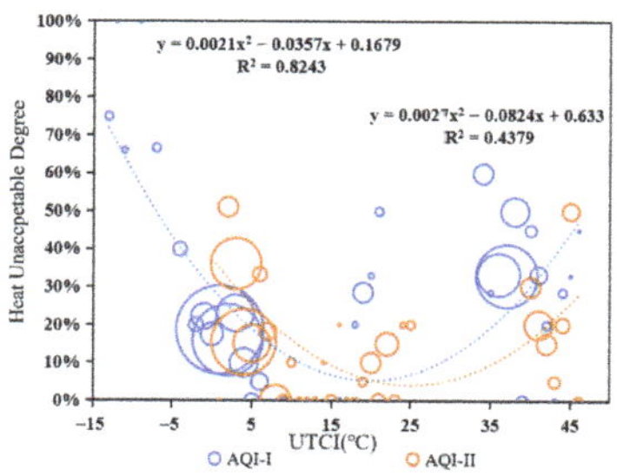

Figure 12. Relationship between thermal unacceptability rate and UTCI for all respondents under different AQI concentration levels.

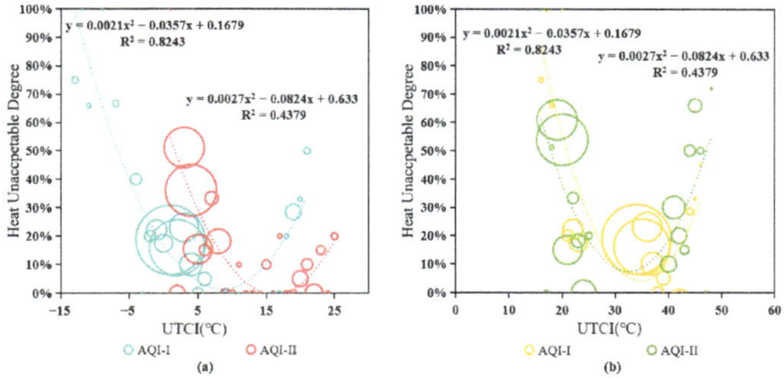

Figure 13. Relationship between thermal unacceptability rate and UTCI for all respondents: (**a**) winter; (**b**) summer.

4.3. Comprehensive Impact of Air Quality and UTCI on Thermal Comfort (TCV)

In this study, UTCI was grouped at 1 °C intervals, and AQI was classified according to the "Ambient Air Quality Standard" (GB 3095-2012) [38]. This grouping clarifies the distribution between UTCI and Thermal Comfort Vote (TCV) and facilitates the identification of goodness-of-fit.

As shown in Figure 14, which presents the variation trend of TCV under different air quality levels and thermal environments, significant differences in human thermal comfort are observed at the same UTCI. Regardless of air quality level changes, TCV increases with UTCI, peaks at approximately 16 °C, and then decreases under overheated conditions. Meanwhile, under identical UTCI conditions, better air quality leads to higher thermal comfort evaluations. That is, improved air quality enhances the perceived comfort of the thermal environment for Wuhan residents, particularly within the 0–10 °C temperature range.

Based on Figures 14 and 15, polynomial fitting was conducted, and the resulting equations are shown in Table 10. A comprehensive analysis reveals that the relationship between UTCI and TCV under different air quality levels conforms to the presented equations. Air quality significantly impacts TCV, and residents provide more sensitive evaluations of thermal comfort under better air quality conditions. Conversely, in cold winter environments, the relationship between air quality and TCV weakens. When residents experience extreme discomfort, the influence of air quality factors on TCV decreases, and UTCI becomes the dominant factor. This phenomenon may occur because residents pay less attention to air quality perception and its impact under extreme cold discomfort, which aligns with empirical expectations. It can thus be inferred that good air quality may enhance mood, thereby improving thermal acceptability. Conversely, poor air quality with severely reduced visibility can negatively affect mood and the evaluation of the thermal environment, consistent with life experience and theoretical expectations. Due to space limitations, this study did not explore the specific physiological pathways through which air quality affects thermal comfort.

Table 10. Regression equations of UTCI and TCV under different seasons and AQI levels.

Air Quality	Season	Thermal Comfort Index UTCI and Average Thermal Comfort Regression Model	
AQI-I	Summer	$TCV = -0.0081UTCI^2 + 0.7564UTCI - 17.155(R^2 = 0.67)$	(6a)
	Winter	$TCV = -0.0015UTCI^2 + 0.0554UTCI - 0.415(R^2 = 0.6572)$	(6b)
	Throughout the year	$TCV = -0.004UTCI^2 + 0.034UTCI - 0.5275(R^2 = 0.3471)$	(6c)
AQI-II	Summer	$TCV = -0.0048UTCI^2 + 0.1552UTCI - 1.2473(R^2 = 0.3273)$	(6d)
	Winter	$TCV = -0.0186UTCI^2 + 1.7321UTCI - 39.758(R^2 = 0.4295)$	(6e)
	Throughout the year	$TCV = -0.003UTCI^2 + 0.0356UTCI - 0.7103(R^2 = 0.3471)$	(6f)

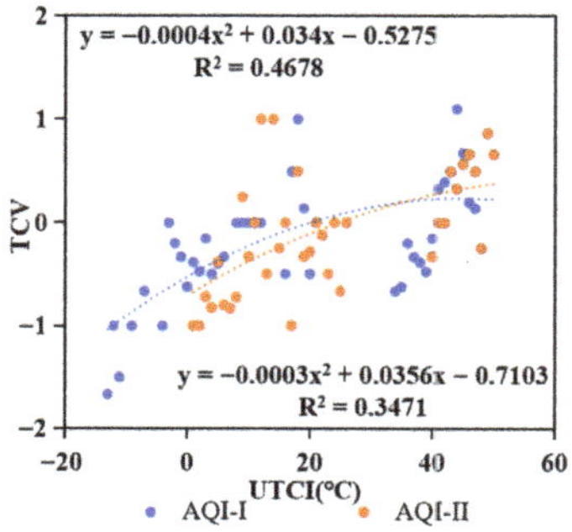

Figure 14. Impact of air quality and thermal environment on TCV for all groups under different AQI levels.

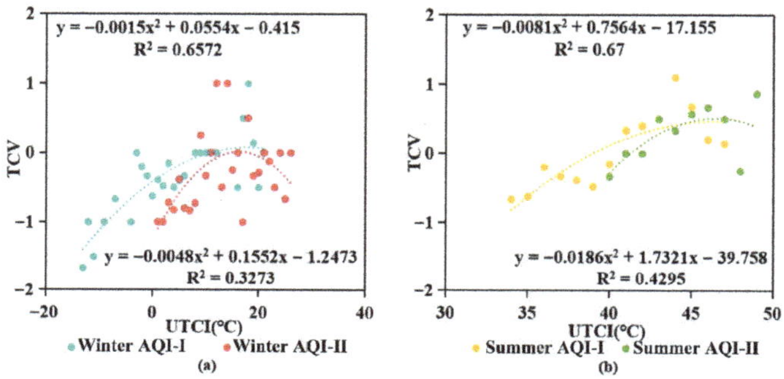

Figure 15. Impact of air quality and thermal environment on TCV: (**a**) summer; (**b**) winter.

5. Discussion

The findings of this study both corroborate and extend existing knowledge on the interplay between thermal comfort and air quality, particularly in the context of high-density residential environments in hot-summer–cold-winter regions.

5.1. Thermal Tolerance and Regional Adaptability

A comparative analysis with Bian et al.'s work [26] in urban parks reveals a distinct adaptive capacity among residential inhabitants. Notably, the climatic and urban environmental contexts of the two studies differ substantially: while Bian et al.'s research was conducted in Tianjin, a city in the cold climate zone of northern China, the present study focuses on Wuhan, located in the hot-summer–cold-winter region of central China. These regional distinctions are characterized by stronger urban heat island effects in Wuhan, especially during the summer, due to its higher background temperatures, humidity levels, and intensive urban development. Specifically, the Thermal Acceptability Range (TAR) thresholds identified in our study extend significantly beyond the conventional neutral UTCI spectrum, suggesting that the habitual exposure and functional requirements characteristic of residential environments—coupled with the region-specific climatic stressors—may foster enhanced thermal tolerance compared to recreational park settings.

5.2. Limitations and Future Research Directions

This study has several limitations that should be considered. First, the quantitative models are specific to Wuhan's climate and cultural context, limiting their generalizability; however, the core finding that the thermal environment and air quality exhibit significant interaction, modulated by seasonal variations and comfort thresholds, provides a valuable conceptual framework for other high-density cities in hot-summer–cold-winter regions. Validation across diverse geographical and climatic zones is necessary.

Furthermore, this study primarily relied on the integrative UTCI to characterize the thermal environment. While this provides a robust physiological assessment, future research with larger datasets could employ advanced statistical modeling, such as multivariable regression with interaction terms or machine learning, to disentangle the complex and potentially non-linear interactions between individual meteorological parameters and air pollutants in shaping occupant perception.

Finally, the underlying physiological and psychological mechanisms by which air quality influences thermal perception were not directly measured. Interdisciplinary studies combining physiological monitoring with environmental surveys are recommended to elucidate these causal pathways.

6. Conclusions

This study selected typical residential areas in Wuhan to conduct research on outdoor thermal comfort in winter in the hot-summer–cold-winter region. Through questionnaire surveys and outdoor physical environment measurements, taking Wuhan as an example, it investigated the impacts of air quality and thermal environment on residents' thermal comfort in this region and further explored the characteristics of outdoor thermal comfort and its correlation with microclimatic factors. Based on the obtained results, the following key conclusions have been drawn:

(1) For every increase of approximately 90 in AQI, satisfaction decreases by one unit for the residence in the studied case. Additionally, this study found that PM2.5 concentration can largely substitute for AQI's role in characterizing air quality for the satisfaction levels.

(2) The explanatory power of the UTCI-MTSV model varies substantially with pollution levels (R2: 0.315–0.776), revealing significant modulation by seasonal and environmental context. Thermal comfort in winter largely adheres to the linear relationship with UTCI, whereas summer exhibits greater vote dispersion due to compounded environmental influences. The lower R2 in summer suggests behavioral adaptations (e.g., seeking shade, adjusting activity), and psychological effects of poor air quality introduce variability not captured by physiological indices alone. These findings highlight the non-linear influence of PM2.5 on thermal perception and underscore the need for integrated models that incorporate microclimatic, pollution, and behavioral data to better predict comfort in complex urban settings.

(3) Residents in Wuhan have a stronger tolerance to changes in the outdoor thermal environment regardless of the AQI levels tested. The lower limit of TAR for AQI-I is 7.1 °C, and that for AQI-II is 8.8 °C, both of which are lower than the lower limits of their respective overall neutral ranges. For the summer, UTCI (30.9 °C) is higher than the upper limits of their respective neutral ranges. However, the fitting between thermal acceptability and UTCI has higher reliability with better air quality, further hinting the increased influence of air quality on the residence's thermal comfort perception.

(4) TCV increases with the increase in UTCI, reaches a peak at approximately 16 °C, and then decreases under overheated conditions. Under the same UTCI conditions, better air quality leads to higher thermal comfort evaluation by people, especially within the temperature range of 0–10 °C.

In summary, this study advances the field of outdoor environmental research through two key contributions. First, it establishes an integrated empirical methodology specifically tailored for high-density residential environments, combining synchronized microclimatic measurements, pollutant monitoring, and subjective comfort surveys. Second, it reveals PM2.5's seasonally modulated interference with thermal perception—a non-linear effect particularly pronounced in summer, where behavioral adaptations and psychological factors amplify the dispersion of thermal sensation votes. Unlike previous studies, this research specifically captures the unique environmental dynamics of high-density urban neighborhoods in hot-summer–cold-winter regions, characterized by strong urban heat island effects and remarkable resident thermal tolerance. These findings provide a new framework for understanding and addressing the combined challenges of heat stress and air pollution in sustainable urban design.

Author Contributions: Conceptualization, W.L. and S.X.; methodology, W.L. and S.X.; software, D.W.; validation, D.W., B.P. and C.Q.; formal analysis, D.W.; investigation, B.P. and C.Q.; resources, S.X.; data curation, D.W., B.P., Y.X. and X.Z.; writing—original draft preparation, D.W.; writing—review and editing, W.L. and Y.X.; visualization, D.W., C.Q., Y.X. and X.Z.; supervision, W.L. and S.X.; project administration, S.X.; funding acquisition, W.L. and S.X. All authors have read and agreed to the published version of the manuscript.

Funding: This research was funded by the National Natural Science Foundation (No. 52378020; No. 52108012), the Program for HUST Academic Frontier Youth Team (No. 2019QYTD10), the Guangdong Natural Science Foundation (No. 2025A1515012125), the Open Foundation of the State Key Laboratory of Subtropical Building and Urban Science (No. 2023KA02), and the Urban Renewal and Transportation Joint Laboratory of Anhui Province (No. 2024CSGX-KF01).

Institutional Review Board Statement: This study was approved by the Institutional Review Board of Guangdong University of Technology (GDUTXS2025022).

Informed Consent Statement: All participants provided informed consent prior to completing the questionnaire.

Data Availability Statement: The data supporting the findings of this study are available within the article and the list of the references.

Conflicts of Interest: Author Dixin Wu was employed by the company China Nuclear Industry 22nd Construction Co., Ltd. Author Bo Pan was employed by the company Wuhan Lingyun Building Decoration Engineering Co., Ltd. Author Congyue Qi was employed by the company China Construction Third Engineering Bureau Group Co., Ltd. The remaining authors declare that the research was conducted in the absence of any commercial or financial relationships that could be construed as a potential conflict of interest.

Appendix A. List of Measuring Instruments and Measurement Point Data

Table A1. Measurement instrument range and accuracy.

Instrument	Testing Items	Instrument Range	Resolution	Accuracy
PC-8/8A mini weather station	Ambient Temperature	−40~80 °C	0.1 °C	±0.2 °C
	Relative Humidity	0~100%	0.1%	±0.2% (≤80%) ±0.5% (>80%)
	PM2.5 Concentration	0–1000 μg/m^3	1 μg/m^3	±10%
	Wind Speed	0~70 m/s	0.1 m/s	±0.3 m/s
	Wind Direction	0~359°	1.0°	±3.0°
TM-188D handheld comprehensive heat index meter	Solar Radiation	0~2000 W/m^2	1.0 W/m^2	<5.0%
	Black Globe Temperature	0~80 °C	0.1 °C	±0.6 °C

Table A2. Illustrative diagram of environmental data from measurement points in the two residential communities.

Measurement Points	**Paris Haoting (1)**					
Name	A1	B1	C1	D1	E1	F1
Orientation of Surrounding Buildings	Open	East–West Orientation	Elevated	Open	East–West Orientation	Open
Underlying Surface Morphology	Hard Brick Pavement	Playground	Concrete Pavement	Hard Brick Pavement	Playground	Hard Brick Pavement
Tree Shading	No	No	Yes	No	Yes	Yes
Green Space Ratio	0.36	0.13	0.07	0.1	0.23	0.18
SVF	0.62	0.4	0.13	0.52	0.33	0.49
SVF map						
measurement points	Jindi Sun City (2)					
Name	A2	B2	C2	D2	E2	F2
Orientation of Surrounding Buildings	Elevated	South–North Orientation	East–West Orientation	East–West Orientation	East–West Orientation	Open
Underlying Surface Morphology	Concrete Pavement	Playground	Lawn	Hard Brick Pavement	Lawn	Hard Brick Pavement
Tree Shading	Yes	No	No	No	Yes	Yes
Green Space Ratio	0	0.2	0.36	0.18	0.28	0.2
SVF	0.25	0.61	0.63	0.59	0.34	0.27
SVF map						

Appendix B.

Base information

Age: ___years old Gender: □Male □Female Height: ___cm Weight: ___kg

1. How long have you been here?
   □Less than 5 minutes □5–15 minutes □More than 15 minutes
2. What were your main activities in the past fifteen minutes?
   □Standing □Sitting □Light physical activity
   Moderate physical activity □Vigorous physical activity
3. What are you wearing today? (Multiple choices)
   Upper garment: □T-shirt □Long-sleeved shirt □Jacket/Coat □Down jacket
   □Sweater □Cardigan □Thin coat □Thick coat
   Lower garment: □Dress □Skirt □Half sleeves □Trousers □Stockings
   □Long socks □Knee-length socks □Ankle socks

Environmental Comfort Survey

4. How do you feel about the current ambient temperature?
   Very cold (−3) | Cold (−2) | slightly cold (−1) | Neutral (0) | slightly warm (1) | Warm (2) | Very hot (3)
5. How would you rate the comfort level of the current ambient climate?
   Very uncomfortable (−2) | Uncomfortable (−1) | Neutral (0) | Comfortable (1) | Very comfortable (2)
6. How do you think the current outdoor air quality is?
   Very good (2) | Good (1) | Neutral (0) | Poor (−1) | Very poor (−2)
7. How acceptable is the current ambient climate to you?
   Completely unacceptable (−2) | Unacceptable (−1) | Neutral (0) | Acceptable (1) | Completely acceptable (2)
8. What is your expectation for the current climate?

Colder	−2	−1	0	1	2	Warmer
Less windy	−2	−1	0	1	2	Windier
Drier	−2	−1	0	1	2	More humid
More shade	−2	−1	0	1	2	More sunshine

9. Regardless of the current weather, which of the following climate changes would make you more willing to go out?
   □Neither too hot nor too cold □Good air quality
10. How do you judge the best air quality?
    □Off-odor (smell) □Visible clarity (vision) □Cold and refreshing air (temperature)

Figure A1. Questionnaire (translated from Chinese to English).

References

1. IPOC Change. *Climate Change 2021: The Physical Science Basis;* Cambridge University Press: Cambridge, UK, 2021.
2. Hu, Y.; Ji, J.S.; Zhao, B. Deaths attributable to indoor $PM_{2.5}$ in urban China when outdoor air meets 2021 WHO air quality guidelines. *Environ. Sci. Technol.* **2022**, *56*, 15882–15891. [CrossRef]
3. Van Hove, L.W.A.; Jacobs, C.M.J.; Heusinkveld, B.G.; Elbers, J.A.; Van Driel, B.L.; Holtslag, A.A.M. Temporal and spatial variability of urban heat island and thermal comfort within the Rotterdam agglomeration. *Build. Environ.* **2015**, *83*, 91–103. [CrossRef]
4. Perini, K.; Magliocco, A. Effects of vegetation, urban density, building height, and atmospheric conditions on local temperatures and thermal comfort. *Urban For. Urban Green.* **2014**, *13*, 495–506. [CrossRef]
5. Shimazaki, Y.; Yoshida, A.; Yamamoto, T. Thermal responses and perceptions under distinct ambient temperature and wind conditions. *J. Therm. Biol.* **2015**, *49*, 1–8. [CrossRef]
6. Abreu-Harbich, L.V.; Labaki, L.C.; Matzarakis, A. Thermal bioclimate as a factor in urban and architectural planning in tropical climates—The case of Campinas, Brazil. *Urban Ecosyst.* **2014**, *17*, 489–500. [CrossRef]
7. Wang, F.; Duan, K.; Zou, L. Urbanization effects on human-perceived temperature changes in the North China Plain. *Sustainability* **2019**, *11*, 3413. [CrossRef]
8. Olesen, B.W.; Parsons, K.C. Introduction to thermal comfort standards and to the proposed new version of EN ISO 7730. *Energy Build.* **2002**, *34*, 537–548. [CrossRef]
9. Wang, Y.; Pan, C.; Huang, Z. Comparison of Applicability of PET and UTCI in Different Seasons in Shanghai. *Build. Sci.* **2020**, *36*, 55–61.
10. Kaewrat, J.; Janta, R.; Sichum, S.; Kanabkaew, T. Indoor air quality and human health risk assessment in the open-air classroom. *Sustainability* **2021**, *13*, 8302. [CrossRef]
11. Sahu, V.; Gurjar, B.R. Spatio-temporal variations of indoor air quality in a university library. *Int. J. Environ. Health Res.* **2021**, *31*, 475–490. [CrossRef] [PubMed]
12. Shen, H.; Hou, W.; Zhu, Y.; Zheng, S.; Ainiwaer, S.; Shen, G.; Chen, Y.; Cheng, H.; Hu, J.; Wan, Y.; et al. Temporal and spatial variation of $PM_{2.5}$ in indoor air monitored by low-cost sensors. *Sci. Total Environ.* **2021**, *770*, 145304. [CrossRef] [PubMed]
13. Chen, M.; Wang, Q.; Yu, H. Analysis of Atmospheric Fine Particulate Matter Sources in Wuhan: A Review. *Environ. Sci. Technol.* **2023**, *46*, 197.
14. Zhang, H.; Wang, Z.; Zhang, W. Exploring spatiotemporal patterns of $PM_{2.5}$ in China based on ground-level observations for 190 cities. *Environ. Pollut.* **2016**, *216*, 559–567. [CrossRef]
15. Yin, Z.; Cao, B.; Wang, H. Dominant patterns of summer ozone pollution in eastern China and associated atmospheric circulations. *Atmos. Chem. Phys.* **2019**, *19*, 13933–13943. [CrossRef]
16. Seco, R.; Peñuelas, J.; Filella, I.; Llusià, J.; Molowny-Horas, R.; Schallhart, S.; Metzger, A.; Müller, M.; Hansel, A. Contrasting winter and summer VOC mixing ratios at a forest site in the Western Mediterranean Basin: The effect of local biogenic emissions. *Atmos. Chem. Phys.* **2011**, *11*, 13161–13179. [CrossRef]
17. Upadhaya, P.; Du, H.; Kommalapati, R.R. Meteorological detrending of ozone at three sites in the Dallas-Fort worth area: Application of KZ filter method. *Atmosphere* **2020**, *11*, 1226. [CrossRef]
18. Li, K.; Liu, X.; Bao, Y. Evaluating the performance of different thermal indices on quantifying outdoor thermal sensation in humid subtropical residential areas of China. *Front. Environ. Sci.* **2022**, *10*, 1071668. [CrossRef]
19. Zhang, L.; Wei, D.; Hou, Y.; Du, J.; Liu, Z.; Zhang, G.; Shi, L. Outdoor thermal comfort of urban park—A case study. *Sustainability* **2020**, *12*, 1961. [CrossRef]
20. Kaushik, A.K.; Arif, M.; Syal, M.M.G.; Rana, M.Q.; Oladinrin, O.T.; Sharif, A.A.; Alshdiefat, A.S. Effect of indoor environment on occupant air comfort and productivity in office buildings: A response surface analysis approach. *Sustainability* **2022**, *14*, 15719. [CrossRef]
21. Xie, X.; Liu, C.H.; Leung, D.Y.C. Impact of building facades and ground heating on wind flow and pollutant transport in street canyons. *Atmos. Environ.* **2007**, *41*, 9030–9049. [CrossRef]
22. Lin, L.; Hang, J.; Wang, X.; Fan, S.; Fan, Q.; Liu, Y. Integrated effects of street layouts and wall heating on vehicular pollutant dispersion and their reentry toward downstream canyons. *Aerosol Air Qual. Res.* **2016**, *16*, 3142–3163. [CrossRef]
23. Mei, S.J.; Liu, C.W.; Liu, D.; Zhao, F.-Y.; Wang, H.-Q.; Li, X.-H. Fluid mechanical dispersion of airborne pollutants inside urban street canyons subjecting to multi-component ventilation and unstable thermal stratifications. *Sci. Total Environ.* **2016**, *565*, 1102–1115. [CrossRef]
24. Miao, C.; Li, P.; Huang, Y.; Sun, Y.; Chen, W.; Yu, S. Coupling outdoor air quality with thermal comfort in the presence of street trees: A pilot investigation in Shenyang, Northeast China. *J. For. Res.* **2023**, *34*, 831–839. [CrossRef]

25. Rui, L.; Buccolieri, R.; Gao, Z.; Ding, W.; Shen, J. The impact of green space layouts on microclimate and air quality in residential districts of Nanjing, China. *Forests* **2018**, *9*, 224. [CrossRef]
26. Bian, G.; Gao, X.; Zou, Q.; Cheng, Q.; Sun, T.; Sha, S.; Zhen, M. Effects of thermal environment and air quality on outdoor thermal comfort in urban parks of Tianjin, China. *Environ. Sci. Pollut. Res.* **2023**, *30*, 97363–97376. [CrossRef]
27. Yilmaz, S.; Sezen, I.; Irmak, M.A.; Külekçi, E.A. Analysis of outdoor thermal comfort and air pollution under the influence of urban morphology in cold-climate cities: Erzurum/Turkey. *Environ. Sci. Pollut. Res.* **2021**, *28*, 64068–64083. [CrossRef] [PubMed]
28. Li, J.; You, W.; Peng, Y.; Ding, W. Exploring the potential of the aspect ratio to predict flow patterns in actual urban spaces for ventilation design by comparing the idealized and actual canyons. *Sustain. Cities Soc.* **2024**, *102*, 105214. [CrossRef]
29. Miao, C.; He, X.; Gao, Z.; Chen, W.; He, B.-J. Assessing the vertical synergies between outdoor thermal comfort and air quality in an urban street canyon based on field measurements. *Build. Environ.* **2023**, *227*, 109810. [CrossRef]
30. Martilli, A.; Nazarian, N.; Krayenhoff, E.S.; Lachapelle, J.; Lu, J.; Rivas, E.; Rodriguez-Sanchez, A.; Sanchez, B.; Santiago, J.L. WRF-Comfort: Simulating microscale variability in outdoor heat stress at the city scale with a mesoscale model. *Geosci. Model Dev.* **2024**, *17*, 5023–5039. [CrossRef]
31. Rui, L.; Buccolieri, R.; Gao, Z.; Gatto, E.; Ding, W. Study of the effect of green quantity and structure on thermal comfort and air quality in an urban-like residential district by ENVI-met modelling. In *Building Simulation*; Tsinghua University Press: Beijing, China, 2019; Volume 12, pp. 183–194.
32. Wuhan Municipal Bureau of Statistics. Wuhan Statistical Yearbook 2022. 2022. Available online: http://tjj.wuhan.gov.cn/ (accessed on 22 November 2025).
33. *ANSI/ASHRAE Standard 55-2023*; Thermal Environmental Conditions for Human Occupancy. American Society of Heating, Refrigerating and Air-Conditioning Engineers: Atlanta, GA, USA, 2023.
34. *ISO 7726:2025*; Ergonomics of the Thermal Environment—Instruments for Measuring Physical Quantities. International Organization for Standardization: Geneva, Switzerland, 2025.
35. National Ambient Air Quality Standards (NAAQS). *40 C.F.R. § 50*; U.S. Environmental Protection Agency: Washington, DC, USA, 2016.
36. Yin, Y.; Luo, W.; Jing, W.; Zhang, J.; Qin, Z.; Zhen, M. Combined effects of Thermal-PM2.5 indicators on subjective evaluation of campus environment. *Build. Environ.* **2022**, *222*, 109381. [CrossRef]
37. Zhang, H.; Arens, E.; Huizenga, C.; Han, T. Thermal Sensation and Comfort Models for Non-Uniform and Transient Environments, Part III: Whole-Body Sensation and Comfort. *Build. Environ.* **2010**, *45*, 399–410. [CrossRef]
38. *GB 3095-2012*; Ambient Air Quality Standards. Ministry of Ecology and Environment of the People's Republic of China: Beijing, China, 2012.

buildings

MDPI

Article

Thermal Performance and Energy Efficiency Evaluation of Thermally Activated Composite Panel for Retrofitted Buildings Across Diverse Climate Zones of Gansu, China

Huajun Zou [1], Ruiwen Chai [1], Zhiqiang Chen [2], Shuying Li [3], Xuefang Bao [2], Sarula Chen [2,*], Kunyu Chen [2] and Yang Yang [3,*]

[1] Gansu Institute of Architectural Design and Research Co., Ltd., Lanzhou 730030, China; zhj-ll@163.com (H.Z.); crwgslz@163.com (R.C.)
[2] College of Architecture and Urban Planning, Anhui Jianzhu University, Hefei 230601, China; 202433102055@stu.ahzju.edu.cn (Z.C.); bxf@ahjzu.edu.cn (X.B.); cky200023@stu.ahzju.edu.cn (K.C.)
[3] College of Architecture and Art, Hefei University of Technology, Hefei 230601, China; 2023110796@mail.hfut.edu.cn
* Correspondence: sarul@tju.edu.cn (S.C.); yangyang@hfut.edu.cn (Y.Y.)

Abstract

This study investigates the use of thermally activated composite panels (TACP) as an energy-efficient retrofitting solution for existing buildings. By utilizing low-grade heat sources, TACP offers significant energy efficiency and carbon emission reduction. To address the challenges posed by high embodied carbon, exterior envelope delamination, and fire safety risks associated with conventional insulation materials, three TACP configurations were proposed: external insulation with externally embedded pipes (EE-TACP), external insulation with internally embedded pipes (EI-TACP), and internal insulation with internally embedded pipes (II-TACP). These were proposed for adapting TACP-walls to the diverse climate zones of Gansu Province. To clarify the thermal performance, energy-saving potential, and reduction effect on conventional insulation thickness of TACP-walls under different climatic conditions and heat source temperatures, dynamic heat transfer models of the TACP-wall were developed and validated. Three energy-saving rate (ε) indicators were introduced to analyze the dynamic thermal behavior in typical cities across the three climate zones of Gansu Province. A comparison was made among the three configurations in terms of instantaneous thermal response characteristics, cumulative internal surface heat load, and ε, relative to both non-insulated walls (NI-walls) and conventional insulated energy-saving walls (CE-walls). The results demonstrate that the application of TACP-walls significantly enhances thermal performance under all three climatic conditions. Irrespective of operating conditions, the ε remains essentially constant corresponding to the four insulation thicknesses, indicating that traditional thickness can be drastically curtailed. EE-TACP, EI-TACP, and II-TACP achieve pronounced ε values of 7.57/7.97/7.50 relative to NI-wall and CE-wall at inlet temperatures of 28–36 °C. Among the three configurations, EI-TACP performs most prominently across all climate zones. The findings provide a valuable reference for the application and optimization of thermally activated technology in energy-efficient retrofits of existing buildings.

Keywords: retrofitting existing buildings; thermally activated composite panel; diverse climate zones; thermal performances; energy-saving rate

1. Introduction

Building energy consumption is a significant contributor to global energy usage and greenhouse gas emissions. Energy efficiency in buildings is not only crucial for improving energy utilization but also an essential strategy for addressing climate change and mitigating global warming [1]. According to statistics, the total floor area of existing buildings in China has far surpassed that of newly constructed buildings. Driven by urban renewal and the growing demand for improved living standards, both national and local governments have introduced policies to promote energy-efficient retrofitting of existing buildings [2]. Among various retrofitting measures, improving the building envelope has become a core focus due to its foundational role, widespread applicability, and significant energy-saving potential [3]. As the key interface for heat exchange between the interior and exterior, the thermal performance of the building envelope directly impacts the building energy consumption and carbon emissions. Therefore, enhancing its insulation properties is a critical means to meet the energy-saving and emission-reduction requirements in different climatic regions [4].

Currently, wall insulation technologies are mainly divided into passive static insulation and active dynamic insulation technologies [5]. The former controls the heat transfer coefficient (K-value) of the envelope structure through the use of insulation materials; however, it faces issues such as high embedded carbon emissions, insufficient durability, high costs, and fire safety risks. In contrast, the latter regulates the temperature difference (ΔT-value) across the envelope, utilizing low-grade renewable energy sources such as air energy and geothermal energy to form a dynamic thermal barrier within the envelope, which reduces heat exchange at its source. Compared to traditional passive methods, active dynamic insulation not only reduces the space occupied by the insulation layer, mitigates fire risks, and extends system life, but also demonstrates greater adaptability and energy-saving potential in the retrofitting of existing buildings [6]. Among various dynamic insulation methods, Thermo-activated Building Systems (TABS) have become a research hotspot both domestically and internationally in recent years [7,8]. TABS embeds hydraulic circulation pipes into the walls, injecting low-grade thermal sources to create an invisible thermal barrier parallel to the interior space, thereby suppressing the heat transfer between the indoor and outdoor environments, achieving both improved building energy efficiency and enhanced indoor environmental quality. Li et al. [9] highlighted that integrating low-grade thermal sources into conventional embedded-pipe wall systems can effectively reduce building thermal loads, which is of significant importance for future low-energy and low-carbon buildings. Yang et al. [10] proposed a novel modular embedded-pipe wall design and, through both experimental and numerical simulations, investigated the effects of pipe cavity dimensions and the thermal conductivity of filler materials on the dynamic thermal behavior and performance metrics of the system. Karanaft et al. [11] further emphasized that using the building framework as a medium for low-grade energy injection and storage represents an effective strategy for achieving both energy savings and occupant comfort.

In recent years, the research on TABS has gradually expanded to the energy retrofitting of existing buildings. Rajput et al. [12] proposed a statistical model of spatial variability of weather for use in building simulation practice. This model effectively simulates the impact of weather variations on the thermal performance of buildings, providing an important reference for optimizing building energy-saving designs using thermally activated technologies. Wu et al. [13] demonstrated significant energy efficiency improvements in retrofitted buildings within dynamic multi-energy building environments by optimizing the resource management system. Especially in cold regions, the heating demand of buildings was reduced by approximately 25%. Zhao et al. [14] further analyzed the application of TABS technology in different climatic regions, noting that in dry and cold areas, TABS can provide a more flexible and efficient energy utilization model compared to traditional static insulation methods. The study revealed that the dynamic thermal barrier of the TABS system not only reduced the intrusion of external heat but also enhanced indoor thermal comfort, particularly during seasonal temperature fluctuations.

However, existing research has mainly focused on areas with single or minimal climatic variation. In regions like Gansu, where there is significant climate diversity and a large geographical span, there remains a lack of systematic performance evaluation and optimization studies. This limits the applicability of existing results in areas with significant seasonal changes and spatial climatic differences. Furthermore, in the retrofitting of existing buildings, structural limitations often prevent the placement of the embedded pipe layer in the middle of the structure, which further restricts the application of TABS in retrofitting projects. The commonly used method of thickening external insulation presents issues such as fire safety risks, high embodied carbon, and excessive thickness that occupy valuable space, while also struggling to meet the demanding retrofitting requirements of cold regions.

To address the above issues, this study integrates TABS technology with existing building envelope structures. It proposes three types of thermally activated composite panel (TACP) (external insulation with external embedded pipes, external insulation with internal embedded pipes, and internal insulation with internal embedded pipes). Unlike previous studies that mainly focused on single climate zones or conventional insulation materials, this research applies thermally activated technology to the energy-efficient retrofitting of existing buildings in Gansu Province, exploring its applicability under multiple climatic conditions. Through numerical simulations, the effects of pipe position, insulation thickness, and heat input temperature on the thermal performance and energy-saving potential are systematically analyzed. Using non-insulated walls and conventional external insulation walls as references, the dynamic thermal characteristics and energy-saving potential of the three TACP configurations are compared across severe cold, cold, and hot summer–cold winter climatic regions, thereby determining the optimal configuration for different climate zones in Gansu.

The paper is structured as follows: Section 2 introduces the research methods and the development of the numerical simulation model; Section 3 investigates the dynamic thermal performance and energy-saving potential of different TACP configurations across the three climate zones of Gansu Province under varying parameters. Specifically, Section 3.1 analyses the effects of the three parameters on TACP thermal performance and determines the optimal scheme. In contrast, Section 3.2 compares the energy-saving potential of TACP with that of two reference walls. Section 4 presents the conclusions and outlook.

2. Methodology

2.1. Working Principle of Thermally Activated Composite Panel

The thermally activated composite panel (TACP) forms an invisible thermal barrier within the wall by circulating low-grade energy available from the building's surroundings through its embedded pipe layer, thereby actively reducing heat transfer between indoor and outdoor environments. Depending on the inlet temperature of the heat source, the TACP operates in three typical modes across different seasons.

The heat transfer principles for three typical winter operational modes are compared between a conventional external insulation wall and TACP walls, using the example of the external insulation with externally embedded pipes TACP (EE-TACP), which are schematically shown in Figure 1. When the inlet temperature lies between the outdoor and indoor air temperatures ($T_e < T_c < T_i$), the system operates in the active insulation mode, which significantly increases the inner surface temperature of the wall and substantially reduces the thermal load (Figure 1b). When the inlet temperature approximates or equals the indoor air temperature ($T_e < T_c = T_i$), the system operates in the neutral mode (Figure 1c), where the negligible temperature difference between the inner wall surface and the indoor environment effectively eliminates heating energy demand caused by envelope heat losses. When the inlet temperature exceeds the indoor air temperature during winter ($T_e < T_i < T_c$), the system operates in the auxiliary heating mode (Figure 1d). In this case, the thermal load of the envelope becomes negative, indicating that the TACP can provide supplementary heating or even serve as a direct heating source for the indoor environment.

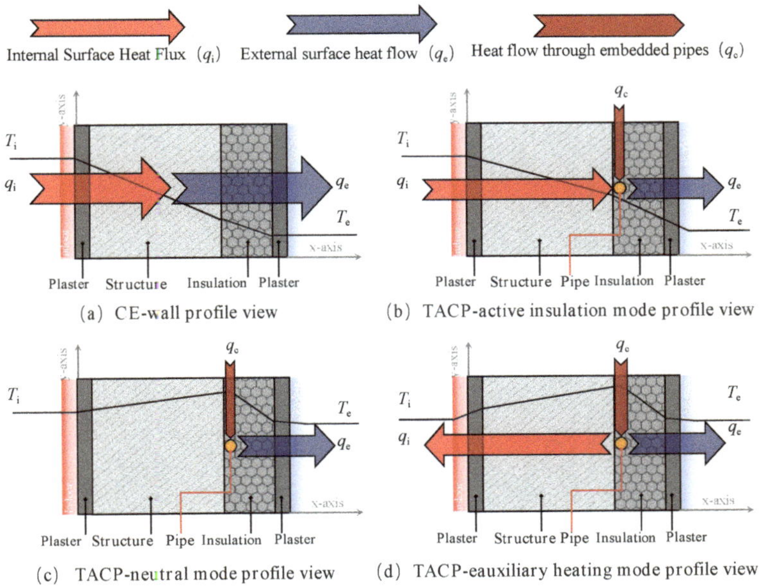

Figure 1. Temperature and heat flux distribution profiles of TACP in different operation modes in winter.

2.2. Physical and Mathematical Models

2.2.1. Geometry Model of TACP

This study uses existing buildings with non-insulated walls (NI-wall) and conventional insulated energy-saving walls (CE-wall) as reference wall types, with their structural and physical models shown in Figure 2a,b. Figure 2c–e illustrate three different configurations of TACP walls: external insulation with externally embedded pipes (EE-TACP), external insulation with internally embedded pipes (EI-TACP), and internal insulation with internally embedded pipes (II-TACP). The NI-wall, lacking an insulation layer, exhibits high heat transfer and poor thermal stability. The CE-wall reduces heat loss through conventional insulation materials; however, it remains a passive thermal control system with limited effectiveness. In contrast, the TACP-wall combines embedded low-temperature water pipes with an insulation layer, enabling active thermal regulation and enhanced heat storage capacity. This configuration significantly improves thermal stability and indoor thermal comfort while achieving higher energy efficiency and carbon-reduction potential. Parametric analysis will be conducted to determine the thermal characteristics and energy-saving potential of these three configurations in the energy retrofitting of existing buildings across three climate zones in Gansu Province, namely, the severe cold climate zones, cold climate zones, and hot summer–cold winter climate zones. The geometric parameters and thermal properties of the five wall types are presented in Table 1. In these configurations, the insulation thickness of the TACP walls (denoted as H_3) is treated as a variable, aimed at comparing the reduction in insulation thickness achieved by TACP relative to the traditional insulation layer. Therefore, the insulation thickness of the CE-wall is set equal to that of the maximum insulation thickness in three TACP configurations, with thicknesses varying according to the energy-saving requirements of each climate zone. In the severe cold climate zones (Jiuquan city), the insulation thickness is set between 60 mm and 120 mm; in the cold climate zones (Lanzhou city), it ranges from 40 mm to 100 mm; and in the hot summer–cold winter climate zones (Wudu city), the thickness ranges from 30 mm to 70 mm [15]. Since most existing buildings in Gansu Province are constructed with brick walls, the structural layer in the model is assumed to be brick, with thicknesses of 370 mm in Jiuquan city and Lanzhou city, and 240 mm in Wudu city.

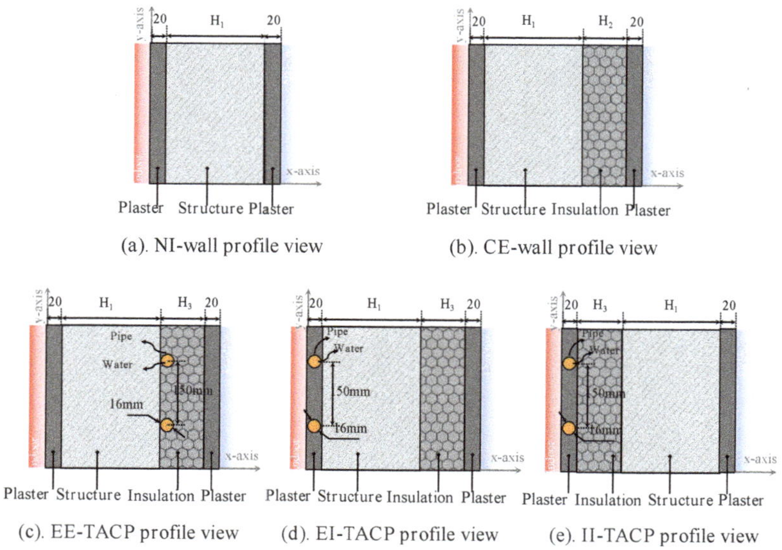

Figure 2. Profiles of two-dimensional geometric structure and model size: (**a**) non-insulated walls; (**b**) conventional insulated energy-saving wall; (**c**) external insulation with externally embedded pipes; (**d**) external insulation with internally embedded pipes; (**e**) internal insulation with internally embedded pipes.

Table 1. Geometric parameters and thermal properties of TABS and two reference walls.

Reference	Layer	Material	Size (mm)	Density ($Kg{\cdot}m^{-3}$)	Cp ($J{\cdot}kg^{-1}{\cdot}{}^{\circ}C^{-1}$)	Thermal Conductivity ($W{\cdot}m^{-1}{\cdot}{}^{\circ}C^{-1}$)
[16,17]	Plaster layers	Brick	20	1400	1050	0.58
[18]	Thermal insulation	XPS	variable	30	1380	0.042
[19]	Embedded pipe	PPR	diameter 16, spacing 150	900	185.205	0.21
[20]	Structure layer	KP1 Clay Brick	370/240	1400	1062.3	0.58
[21]	Thermal grease	Laird Tflex 600	/	2700	1600	3.5

2.2.2. Meshing Method

All numerical simulations in this study were performed using the commercial finite element analysis software ANSYS Workbench 2023 R1 (Ansys, Inc., Canonsburg, PA, USA). First, the geometry modeling for all cases was completed in the DesignModeler module, and mesh generation and boundary naming were further carried out after importing them into the Mesh module. Next, all simulation operations were conducted in the Fluent solver, including setting up the energy and viscosity models, specifying material properties and operating parameters, initializing the computational model, configuring computational settings, running the simulation, and ensuring that the computational model met the convergence criteria.

To avoid the adverse effects of grid quantity and division methods on the simulation results, all cases were generated using the same meshing strategy. The grids were optimized under the prerequisite of ensuring mesh quality, thereby balancing accuracy with computational efficiency. For all five wall types, a structured hexahedral meshing method was applied to the computational domains. This process comprised two core steps: firstly, the radial boundary ends of each domain were divided proportionally to generate high-quality surface meshes; secondly, the radial boundaries in the depth direction were further subdivided proportionally, followed by the sweeping method to create the volume meshes. The mesh quality of all simulation cases was maintained above 0.8, with no non-physical harmful volume elements, ensuring computational stability and reliable results. As illustrated in Figure 3, the examples include the mesh division results for the NI-wall and CE-wall with a 70 mm insulation layer, as well as three types of TACP walls (EE-TACP, EI-TACP, II-TACP), based on a case study in Wudu City, a representative region with hot summers–cold winters.

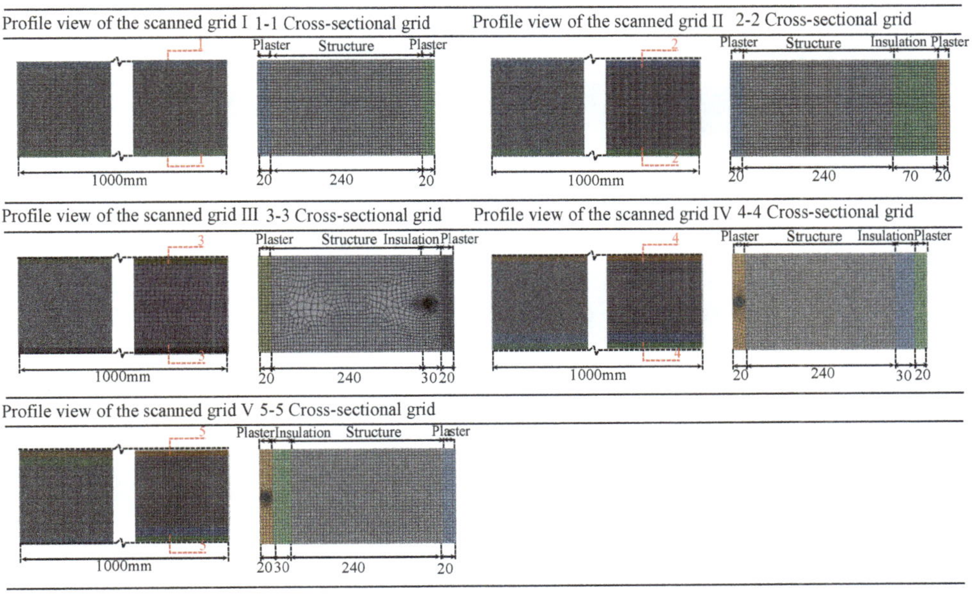

Figure 3. Meshing method of TACP wall profiles and meshing results of 5 different wall configurations.

2.2.3. Boundary and Initial Conditions

This study selected the hourly meteorological data for the coldest month of January from the meteorological stations in three typical cities of Gansu Province, namely Jiuquan,

Lanzhou, and Wudu, as the external boundary conditions for the simulations. All cases were transiently simulated based on these observed meteorological data, with the corresponding hourly data shown in Figure 4.

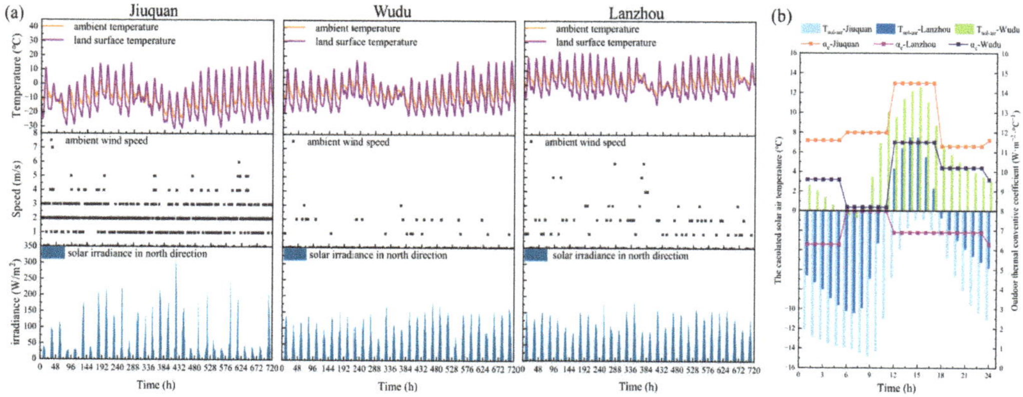

Figure 4. (**a**) Variation in outdoor meteorological parameters over time for typical cities in three climatic regions of Gansu Province in January; (**b**) Hourly average values of outdoor composite temperature and external surface heat transfer coefficient.

To simplify the mathematical model, the following necessary basic assumptions are made: (a) The fluid water injected into the embedded pipe and the air that undergoes convective heat transfer with the inner surface and outer surfaces of the wall are treated as incompressible fluids with constant density, and their thermal properties are assumed to be constant. (b) All solid materials involved in the model (Table 1) are considered homogeneous and isotropic in terms of thermal properties. (c) It is assumed that there are no gaps between different material layers inside the wall, and additional thermal resistance due to poor contact is neglected. (d) The arrangement of embedded tubes inside the wall follows a parallel configuration; thus, the interface between adjacent pipes, due to its symmetrical heat transfer behavior, can be treated as a symmetry boundary condition.

Based on the above assumptions, the energy equation for the solid region is as follows:

$$\left(\rho_{\mathrm{k}} c_{\mathrm{p,k}}\right)/\lambda\,|_{\mathrm{k}} \partial T/\partial \tau = \partial^2 T/\partial x^2 + \partial^2 T/\partial y^2 + \partial^2 T/\partial z^2 + S_{\mathrm{m}} \quad \mathrm{k} = 1,\ 2,\ 3,\ 4 \tag{1}$$

In the formula, ρ_{k} (kg·m^{-3}), t (s), $c_{\mathrm{p,k}}$ (J·kg^{-1}·°C^{-1}), λ (W·m^{-1}·°C^{-1}), T (°C), S_{m} (/) are the density of solid region, time, specific heat, thermal conductivity, temperature, and source term of momentum, respectively; and k = 1 to 4 denotes the four solid layers of TACP, exterior plaster, insulation layer, structural layer, pipe-embedded layer, and interior plaster, respectively.

The energy equation for the liquid region is as follows:

$$\frac{\partial(\rho_{\mathrm{x}} c_{\mathrm{p,x}} F)}{\partial t} + \nabla(\rho_{\mathrm{x}} U F) = \nabla(\Gamma_{\varphi} \mathrm{grad} F) + S_{\varphi} \tag{2}$$

In the formula, ρ_{x} (kg·m^{-3}), $c_{\mathrm{p,x}}$ (J·kg^{-1}·°C^{-1}), F (°C), t (s), U (/), Γ_{Φ} (/), and S_{Φ} (/) are the density of working fluid (water), specific heat, temperature, time, velocity vector, diffusion coefficient, and source term of energy conservation equation, respectively.

The heat transfer boundary at the inner surface of the TACP wall is as follows:

$$\lambda_0 \frac{\partial T}{\partial x} \Big|_{x=0} = \alpha_i (T_i - T_{x=0}), \tag{3}$$

In the formula, λ_0 (W·m^{-1}·°C^{-1}), α_i (W·m^{-2}·°C^{-1}), T_i (°C), $T_{x=0}$ (°C) represent the thermal conductivity of inner plaster layer, the convective heat transfer coefficient of room side (α_i = 8.7 W·m^{-2}·°C^{-1}) [22], room set point (T_i = 24 °C), and inner surface temperature, respectively.

The heat transfer boundary at the outer surface of the TACP wall is as follows:

$$\lambda_L \frac{\partial T}{\partial x} \Big|_{x=L} = \alpha_e (T_a - T_{x=L}) + \rho_s I - R_{es} \tag{4}$$

In the formula, λ_L (W·m^{-1}·°C^{-1}), α_e (W·m^{-2}·°C^{-1}), T_a (°C), $T_{x=L}$ (°C), ρ_s (m^2), I (W·m^{-2}), R_{es}(K·W^{-1}) represent the thermal conductivity of exterior plaster layer, the convective heat transfer coefficient of outer side, outer surface temperature, radiative temperature of the outdoor environment, absorption coefficient of outer surface (ρ_s = 0.65), solar radiation at the outer surface, and resistance heating, respectively. Among them, the α_e calculation method is as follows [23,24]:

$$\alpha_e = 5.62 + 3.9v \tag{5}$$

In the formula, v (m·s^{-1}) is the air velocity of the outdoor environment.

The solar radiation absorption on the outer surface of the TACP wall and the convective heat transfer between the wall and the outdoor environment are based on the outdoor air composite temperature, as given by the following equation [23,24]:

$$T_0 = \frac{I \cdot \rho_s}{\alpha_e} + T_a \tag{6}$$

In the formula, T_0 (°C) is the comprehensive outdoor air temperature.

The numerical model in this study employs a Pressure-Based Segregated Solver for fluid–structure interaction calculations. The coupling between pressure and velocity is achieved through the SIMPLE algorithm, a type of pressure correction method. This algorithm iteratively solves the momentum and pressure correction equations to ensure that the velocity field satisfies both mass conservation and momentum conservation. The convective terms in all governing equations are discretized using a second-order upwind scheme. This scheme interpolates between the nearest upstream node and the next upstream node, offering higher accuracy compared to the first-order upwind scheme, which helps reduce numerical diffusion.

The convergence of the calculations is determined by the residuals of the governing equations. The convergence criteria for the residuals are set as follows:

(1). The residual for the mass conservation (continuity) equation is set to 1×10^{-5};

(2). The residual for the momentum equation is set to 1×10^{-6};

(3). The residual for the energy equation is set to 1×10^{-8}.

In this study, the temperature of the injected heat in the embedded pipe is set between 24 °C and 36 °C, representing low-grade hot water. The inlet velocity is set to 0.5 m·s^{-1}, while the indoor air temperature is maintained at a constant 24 °C.

2.3. Model Validation

The dynamic heat transfer model used in this study is an improvement of the previous research on the concrete traditional thermally activated wall (CTAW), based on the structural features and heat transfer characteristics of the TACP. The fundamental assumptions and validity of the model have been fully verified in prior studies [25]. The primary difference between TACP and CTAW lies in the placement of the embedded pipe layer: TACP positions the pipe layer adjacent to the structural layer, either within the plaster or insulation layer, whereas CTAW places the pipe layer within the concrete structural layer. By adjusting the layout of the embedded pipes, this study aims to explore the optimal wall structure of TACP in the context of various climate zones in Gansu. Furthermore, in the heat transfer simulation of the building envelope, replacing materials that do not undergo phase changes typically does not affect the reliability of the heat transfer model itself. This ensures the consistency and credibility of the results in this study.

2.4. Simulation Scheme and Performance Evaluation Methods

2.4.1. Research Variables and Simulation Scheme

For TACP, the key factors influencing the thermal performance and energy-saving potential of existing building envelope structures are the pipe position, insulation thickness, and injected heat source temperature. Therefore, this study designs simulation scenarios based on these three parameters (Table 2). Using the NI-wall of existing buildings and the CE-wall as reference walls, the study compares the impact of these parameters on the thermal performance and energy-saving potential of TACP. Section 3.1 analyses the thermal performance of TACP by comparing the hourly thermal response of cumulative heat load on the inner and outer surfaces under the influence of the three parameters. Section 3.2 evaluates the impact of the three parameters on TACP energy-saving potential by comparing it with the reference walls.

Table 2. The variable range and default values adopted under different TACP configurations.

TACP Configurations	Insulation Thickness (mm)	Climate Zones/Typical City	Inlet Temperature (°C)	Inlet Velocity (m/s)
external insulation with externally embedded pipes	60 80 100 120	severe cold climate zones, Jiuquan city	24, 28, 32, 36	0.5
	40 60 80 100	cold climate zones, Lanzhou city		
	30 40 55 70	hot summer–cold winter climate zones, Wudu city		
external insulation with internally embedded pipes	60 80 100 120	severe cold climate zones, Jiuquan city		
	40 60 80 100	cold climate zones, Lanzhou city		
	30 40 55 70	hot summer–cold winter climate zones, Wudu city		
internal insulation with internally embedded pipes	60 80 100 120	severe cold climate zones, Jiuquan city		
	40 60 80 100	cold climate zones, Lanzhou city		
	30 40 55 70	hot summer–cold winter climate zones, Wudu city		

2.4.2. Performance Evaluation

This study evaluates the thermal performance and energy-saving potential of wall structures by comparing the cumulative heat load and energy-saving rate (ε) indicators of TACP and two reference wall types. Thermal performance is assessed using the temperature distribution and heat flux of the internal/external surface of the wall and embedded pipe. The energy-saving rate, ε, is calculated using the following formula. When the cumulative heat load on the inner surface of the wall is negative, it indicates that no additional heat load is required indoors, with the wall providing auxiliary heating. Moreover, a higher ε value signifies greater energy-saving potential for the building. In Equation (7), ε_1 represents the calculation formula for the energy-saving potential of TACP relative to NI-wall, while in Equation (8), ε_2 represents the energy-saving potential of TACP relative to CE-wall. In addition, in Equation (9), ε_3 represents the energy-saving potential of the NI-wall relative to the CE-wall:

$$\varepsilon_1 = \frac{Q_{ni} - Q_{tacp}}{Q_{ni}} \tag{7}$$

$$\varepsilon_2 = \frac{Q_{ce} - Q_{tacp}}{Q_{ce}} \tag{8}$$

$$\varepsilon_3 = \frac{Q_{ni} - Q_{ce}}{Q_{ni}} \tag{9}$$

In the formula, Q_{ni} ($kW \cdot h \cdot m^{-2}$), Q_{tacp} ($kW \cdot h \cdot m^{-2}$), Q_{ce} ($kW \cdot h \cdot m^{-2}$) represent the daily cumulative heat load per unit area on the inner surface of the wall for existing buildings with no insulation, the daily cumulative heat load per unit area on the inner surface of the wall for the TACP, the daily cumulative heat load per unit area on the inner surface of the wall for the conventional insulation-retrofitted wall, respectively.

3. Results and Discussions

3.1. Thermal Performance of TACP-Wall Configurations in Different Climate Zones

3.1.1. Analysis of Wall Thermal Response Characteristics in Three Climate Zones

Figure 5 illustrates the variation in the exterior surface temperature of the EI-wall over time under different inlet temperatures (T_i) for representative cities (Jiuquan, Lanzhou, Wudu) across the climate zones of Gansu Province. It can be observed that the exterior surface temperatures differ in magnitude across the three climate zones due to differences in ambient temperatures. Nevertheless, their overall variation pattern remains identical during stable TACP operation. For all three TACP configurations, the exterior surface temperature is lower than that of the NI-wall when the outdoor temperature is relatively low during the day, and higher than that of the NI-wall when the sol-air temperature is high. This indicates that the heat loss through the energy-efficient wall assemblies is consistently lower than that of the original walls prior to retrofitting. As T_i increases, the exterior surface temperature of the EI-wall generally rises relative to the CE-wall, though the increase is modest. Taking a 60 mm insulation thickness in Jiuquan city as an example, the average increases in exterior surface temperature at T_i = 24 °C, 28 °C, 32 °C, and 36 °C are 0.44 °C, 0.45 °C, 0.51 °C, and 0.58 °C, respectively. This suggests that raising T_i does not significantly affect outward heat loss. Furthermore, under the same climatic zone and T_i condition, increasing the insulation thickness leads to a decrease in the exterior surface temperature. For instance, in Lanzhou at T_i = 28 °C, the average exterior surface temperatures of the EI-wall are −2.87 °C (60 mm), −3.30 °C (80 mm), −3.42 °C (100 mm), and −3.43 °C (120 mm). This implies that greater insulation thickness reduces heat loss through the exterior surface during winter, as the heat transfer from the pipe-embedded layer to the outdoors is diminished.

Figure 6 illustrates the variation in heat flux over time at the exterior surface and the pipe-embedded surface of the EI-wall under different inlet temperatures (T_i) for representative cities (Jiuquan, Lanzhou, Wudu) across the climate zones of Gansu Province. It can be seen that the heat flux at the pipe-embedded surface remains highly stable during operation, showing little influence from climatic variations or outdoor temperature fluctuations, while increasing significantly with rising T_i. In contrast, the exterior surface heat flux exhibits periodic variations in response to changes in outdoor temperature. Although the sol-air temperature in the severe cold climate zones is significantly lower than that in the hot summer–cold winter climate zones, leading to greater heat loss through the exterior wall surface, the heat flux on the surface of the embedded pipe remains unaffected. This does not affect the heat flux at the pipe-embedded surface. This further demonstrates that although the TACP retrofitting scheme increases heat loss at the exterior surface due to the active heat injection, the increase is not significant. For example, in Jiuquan, with an insulation thickness of 60 mm, the heat flux at the exterior and pipe-embedded surfaces of the EI-wall increases relative to the NI-wall/CE-wall by 23.82/−6.16 W/m^2 (at 24 °C), 22.70/−7.30 W/m^2 (at 28 °C), 21.57/−8.41 W/m^2 (at 32 °C), and 20.44/−9.54 W/m^2 (at 36 °C), respectively. Additionally, for every 4.0 °C increase in T_i, the heat flux at the pipe-embedded surface decreases by approximately 1.1, 2.2 W/m^2, and 3.3 W/m^2, indicating that the heating performance of the EI-wall is largely unaffected by the sol-air temperature. The results also show that variations in insulation thickness have an insignificant impact on the heat flux density at the pipe-embedded surface. For instance, at T_i = 36 °C, when the insulation thickness increases from 60 mm to 120 mm, the maximum increase in heat flux density is only about 7.36 W/m^2 (in Jiuquan), 10.40 W/m^2 (in Lanzhou), and 9.04 W/m^2 (in Wudu). Therefore, while the insulation layer mitigates heat loss from the pipe-embedded layer, additional increments in thickness confer negligible further benefit to the overall energy performance of the wall. From the perspective of reducing the embodied carbon of the insulation material, a thicker layer is, in fact, counterproductive.

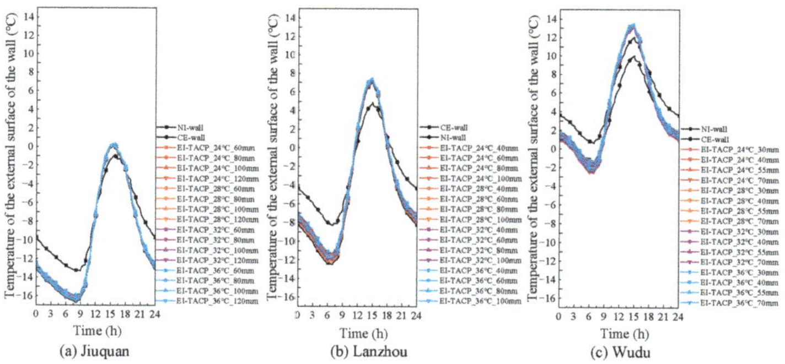

Figure 5. The variation in the exterior surface temperature of the EI-TACP wall over time under the same configuration scheme in different climate zones.

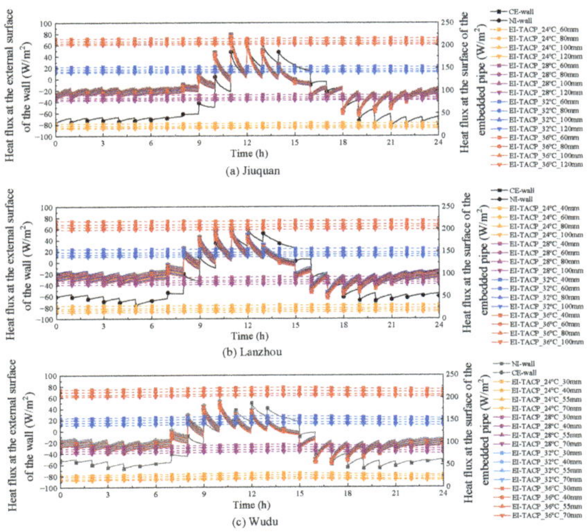

Figure 6. The heat flux variation at the exterior surface and the pipe-embedded surface of the EI-wall over time under the same configuration scheme in different climate zones.

Figure 7 illustrates the variation in the interior surface temperature of the EI-wall over time under different inlet temperatures (T_i) for representative cities (Jiuquan, Lanzhou, Wudu) across the climate zones of Gansu Province. In contrast to the exterior surface temperature, the interior surface temperature under the energy-efficient retrofitting remains highly stable, whereas that of the non-retrofitted wall not only fluctuates but is also significantly lower than the interior surface temperatures of the other wall. Under various operating conditions in the three climate zones, the interior surface temperatures of the three TACP-wall configurations are consistently higher than those of the CE-wall, and increase with rising T_i. As T_i increases, the TACP transitions from a load-reduction mode to an auxiliary heating mode. For instance, at T_i = 28 °C, the interior surface temperature reaches 25.76 °C, indicating that with higher T_i, the interior surface temperature of the EI-wall can be further stabilized and approach the indoor setpoint temperature. Figure 7 also shows that the interior surface temperature is largely unaffected by meteorological conditions in different climate zones. For example, at T_i = 36 °C, the average interior surface temperatures of the EI-wall with insulation thicknesses of 60, 80, 100, and 120 mm are 30.13 °C, 30.14 °C, 30.37 °C, and 30.44 °C in Jiuquan; 30.06 °C, 30.24 °C, 30.36 °C, and 30.47 °C in Lanzhou; and 30.02 °C, 30.16 °C, 30.30 °C, and 30.40 °C in Wudu, respectively. This demonstrates that during winter, the EI-wall enhances the thermal performance of the building envelope, maintaining the interior surface temperature at a consistently high level. This not only substantially reduces heat loss through the wall but also contributes to improved indoor thermal comfort.

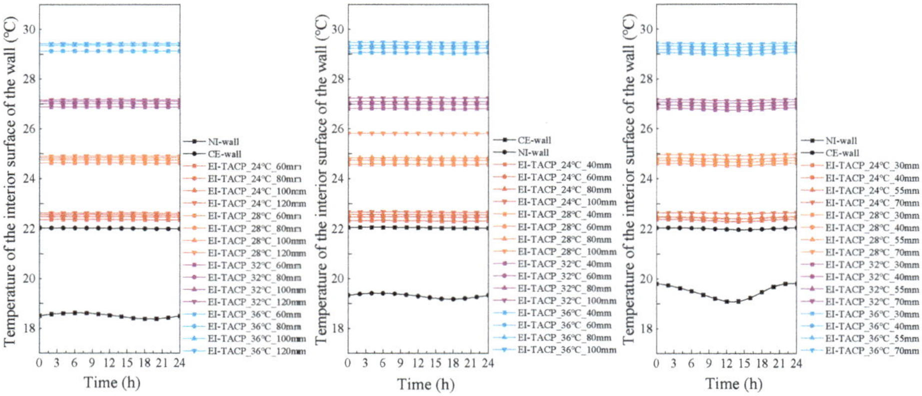

(a) Jiuquan (b) Lanzhou (c) Wudu

Figure 7. The variation in the interior surface temperature of the EI-TACP wall over time under the same configuration scheme in different climate zones.

Figure 8 illustrates the variation in heat flux at the interior surface of the EI-wall over time under different inlet temperatures (T_i) for representative cities (Jiuquan, Lanzhou, Wudu) across the climate zones of Gansu Province. In sharp contrast to the heat flux at the exterior surface, the interior surface heat flux of the three TACP configurations is negative at all heat injection temperatures except T_i = 24 °C, indicating that the TACP system not only prevents indoor heat loss to the outdoors but also provides additional heat to the interior space. In comparison, the interior surface heat flux of the non-insulated wall reaches approximately 30 W/m^2 in all climate zones, with obvious fluctuations throughout the day. Within the same climatic zone, the heat flux released to the interior space increases notably with higher heat source temperatures. Furthermore, under different climatic conditions, the changes in interior surface heat flux for the EI-wall with a 60 mm insulation thickness at T_i = 24 °C relative to the NI-wall/CE-wall are 33.54/3.03 W/m^2 (Jiuquan), 26.23/2.37 W/m^2 (Lanzhou), and 24.97/2.88 W/m^2 (Wudu). This demonstrates that the interior surface heat flux is strongly influenced by the heat source temperature, while the impact of outdoor climate conditions is negligible. Additionally, when the insulation thickness is fixed, for example, at 100 mm for the EI-wall, the interior surface heat flux decreases by approximately 19.75 W/m^2 (Jiuquan), 19.70 W/m^2 (Lanzhou), and 19.82 W/m^2 (Wudu) for every 4.0 °C increase in T_i. This further confirms that the interior surface heat flux of the EI-wall is consistently governed by the heat source temperature under winter conditions. The insulation thickness also has a discernible influence on the interior surface heat flux. For instance, in Wudu, at T_i = 36 °C, the interior surface heat fluxes of the EI-wall relative to the NI-wall/CE-wall are 83.0/60.9 W/m^2 (60 mm), 84.3/62.2 W/m^2 (80 mm), 85.5/63.4 W/m^2 (100 mm), and 86.4/64.3 W/m^2 (120 mm). These results indicate that variations in insulation thickness have only a minor effect on the interior surface heat flux.

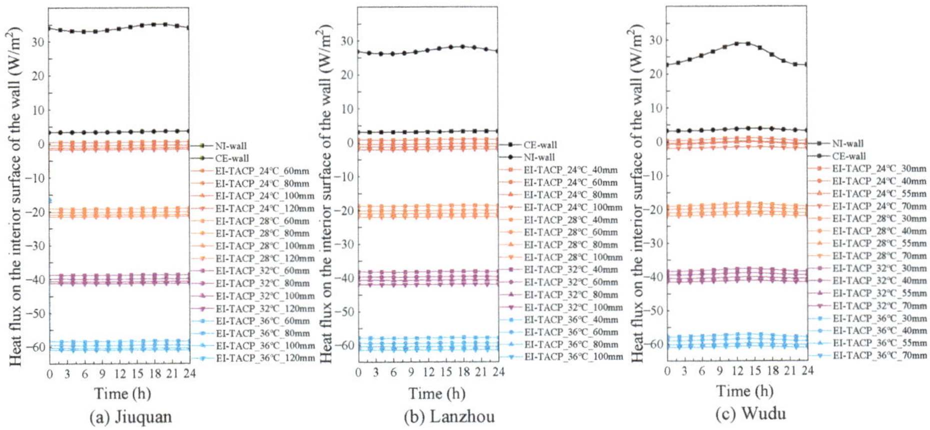

Figure 8. The heat flux variation at the interior surface of the EI-wall over time under the same configuration scheme in different climate zones.

3.1.2. Performance Evaluation of Three TACP Configurations in the Severe-Cold Climate Zones

Figures 9 and 10 present the variations in interior surface temperature and cumulative interior surface heat load for the three TACP configurations, EE-TACP, EI-TACP, and II-TACP, under different inlet temperatures (T_i = 24, 28, 32, and 36 °C) and insulation thicknesses (60, 80, 100, and 120 mm), relative to two types of reference walls. As observed in Figure 9a–d, when the T_i is 24 °C, the interior surface temperatures of all walls except the non-insulated wall are relatively similar. However, as T_i increases from 28 °C to 36 °C, the differences in interior surface temperature among the three types of TACP-walls compared to the CE-wall, as well as among the EE-TACP, EI-TACP, and II-TACP systems, become more pronounced. Among these, the interior surface temperature of EI-TACP is significantly higher than that of EE-TACP and II-TACP. For example, with an insulation thickness of 120 mm, the interior surface temperature of EI-TACP increases relative to the NI-wall/CE-wall by 6.38/2.87 °C (at 28 °C), 8.66/5.15 °C (at 32 °C), and 10.93/7.42 °C (at 36 °C). Compared to the EE-TACP/II-TACP walls, the increases are 1.67/0.22 °C (28 °C), 3.49/0.40 °C (32 °C), and 5.31/0.58 °C (36 °C), respectively. Furthermore, under the studied operating conditions, increasing the insulation thickness has an insignificant effect on the interior surface temperature. For instance, at T_i = 36 °C, increasing the insulation thickness of the EI-TACP wall from 60 mm to 120 mm raises the temperature by only 0.34 °C. Figure 10 clearly shows that the cumulative interior surface heat load of the three TACP-walls nearly approaches zero at a heat source temperature of 24 °C and turns increasingly negative as the injection temperature rises. This confirms that the TACP-walls are supplying heat to the indoor space. Moreover, under all four T_i conditions, the performance of EI-TACP is significantly superior to that of EE-TACP and II-TACP.

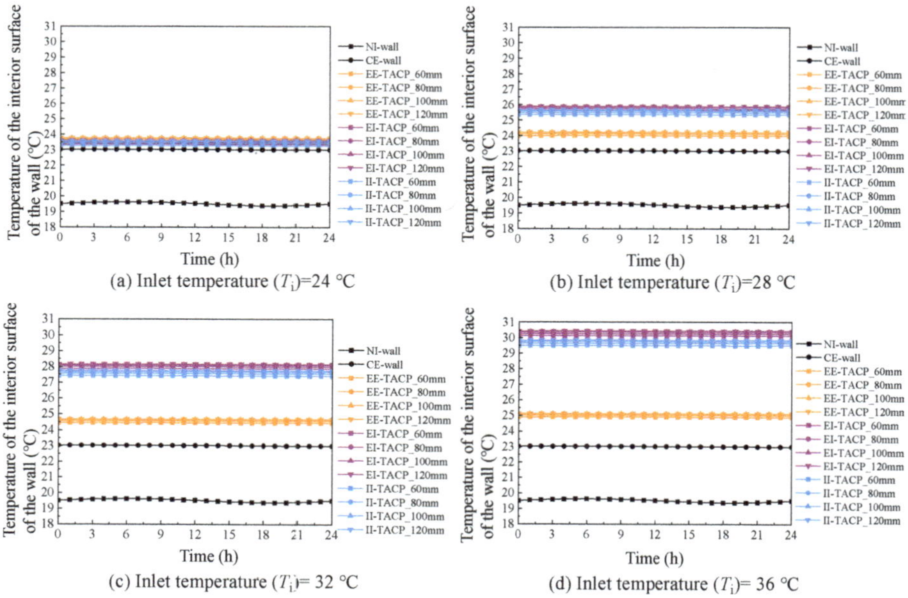

Figure 9. Variation in interior surface temperature among EE-TACP, EI-TACP, II-TACP, NI-wall, and CE-wall under varying inlet temperatures and insulation thicknesses in the severe cold climate zones.

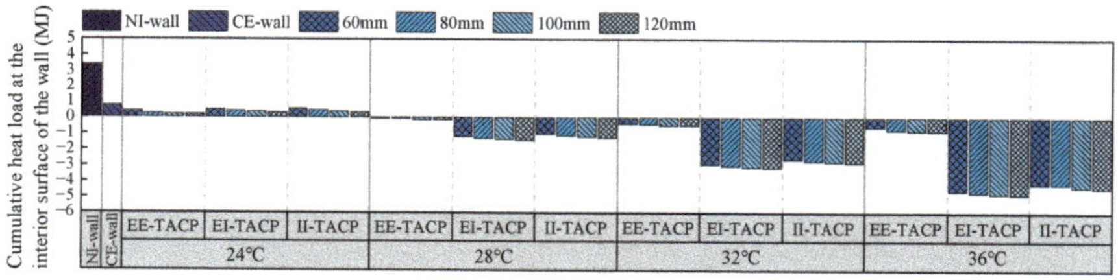

Figure 10. Cumulative heat load at the interior surface among EE-TACP, EI-TACP, II-TACP, NI-wall, and CE-wall under varying inlet temperatures and insulation thicknesses in the severe cold climate zones.

3.1.3. Performance Evaluation of Three TACP Configurations in the Cold Climate Zones

Figures 11 and 12 present the variations in interior surface temperature and cumulative interior surface heat load for the three TACP configurations, EE-TACP, EI-TACP, and II-TACP, under different inlet temperatures (T_i = 24, 28, 32, and 36 °C) and insulation thicknesses (40, 60, 80, and 100 mm), relative to the non-insulated wall (NI-wall) and the conventional energy-saving wall (CE-wall). Consistent with the patterns observed in the severe cold climate zones, when the T_i is 24 °C, the interior surface temperatures of all wall types, except the NI-wall, are relatively similar, as shown in Figure 11a–d. However, as T_i increases from 28 °C to 36 °C, the differences in interior surface temperature between the three types of TACP-walls and the CE-wall become more pronounced. In particular, the interior surface temperature of EI-TACP increases significantly compared to that of EE-TACP as T_i rises, while the difference relative to II-TACP remains minimal. Furthermore, at T_i = 24 °C, the interior surface temperatures of the EE-TACP, EI-TACP, and II-TACP walls already approach the indoor setpoint temperature and increase noticeably with higher heat injection temperatures. For example, at T_i = 36 °C and an insulation thickness of 40 mm, the interior surface temperatures of EE-TACP, EI-TACP, and II-TACP increase by 4.30/1.56 °C, 9.75/7.01 °C, and 9.13/6.38 °C relative to the NI-wall and CE-wall, respectively. This clearly demonstrates that the invisible thermal barrier formed by the low-grade heat injected into the wall can effectively enhance the dynamic thermal behavior of the building envelope, with EI-TACP exhibiting the most favorable performance. As also observed in Figure 12, in the cold climate zones, the cumulative interior surface heat load of the three types of TACP-walls approaches zero at a heat source temperature of 24 °C and increases with higher injection temperatures, indicating that the TACP-walls are supplying heat to the indoor space. Under all four inlet temperature conditions, II-TACP outperforms EE-TACP, while EI-TACP delivers superior performance compared to II-TACP. When the insulation thickness increases from 40 mm to 100 mm, the maximum increase in cumulative interior surface heat load for EI-TACP is 0.23 W/m^2 (at 24 °C), 0.28 W/m^2 (at 28 °C), 1.97 W/m^2 (at 32 °C), and 0.73 W/m^2 (at 36 °C).

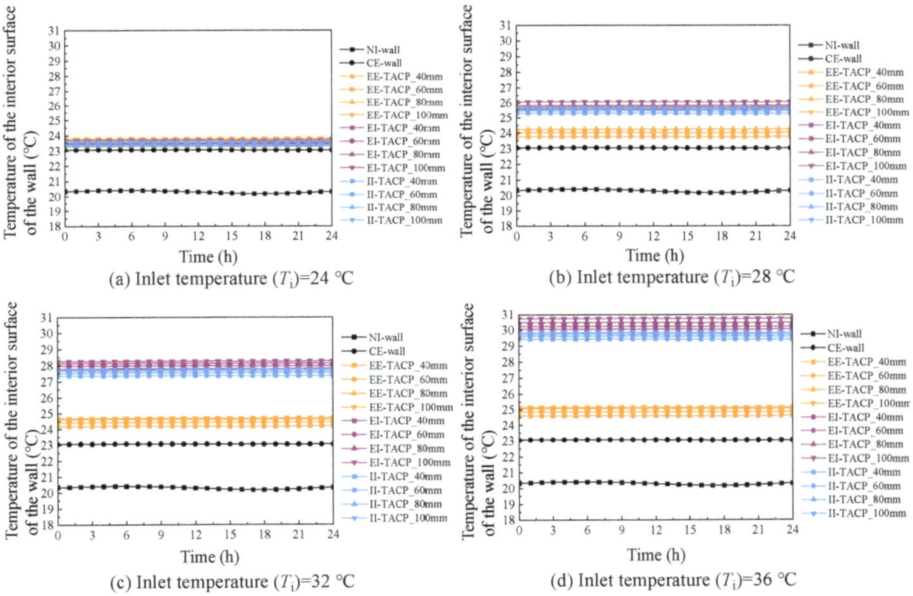

Figure 11. Variation in interior surface temperature among EE-TACP, EI-TACP, II-TACP, NI-wall, and CE-wall under varying inlet temperatures and insulation thicknesses in the cold climate zones.

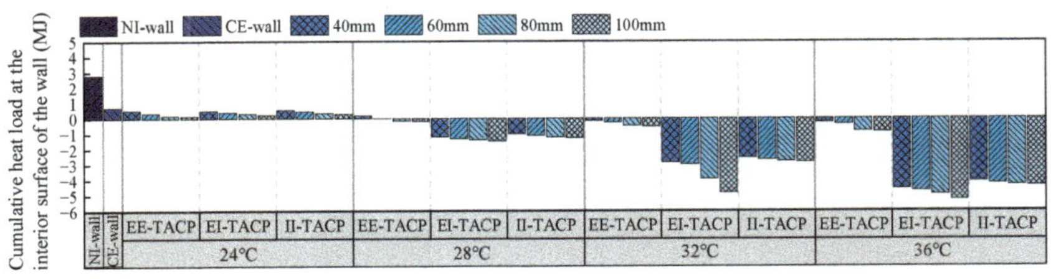

Figure 12. Cumulative heat load at the interior surface among EE-TACP, EI-TACP, II-TACP, NI-wall, and CE-wall under varying inlet temperatures and insulation thicknesses in the cold climate zones.

3.1.4. Performance Evaluation of Three TACP Configurations in the Hot Summer–Cold Winter Climate Zones

Figures 13 and 14 present the variations in interior surface temperature and cumulative interior surface heat load for the three wall configurations, EE-TACP, EI-TACP, and II-TACP, under different inlet temperatures (T_i = 24, 28, 32, and 36 °C) and insulation thicknesses (30, 40, 55, and 70 mm), relative to the NI-wall and the CE-wall. As shown in Figure 13a–d, the interior surface temperature of the NI-wall exhibits more pronounced fluctuations compared to those observed in the severe cold and cold climate zones, which may be attributed to the thinner insulation layers typically used in the hot summer–cold winter climate zones. In addition, as T_i increases, the interior surface temperatures of the EE-TACP, EI-TACP, and II-TACP walls are significantly elevated relative to both reference walls. However, a distinct pattern emerges at T_i = 32 °C that differs from the other two climate zones: under insulation thicknesses of 55 mm and 70 mm, the interior surface temperature of the II-TACP wall slightly exceeds that of the EI-TACP wall. Under the other three T_i conditions, the differences in interior surface temperature and heat flux between the II-TACP and EI-TACP walls are negligible. For example, at T_i = 36 °C, the maximum increases in cumulative interior surface heat load for EI-TACP relative to the NI-wall and CE-wall are 7.18/5.27 W/m^2 (30 mm), 7.29/5.38 W/m^2 (40 mm), 7.40/5.49 W/m^2 (55 mm), and 7.48/5.57 W/m^2 (70 mm). Similarly, the corresponding values for II-TACP are 8.91/6.37 W/m^2 (30 mm), 9.07/6.53 W/m^2 (40 mm), 9.23/6.70 W/m^2 (55 mm), and 7.42/6.88 W/m^2 (70 mm). Therefore, in the hot summer–cold winter climate zones, based on the thermal response characteristics and cumulative interior surface heat load of the TACP-walls, both EI-TACP and II-TACP represent viable retrofitting options, depending on the heat source temperature applied to the pipe-embedded layer.

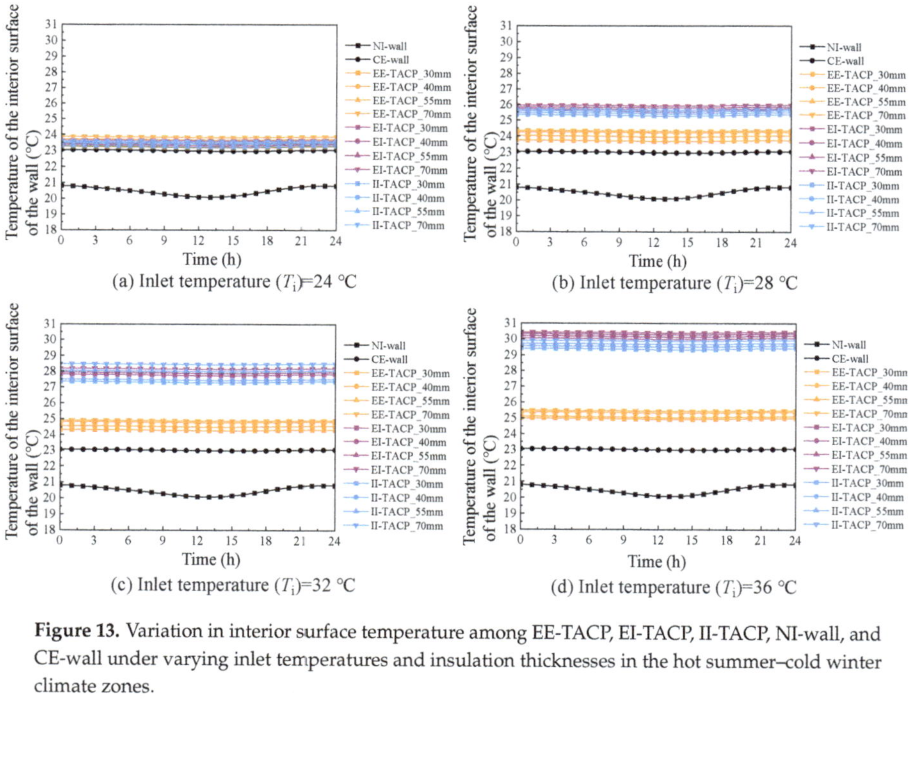

Figure 13. Variation in interior surface temperature among EE-TACP, EI-TACP, II-TACP, NI-wall, and CE-wall under varying inlet temperatures and insulation thicknesses in the hot summer–cold winter climate zones.

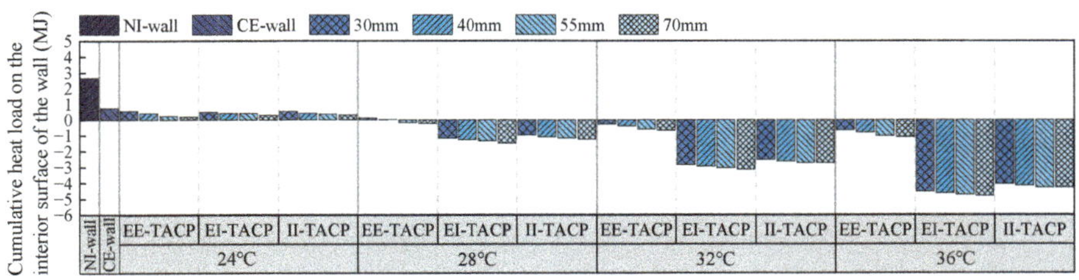

Figure 14. Cumulative heat load at the interior surface among EE-TACP, EI-TACP, II-TACP, NI-wall, and CE-wall under varying inlet temperatures and insulation thicknesses in the hot summer–cold winter climate zones.

3.2. *Energy-Saving Potential of Alternative TACP-Wall Configurations in Different Climate Zones*

This section presents a comparative analysis of the energy-saving potential of three types of thermally activated composite panel wall (TACP-wall) configurations, i.e., external insulation with externally embedded pipes (EE-TACP), external insulation with internally embedded pipes (EI-TACP), and internal insulation with internally embedded pipes (II-TACP), under varying injected heat temperatures (T_i = 24, 28, 32, and 36 °C) and insulation thicknesses (severe cold climate zones: 60–120 mm; cold climate zones: 40–100 mm; hot summer–cold winter climate zones: 30–70 mm), relative to both non-insulated wall (NI-wall) and conventional insulated energy-saving wall (CE-wall). The energy savings of the CE-wall relative to the NI-wall are also assessed to identify optimal TACP-wall retrofitting strategies for each climatic zone of Gansu Province.

3.2.1. Energy-Saving Potential of TACP-Wall Configurations in Severe Cold Climatic Zone

Figure 15 illustrates the variations in energy-saving potential of retrofitted walls using EE-TACP, EI-TACP, and II-TACP relative to both the non-insulated (NI) and the conventional insulated (CE) reference walls in Jiuquan city (severe-cold climate zones), under different inlet temperatures (T_i = 24, 28, 32, and 36 °C) and insulation layer thicknesses (60, 80, 100, and 120 mm). The results indicate that in the severe cold climate zones, the energy-saving rates of the three retrofitting schemes range from 0.28 to 7.57 across the different inlet temperatures (T_i). Furthermore, as shown in Figure 15a–d, under the same T_i, the improvement in energy-saving rate due to changes in insulation thickness is not significant for a given wall type. For example, for the EE-TACP configuration at T_i = 24 °C, when the insulation thickness increases from 60 mm to 120 mm, the energy-saving rate increases by only 0.026 (compared to the NI-wall) to 0.12 (compared to the CE-wall). Therefore, from the perspectives of economy and embodied carbon emissions, using the minimum insulation thickness can adequately meet energy-saving requirements. Additionally, except under the T_i = 24 °C condition, the energy-saving rates of EI-TACP and II-TACP are significantly higher than that of EE-TACP, with the difference becoming more pronounced as the inlet temperature increases. At T_i = 36 °C, the energy-saving rates of EI-TACP and II-TACP increase markedly compared to EE-TACP, with EI-TACP performing particularly well, showing improvements of 2.37–2.44 and 7.26–7.57 relative to the NI-wall and CE-wall, respectively. This study further reveals that, under conditions where the CE-wall is more energy-efficient than the NI-wall, the energy-saving rates of the three types of TACP-wall configurations relative to the CE-wall are substantially higher than those relative to the NI-wall, with the disparity increasing at higher inlet temperatures. This suggests that the reduction in heat loss to the outdoors is far greater in TACP-walls with insulation layers than in CE-walls. In summary, considering energy-saving performance, economic feasibility, and carbon emissions, the EI-TACP configuration with the minimum insulation thickness is identified as the most suitable retrofitting option among the three TACP-wall types in the severe cold climate zones.

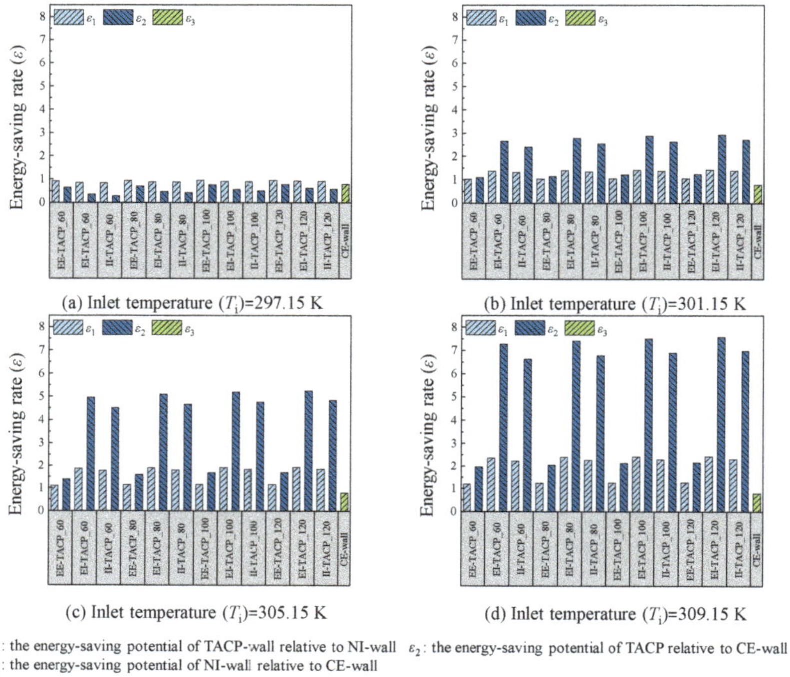

Figure 15. Energy-saving potential (ε) of different TACP wall configurations under different inlet temperatures and insulation thicknesses compared to NI-wall and CE-wall in severe cold climate zones, Jiuquan City.

3.2.2. Energy-Saving Potential of TACP-Wall Configurations in Cold Climatic Zone

Figure 16 presents the variations in energy-saving potential of retrofitted walls using EE-TACP, EI-TACP, and II-TACP active energy-saving schemes in Lanzhou city under different inlet temperatures (T_i = 24, 28, 32, and 36 °C) and insulation thicknesses (40, 60, 80, and 100 mm), relative to two types of reference walls. Figure 16a–d showed that the energy-saving rate of the three retrofitting schemes increases with rising T_i in the cold climate zones. Compared to the NI-wall/CE-wall, the improvements in energy-saving rate for EE-TACP, EI-TACP, and II-TACP range from 0.81–0.95/0.27–0.79 (T_i = 24 °C), 0.82–0.93/0.29–0.67 (T_i = 28 °C), and 0.81–0.95/0.27–0.79 (T_i = 32 °C), to 0.80–0.89/0.21–0.58 (T_i = 36 °C). Moreover, within any single T_i scenario, the increase in energy-saving rate resulting from variations in insulation thickness remains insignificant for each wall type. For instance, when the insulation thickness of EE-TACP increases from 40 mm to 100 mm, the improvement relative to the NI-wall/CE-wall is only 0.13/0.45 (T_i = 24 °C), 0.13/0.50 (T_i = 28 °C), 0.14/0.54 (T_i = 32 °C), and 0.15/0.57 (T_i = 36 °C). Therefore, considering both economic efficiency and embodied carbon emissions, the use of the minimum insulation thickness is sufficient to meet energy-saving requirements. The result further reveals that, except under the T_i = 24 °C condition, the energy-saving rates of EI-TACP and II-TACP are significantly higher than that of EE-TACP as T_i increases, with EI-TACP demonstrating particularly outstanding performance. For instance, at T_i = 36 °C, EI-TACP exhibits improvements relative to the NI-wall/CE-wall of 2.64/7.40 (40 mm), 2.70/7.59 (60 mm), 2.83/7.74 (80 mm), and 3.05/7.97 (100 mm). Consequently, integrating thermal performance, energy-saving magnitude, cost-effectiveness, and carbon footprint, the EI-TACP configuration with the minimal insulation thickness emerges as the optimal retrofit solution among the three TACP-wall variants in cold-climate regions.

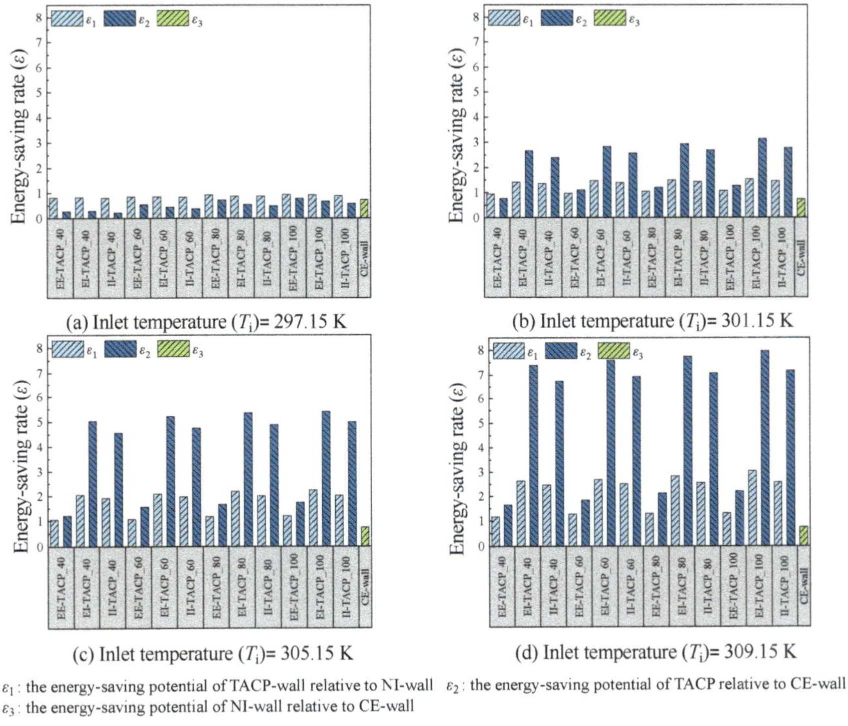

Figure 16. Energy-saving rate (ε) of different TACP-wall configurations under different inlet temperatures and insulation thicknesses compared to NI-wall and CE-wall in cold climate zones, Lanzhou City.

3.2.3. Energy-Saving Potential of TACP-Wall Configurations in Hot Summer–Cold Winter Climatic Zone

Figure 17 quantifies the energy-saving potential of retrofitted existing building walls using EE-TACP, EI-TACP, and II-TACP schemes in the Wudu area, evaluated under different inlet temperatures (T_i = 24, 28, 32, and 36 °C) and insulation layer thicknesses (30, 40, 55, and 70 mm), relative to non-insulated walls (NI-walls) and conventional insulated energy-saving walls (CE-walls). The results showed that in the hot summer–cold winter climate zones, the energy-saving rate (ε) of the three retrofitting schemes varies considerably with T_i. At T_i = 24 °C, the values of ε_1 and ε_2 differ significantly from those at T_i = 28, 32, and 36 °C, with a wide range of 0.33–7.50. Overall, ε exhibits an increasing trend as T_i rises, with particularly notable performance under high-temperature conditions of T_i = 36 °C. Furthermore, as observed in Figure 17a–d, increasing the insulation thickness under the same injected heat temperature has a minimal influence on ε. Taking the EE-TACP scheme at T_i = 24 °C as an example, ε relative to the NI-wall and CE-wall increases only marginally across the four insulation thicknesses: 0.79/0.33 (30 mm), 0.88/0.43 (40 mm), 0.93/0.67 (55 mm), and 0.94/0.74 (70 mm). It can be seen that variations in ε due to changes in insulation thickness are negligible when compared to the NI-wall, while the increase relative to the CE-wall remains modest, ranging from 0.10 to 0.41, a trend consistent across all T_i levels.

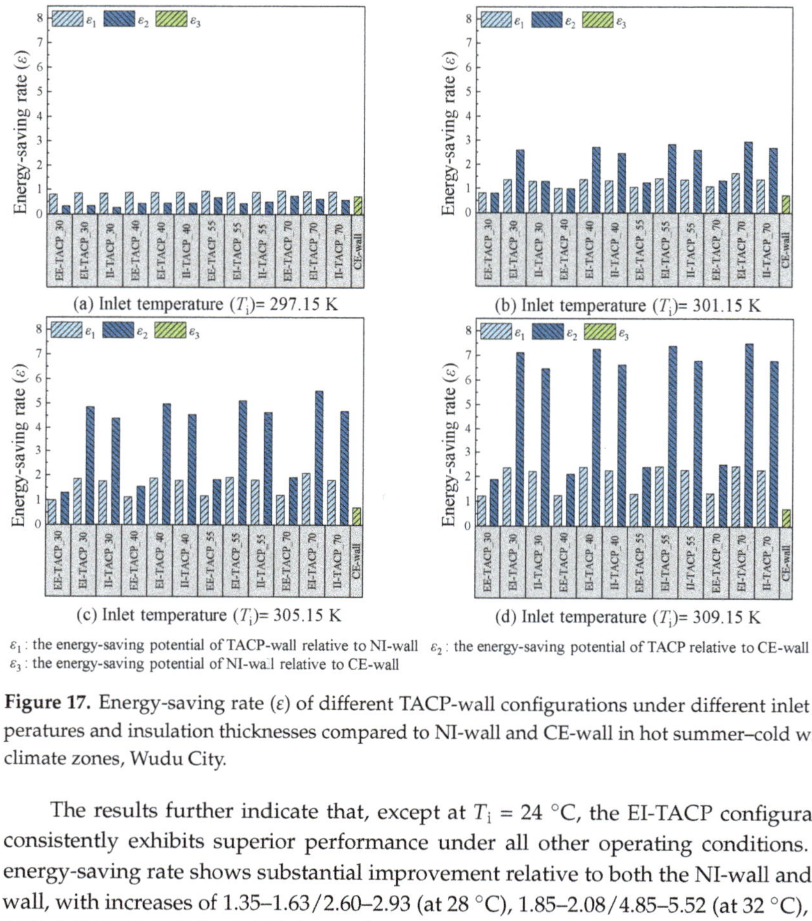

Figure 17. Energy-saving rate (ε) of different TACP-wall configurations under different inlet temperatures and insulation thicknesses compared to NI-wall and CE-wall in hot summer–cold winter climate zones, Wudu City.

The results further indicate that, except at T_i = 24 °C, the EI-TACP configuration consistently exhibits superior performance under all other operating conditions. Its energy-saving rate shows substantial improvement relative to both the NI-wall and CE-wall, with increases of 1.35–1.63/2.60–2.93 (at 28 °C), 1.85–2.08/4.85–5.52 (at 32 °C), and 2.34–2.43/7.10–7.50 (at 36 °C). In conclusion, based on a comprehensive consideration of the thermal performance results presented in Section 3.1 and the energy-saving potential analyzed in this section, the EI-TACP scheme with the minimum insulation thickness emerges as the most economically efficient and low-carbon option among the three TACP-wall retrofitting strategies in the hot summer–cold winter climate zones.

4. Conclusions

This study presents three configurations of thermally activated composite panels (TACP) as effective solutions for improving energy efficiency and reducing carbon emissions in existing buildings, particularly in areas with diverse climates like Gansu Province. The configurations of thermally activated composite panels (TACP) are external insulation with externally embedded pipes (EE-TACP), external insulation with internally embedded pipes (EI-TACP), and internal insulation with internally embedded pipes (II-TACP). To elucidate the performance of the three configurations under varying conditions, a systematic analysis was conducted on the effects of insulation thickness and injection temperature on thermal performance and energy-saving potential in the severe cold, cold, and hot summer–cold winter climate zones. The key findings of this study can be summarized as follows:

(1) The three TACP-wall configurations exhibit superior thermal performance and economic efficiency compared with non-insulated walls (NI-wall) and conventional insulated energy-saving walls (CE-wall) under all operating conditions. These results demonstrate the feasibility of the proposed retrofit solution across different climate zones in Gansu Province and provide a practical technical pathway for energy-efficient renovation of existing buildings.

(2) Increasing the insulation thickness has a negligible effect on the interior surface temperature, interior heat flux, and energy-saving rate of the TACP-wall. However, due to the presence of the insulation layer, the heat loss from the TACP-wall to the outdoors is significantly smaller relative to the CE-wall than to the NI-wall. Hence, under identical heat injection temperatures and outdoor climatic conditions, the energy-saving rate ε_2 is considerably greater than ε_1 for the same TACP configuration.

(3) The energy-saving rates of the three configurations, EE-TACP, EI-TACP, and II-TACP, all increased with rising heat injection temperature, reaching maximum values of up to 7.57 in the severe cold climate zones, 7.97 in the cold climate zones, and 7.50 in the hot summer–cold winter climate zones. When the injection temperature exceeded the room temperature, the absolute value of the cumulative heat load at the interior surface also increased significantly with higher injection temperatures.

(4) Under identical operating conditions across the three climate zones, the performance of both EI-TACP and II-TACP walls is significantly superior to that of the EE-TACP wall. In the severe cold and cold climate zones, the overall performance of the EI-TACP wall exceeds that of the II-TACP wall. However, in the hot summer–cold winter zones, the performance of the EI-TACP and II-TACP walls is comparable, with the II-TACP wall performing slightly better than the EI-TACP wall, particularly at a heat injection temperature of 24 °C. Therefore, based on a comprehensive evaluation, the most suitable scheme for the severe cold zones is EI-TACP. For the cold zones, EI-TACP is the primary recommendation, followed by II-TACP as a secondary option. In the hot summer–cold winter zones, if only the energy-saving rate is considered, EI-TACP is recommended, yet II-TACP may be selected when retrofit constraints and whole winter operation are also taken into account.

5. Research Prospect

Future research will focus on implementing continuous on-site monitoring in retrofitted buildings across different climatic zones of Gansu Province to verify the long-term thermal performance and energy-saving effectiveness of TACP under real meteorological conditions. In addition, efforts will be made to optimize economic performance, enhance material sustainability, and improve system integration, providing support for the large-scale application of TACP technology in energy-efficient retrofitting of existing buildings.

Author Contributions: H.Z.: Project administration; Resources; Funding acquisition. R.C.: Project administration; Resources. Z.C.: Validation; Formal analysis; Data curation; Writing—original draft. S.L.: Investigation; Data curation; Writing—review and editing. X.B.: Visualization; Funding acquisition; Data curation. S.C.: Methodology; Investigation; Writing—original draft; Supervision.

K.C.: Formal analysis; Data curation; Visualization. Y.Y.: Methodology; Supervision; Software; Validation; Writing—original draft. All authors have read and agreed to the published version of the manuscript.

Funding: The research was supported by Research on Planning and Aesthetic Renewal Technologies for Existing Urban Districts (NO. JK2023-20; NO. KY2023-05), the National Natural Science Foundation of China (NO. 52208103), and Anhui Provincial University Research Project (NO. 2024AH050240).

Institutional Review Board Statement: Not applicable.

Informed Consent Statement: Not applicable.

Data Availability Statement: The data presented in this study are available upon request from the corresponding author.

Conflicts of Interest: Authors Huajun Zou and Ruiwen Chai were employed by the company Gansu Institute of Architectural Design and Research Co., Ltd. The remaining authors declare that the research was conducted in the absence of any commercial or financial relationships that could be construed as a potential conflict of interest.

Nomenclature

V/v	Speed, m/s	Abbreviation	
Q	Heat exchange intensity, MJ·m^{-2}	TACP	Thermally activated composite panel
H	Length size, mm	EE-TACP	Externally embedded pipes
q	Heat flux, W·m^{-2}	EI-TACP	External insulation with internally embedded pipes
T	Temperature,°C	II-TACP	Internal insulation with internally embedded pipes
c_p	Specific heat capacity, J·kg^{-1}·°C^{-1}	NI-walls	Non-insulated walls
g	Gravitational acceleration, m·s^{-2}	CE-walls	Conventional insulated energy-saving walls
h	Thermal convective coefficient, W/(m^2·°C^{-1})	CTAW	Concrete traditional thermally activated wall
t	Simulation time, s		
D	Pipe diameter, mm	Subscripts	
I	Solar radiance, W·m^{-2}	i	Interior surface
		e	Exterior surface
Greek symbols		ab	Ambient air
ρ	Mass density, kg·m^{-3}	c	Water pipe
ε	The energy-saving rate	ins	Thermal insulation layer
λ	Heat conductivity factor. W/(m.°C)	id	Indoor set
α	Heat transfer coefficient, W/(m^2·°C^{-1})	p	Plaster layer

References

1. Chen, X.; Vand, B.; Baldi, S. Challenges and strategies for achieving high energy efficiency in building districts. *Buildings* **2024**, *14*, 1839. [CrossRef]
2. *GB 50189-2015*; Building Energy Efficiency Design Standard. Architecture & Building Press: Beijing, China, 2015. (In Chinese)
3. Zhang, Y.; Hu, S.; Guo, F.; Mastrucci, A.; Zhang, S.; Yang, Z.; Yan, D. Assessing the potential of decarbonizing China's building construction by 2060 and synergy with industry sector. *J. Clean. Prod.* **2022**, *359*, 132086. [CrossRef]
4. Yang, Y.; Chen, S. Comprehensive investigation on dynamic energy performances of pipe-embedded enclosure structures with thermal anisotropic injection and diffusion features. *Energy Convers. Manag.* **2025**, *343*, 120261. [CrossRef]
5. Yang, Y.; Chen, S. Thermal insulation solutions for opaque envelope of low-energy buildings: A systematic review of methods and applications. *Renew. Sustain. Energy Rev.* **2022**, *167*, 112738. [CrossRef]
6. Zhao, H.; Wu, Y.; Jiang, G.; Zhong, M.; Sun, H.; Lin, B. Material and Structural Optimization of Novel Phase-Change Thermal Diode for Dynamic Building Envelope. *Engineering*, 2025; *in press*. [CrossRef]

7. Chen, S.; Yang, Y.; Olomi, C.; Zhu, L. Numerical study on the winter thermal performance and energy saving potential of thermo-activated PCM composite wall in existing buildings. *Build. Simul.* **2020**, *13*, 237–256. [CrossRef]
8. Dharmasastha, K.; Samuel, D.; Nagendra, S.; Maiya, M.P. Thermal comfort of a radiant cooling system in glass fiber reinforced gypsum roof-An experimental study. *Appl. Therm. Eng.* **2022**, *214*, 118842. [CrossRef]
9. Jiang, S.; Li, X.; Lyu, W.; Wang, B.; Shi, W. Numerical investigation of the energy efficiency of a serial pipe-embedded external wall system considering water temperature changes in the pipeline. *J. Build. Eng.* **2020**, *31*, 101435. [CrossRef]
10. Yang, Y.; Chen, S.; Zhang, J. A comprehensive study on transient thermal behaviors and performances of the modular pipe-embedded energy wall system under intermittent operation conditions. *Energy* **2023**, *280*, 128083. [CrossRef]
11. Karanafti, A.; Theodosiou, T.; Tsikaloudaki, K. Assessment of buildings' dynamic thermal insulation technologies—A review. *Appl. Energy* **2022**, *326*, 119985. [CrossRef]
12. Rajput, M.; Reisi Gahrooei, M.; Augenbroe, G. A statistical model of the spatial variability of weather for use in building simulation practice. *Build. Environ.* **2021**, *206*, 108331. [CrossRef]
13. Wu, H.; Qiu, D.; Zhang, L.; Sun, M. Adaptive multi-agent reinforcement learning for flexible resource management in a virtual power plant with dynamic participating multi-energy buildings. *Appl. Energy* **2024**, *374*, 123998. [CrossRef]
14. Zhao, X.; Zhang, Z.; Li, W. Performance analysis of TABS in retrofitting buildings in cold and arid regions. *J. Build. Perform.* **2021**, *12*, 1402. [CrossRef]
15. Zhao, Y.; Li, X. Energy efficiency and economic performance of a low-temperature heating system combining double-layer pipe-embedded wall and ground source heat pump. *Renew. Energy* **2025**, *239*, 122087. [CrossRef]
16. Chen, Y.; Sun, Y.; Yang, J.; Liu, Y.; Gao, D.-C. Demand response with PCM-based pipe-embedded wall in commercial buildings: Combined passive and active energy storage in envelopes. *Energy* **2024**, *308*, 132980. [CrossRef]
17. Arroyo, J.; Spiessens, F.; Helsen, L. Comparison of Model Complexities in Optimal Control Tested in a Real Thermally Activated Building System. *Buildings* **2022**, *12*, 539. [CrossRef]
18. Yang, Y.; Chen, S.; Chang, T.X.; Ma, J.; Sun, Y. Uncertainty and global sensitivity analysis on thermal performances of pipe embedded building envelope in the heating season. *Energy Convers. Manag.* **2021**, *244*, 114509. [CrossRef]
19. Jung, S.; Yoon, Y.; Im, P.; Salonvaara, M.; Dong, J.; Cui, B.; Lapsa, M. Peak cooling load shift capability of a thermal energy storage system integrated with an active insulation system in US climate zones. *Energy Build.* **2022**, *277*, 112484. [CrossRef]
20. Atkins, C.; Hun, D.; Im, P.; Post, B.; Slattery, B.; Iffa, E.; Cui, B.; Dong, J.; Barnes, A.; Vaughan, J.; et al. Empower Wall: Active insulation system leveraging additive manufacturing and model predictive control. *Energy Convers. Manag.* **2022**, *266*, 115823. [CrossRef]
21. Hwang, Y.; Wasti, A.; Borca-Tasciuc, T.; Shultz, J.; Vanfretti, L.; Tsamis, A. Modeling, Simulation and evaluation of a double-sided hydronic layer embedded opaque climate-adaptive building envelope. In Proceedings of the Building Performance Analysis Conference and SimBuild, Chicago, IL, USA, 14–16 September 2022. [CrossRef]
22. *GB/T 50378-2024*; Green Building Evaluation Standard. Architecture & Building Press: Beijing, China, 2024. (In Chinese)
23. Zhu, L.; Yang, Y.; Chen, S.; Sun, Y. Numerical study on the thermal performance of lightweight temporary building integrated with phase change materials. *Appl. Therm. Eng.* **2018**, *138*, 35–47. [CrossRef]
24. *DB13(J)185-2020*; Energy-Saving Design Standards for Residential Buildings (Energy Saving 75 %). China Architecture & Building Press: Beijing, China, 2020. (In Chinese)
25. Yan, Y.; Chen, S. Global sensitivity analysis on thermal performances of pipe-embedded thermo-activated composite wall. *Acta Energiae Solaris Sin.* **2023**, *44*, 248–255. (In Chinese)

buildings

MDPI

Article

Research on the Factors Influencing Broken-Bridge Aluminum Exterior Windows with the Goal of Thermal Performance Improvement

Feining Yang [1], Yu Sun [1], Lei Li [1,*], Yu Zhang [1], Lingyun Sun [1], Dong Wang [1], Fengjun Sun [1] and Lin Liu [2]

1 China MCC5 Group Corp. Ltd., Chengdu 610063, China; yangfeining@mcc5.com.cn (F.Y.); sunyu@mcc5.com.cn (Y.S.); zhangyu19981228@163.com (Y.Z.); sunlingyun@mcc5.com.cn (L.S.); wangdong@mcc5.com.cn (D.W.); sunfengjun@mcc5.com.cn (F.S.)
2 School of Civil and Transportation Engineering, Guangdong University of Technology, Guangzhou 510006, China; liulinhit@163.com
* Correspondence: wstone_ll@163.com

Abstract: The enhancement of the thermal performance of transparent envelopes has been demonstrated to be an effective measure for reducing building energy consumption. To achieve enhanced performance in the domain of building windows, considerable economic investments are frequently necessary. The present study focuses on the 70 series of broken-bridge aluminum windows, which are widely used in China. The investigation explores the potential factors affecting the thermal performance of these windows, mainly considering their glass and profiles. The study also explores the influencing characteristics of various factors and their optimal combinations for their combined effect by comprehensively applying the single-variable method, orthogonal test method, and a numerical simulation. The findings of the study indicate a substantial reduction in the heat transfer coefficient of the 70 series of broken-bridge aluminum windows, achieving a decrease of 25.98%. Additionally, the heat transfer coefficient of the frame and edge is reduced by 29–35%, and the heat transfer coefficient of the entire window is reduced by 28.34%. These findings suggest that the energy-saving requirements for green and low-carbon buildings in most regions of China can be met. Furthermore, the product offers a significant economic advantage over similar market offerings with an equivalent performance. Concurrently, the performance enhancement measures derived from the study are applicable to other types of aluminum windows.

Keywords: broken-bridge aluminum alloy exterior windows; thermal performance; heat transfer coefficient; performance improvement; influencing factors

1. Introduction

A global consensus has emerged in support of reducing energy consumption. According to the U.S. Energy Information Administration (EIA), the world's energy consumption is predicted to increase by approximately 50% by the year 2050 [1]. It is widely acknowledged that the construction, industry, and transportation sectors are of paramount importance in terms of global energy consumption. When the entire life cycle of buildings is taken into consideration, it is estimated that the building sector accounts for approximately 37% of global energy consumption [2]. Projections indicate that the global building area will expand by 75% by the year 2050 [3]. The imperative to curtail building energy consumption has garnered global attention, a pivotal strategy for which is the implementation of green and energy-efficient building technology [4]. Among these approaches, passive building design has emerged as a critical facet of green energy-saving building technology, leveraging the natural environment to ensure indoor environmental comfort and curtail energy consumption [5]. As indicated by the extant literature, envelope energy consumption is the most significant component of a building's energy consumption, given its direct impact on safeguarding a building from external environmental influences [6]. It has been determined that, among the four major enclosure structures, external doors and windows, external walls, roofs, and floors, the energy consumption of door and window systems is approximately four times that of wall systems, five times that of roof systems, and twenty times that of floor systems. This accounts for approximately 40% to 50% of the total energy consumption of a building enclosure [6,7]. As the surface area share of windows and doors in an envelope system increases, their energy consumption will increase accordingly.

Presently, broken-bridge aluminum exterior windows constitute the primary products in energy-saving building envelopes. The China Windows and Doors Industry Report 2024 indicated a considerable surge in the demand for such windows and doors in recent years, which has exhibited a consistent annual growth rate of over 20% and currently holds the dominant market share. However, the industry is confronted with two significant challenges: First, the implementation of the Chinese regulation GB 55015-2021, also known as the "General Specification for Building Energy Saving and Renewable Energy Utilization", has led to a significant upgrade to the country's energy-saving requirements. This has created a challenge for the thermal performance of conventional broken-bridge aluminum exterior windows to meet these new and elevated standards, particularly with regard to energy savings. Additionally, there are issues related to condensation on the interior surfaces of these windows and their poor acoustic performance. Secondly, the economic burden associated with high-performance broken-bridge aluminum exterior windows and doors is often prohibitive. In general, the cost of high-performance aluminum window products is two to three times higher than that of conventional aluminum windows. Additionally, their scope of application is constrained by financial limitations. Consequently, the focal point of research endeavors concerning door and window systems should center on enhancing the thermal performance of conventional aluminum alloy doors and windows, with the objective of aligning them with emerging energy-saving requirements and ensuring their suitability for large-scale use.

Exterior window systems consist of glass and profiles. The surface area percentage of glass is significantly higher than that of profiles, a finding that has garnered attention in research circles. S. Grynning et al. highlighted that glass accounts for a substantial portion of a building's energy consumption, underscoring the significance of glass optimization as a pivotal strategy for enhancing the thermal performance of exterior windows [7]. Presently, the field of research on glass is concentrated on the development of novel glass types, including vacuum glass, silica aerogel glass, thermochromic glass, and electrochromic glass. Vacuum glass has been demonstrated to exhibit minimal heat loss and high visible-light transmission in thin window products [8]; it can be combined with photovoltaic glass to form photovoltaic-combined vacuum glazing [9–11], and with water-flow windows to form vacuum-glazed insulated water-flow windows [12]. The incorporation of silicon dioxide aerogel glass into the interglazing results in the enhanced thermal performance of the glass due to the presence of the highly insulating aerogel [13,14]. Thermochromic/electrochromic glass is a material that has been developed to regulate light and heat. This is achieved by overlaying thermo-responsive and electrochromic materials between the glass layers [15]. Thermo-responsive materials rely on the thermal reaction of the material itself [16], while electrochromic materials undergo a color change through a redox reaction when an external alternating voltage is applied [17,18]. Moreover, experimental evidence has demonstrated that the incorporation of advanced spacers and edge seals can lead to substantial enhancements in the thermal performance of the entire window system [19]. In comparison with metal spacers, composite spacers have been shown to exhibit superior thermal performance and demonstrate a greater application potential [20]. Given that these new glasses have not yet reached production scale and that their application in practice is not common, this study will focus on the performance enhancement potential of double-insulating glass, which is the most widely used type.

Profiles have been identified as a critical weak link in the context of external window systems. Profiles, which encompass a wide range of materials—including wood, aluminum, and polymers—have undergone significant transformations in both their composition and their design [21]. Several studies have indicated the potential for future advancements in exterior window profiles, with particular emphasis on the development of glass fiber-reinforced polyurethane (GFRP) profiles and high-performance bridged aluminum profiles [22]. Presently, glass fiber-reinforced polyurethane (GFRP) profiles are utilized less frequently due to their structural performance, construction maturity, and other concerns [23]. Broken-bridge aluminum profiles have gained popularity due to their lightweight nature, ease of installation, and aesthetic appeal [24]. The present study is an investigation into the enhancement of the thermal performance of broken-bridge aluminum profiles. Previous investigations have identified that the implementation of materials with reduced thermal conductivity, in conjunction with altering the geometry of the profiles, has the potential to achieve a substantial reduction in the heat transfer coefficient [21,25,26]. The measures primarily consist of the following: the utilization of thermal break with reduced thermal conductivity [19], the application of insulation materials within the broken-bridge cavities [21,27], the augmentation of the sectional dimensions of the profiles [28], and the extension length of the thermal break [3]. Furthermore, it has been posited that the implementation of long-tailed rubber strips within the glazing channel [29], the integration of multiple seals between the window frame and the sash [30], and a linear isotherm design [31] could also contribute to the enhancement of the thermal performance of the profiles. The aforementioned conclusions were derived from the control variable method, which was employed to investigate the impact of a single factor on the experimental outcomes. This methodological approach does not encompass a comprehensive analysis from the standpoint of a multifactor implementation. Consequently, these previous studies were not able ascertain the relative influence of each factor on the performance of the profiles. Additionally, they were not able to determine whether the optimal solution was obtained following a comprehensive analysis.

In light of the aforementioned findings, this study aspires to investigate the thermal performance of external windows by focusing on conventional broken-bridge aluminum alloy profiles with double-layer insulating glass. The study aims to identify the various factors that influence the thermal performance and to determine the optimal combination of parameters through numerical simulation, single-variable and orthogonal test methods, and other research methods.

2. Methodology

2.1. Research Design

As demonstrated in Figure 1, the research focuses on broken-bridge aluminum exterior windows, which are currently the most prevalent type of window in China. Secondly, drawing upon the extant research and related practices, the potential factors affecting the thermal performance of external windows are identified and summarized. The influence of each factor on the thermal performance of external windows is then analyzed using the single-variable method and the orthogonal test method, according to the properties of each variable. Finally, a numerical simulation is employed to derive the optimal combination of the multiple variables for the typical thermal performance enhancement of external windows.

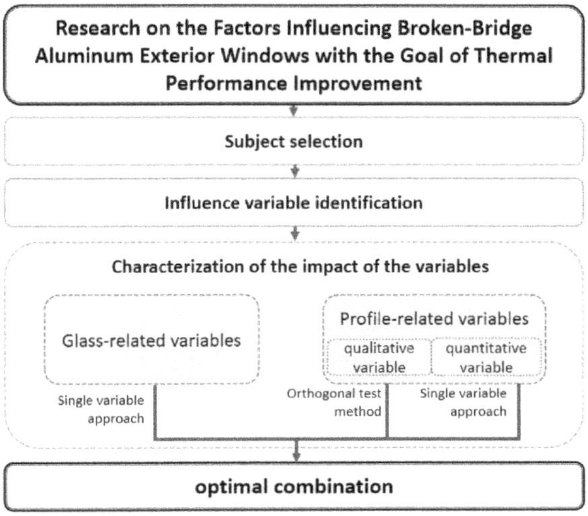

Figure 1. Diagram showing the flow of this research study.

2.2. Subject Selection

At present, there are a considerable number of broken-bridge aluminum alloy window products available, but the node structure of each is similar. To enhance the generalizability of the research outcomes, this study selects the nodes from the China Standard Atlas 22J603-1 [32], entitled "Aluminium alloy windows and doors" (henceforth referred to as "atlas").

The Atlas range of aluminum exterior window models encompasses a spectrum from the 50 series to the 112 series, thereby offering a comprehensive selection that meets diverse architectural needs. According to the economic criteria, this study selects the 70 series of broken-bridge aluminum internal casement windows, with 5 mm (low-E) + 12 A + 5 mm insulating glass, as the typical object (Figure 2). The heat transfer coefficient of an external window is contingent upon the area ratio of the window frame and window opening. An increase in the area ratio of the window frame and window opening results in an increase in the heat transfer coefficient of the entire window (Table 1). According to GB 55015-2021, "General Specification for Energy Saving and Renewable Energy Utilization in Buildings", it is evident that the performance requirements for external windows are exclusively applicable to residential buildings in climates characterized by hot summers and cold winters, hot summers and warm winters, and mild regions, as well as a limited number of public buildings with low window-to-wall ratios [33]. Presently, the architectural landscape is undergoing a transition, with a growing inclination toward buildings with elevated window-to-wall ratios [34–36]. This trend portends a diminution in the viability of these windows.

Table 1. Thermal performance of the 70 series aluminum tilt-and-turn windows [32].

Model Number	Extruded Profile	Glass Configuration	Glass U-Value	Whole Window Heat Transfer Coefficient U-Value		
				Ratio of Area of Window Frames and Openings		
				15%	20%	30%
70 series aluminum tilt-and-turn windows	Heat-insulating aluminum alloy profiles	5mm(low-E) + 12 A + 5 mm	1.75	1.88	1.96	2.14

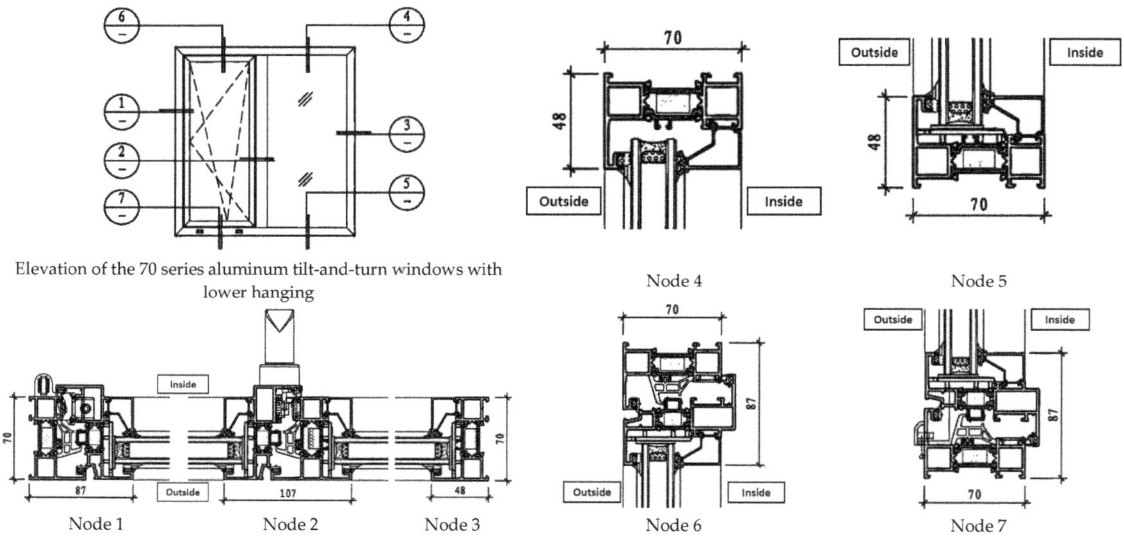

Figure 2. The node details of the 70 series broken-bridge aluminum tilt-and-turn windows [32]. Note: 'Node' means the profile construction of the window frame and glazing.

2.3. Variable Identification

Combined with the existing research and practical experience, this study systematically analyzes the key factors affecting the thermal performance of broken-bridge aluminum exterior windows, focusing on the roles played by glass and aluminum alloy profiles. The primary research variables are elucidated in Table 2 and Figure 3.

In the context of glass, three factors must be considered: the configuration of the glass itself, and the type and thickness of the cavity. The thermal performance of glass is contingent upon its thickness, the number of sheets, the manufacturing process, the type of coating material, the number of layers, and the layers' positions. These factors can even influence the performance of the glass, depending on the manufacturer [37]. In accordance with the principles of economy and control variables, the configuration of the glass itself is considered a fixed variable in this study, and only two factors are considered to change the type and thickness of the cavity. The gas within the cavity is typically air, and the thermal performance of the glass can be enhanced by filling it with inert gases, such as argon or krypton [38]. The thickness of the cavity is typically between 9 mm and 12 mm. To more accurately reflect the effects of different cavity thicknesses on the glass properties, the study will simulate and analyze the three gases, namely air, argon, and krypton, one by one, in varying cavities sized between 6 mm and 20 mm.

The profiles are regarded as the constraining element that hinders the enhancement of the thermal performance of an entire window. This is due to their elevated heat transfer coefficient in comparison to that of glass [19]. The present study will explore potential performance enhancement factors from the perspective of node construction, including the following eight factors: the number of profile cavities, the length of the thermal break, the number of sealing channels between the window frame and sash, the shape of the sealing strips, whether or not to set gaskets under the glass, whether or not to fill the insulation material in the broken-bridge cavities, the isothermal design, and the type of glazing spacer strips. The categorization of factors is predicated on their divergent characteristics, which are then classified as either quantitative or qualitative variables.

The correlation between the quantity of cavities in a given profile and its respective profile model is a matter of established significance. The higher the model, the greater the number of cavities, and thus the superior thermal performance and strength of the entire window. The 70 series of broken-bridge aluminum alloy profiles, selected for the present study, have three cavities, which are considered a fixed variable in accordance with the economic factors.

Table 2. Factors potentially influencing the thermal performance of broken-bridge aluminum exterior windows.

Typology	Potential Factor	Variable Classification	Specific Measures	Whether to Study
Fiberglass (Figure 3a)	Configuration of the glass itself	Fixed variable	China South Glass 5 mm (low-e) + 5 mm insulating glass	No
	Cavity gas types	Qualitative variable	Air/argon/krypton	Yes
	Cavity thickness	Quantitative variable	6~20 mm one-by-one simulation	Yes
Extruded profile	Number of profile cavities	Fixed variable	70 Series 3-Cavity	No
	Length of the thermal break (Figure 3b)	Quantitative variable	25 mm, 28 mm, 30 mm, 32 mm, 35 mm, 40 mm	Yes
	Number of sealing channels between window frame and window sash (Figure 3c)	Qualitative variable	Two seals/three seals	Yes
	Shape of sealing tape (Figure 3d)	Qualitative variable	Tower tape/Long-tail tape	Yes
	Whether or not to set the gasket under the glass (Figure 3e)	Qualitative variable	Set shim/no shim	Yes

Table 2. *Cont.*

Typology	Potential Factor	Variable Classification	Specific Measures	Whether to Study
Extruded profile	Whether the cavity is filled with insulation material (Figure 3f)	Qualitative variable	Filled insulation/unfilled insulation	Yes
	Isothermal design (Figure 3g)	Qualitative variable	Linear isotherm design/non-linear isotherm design	Yes
	Spacer types (Figure 3h)	Qualitative variable	Aluminum spacer/composite spacer	Yes

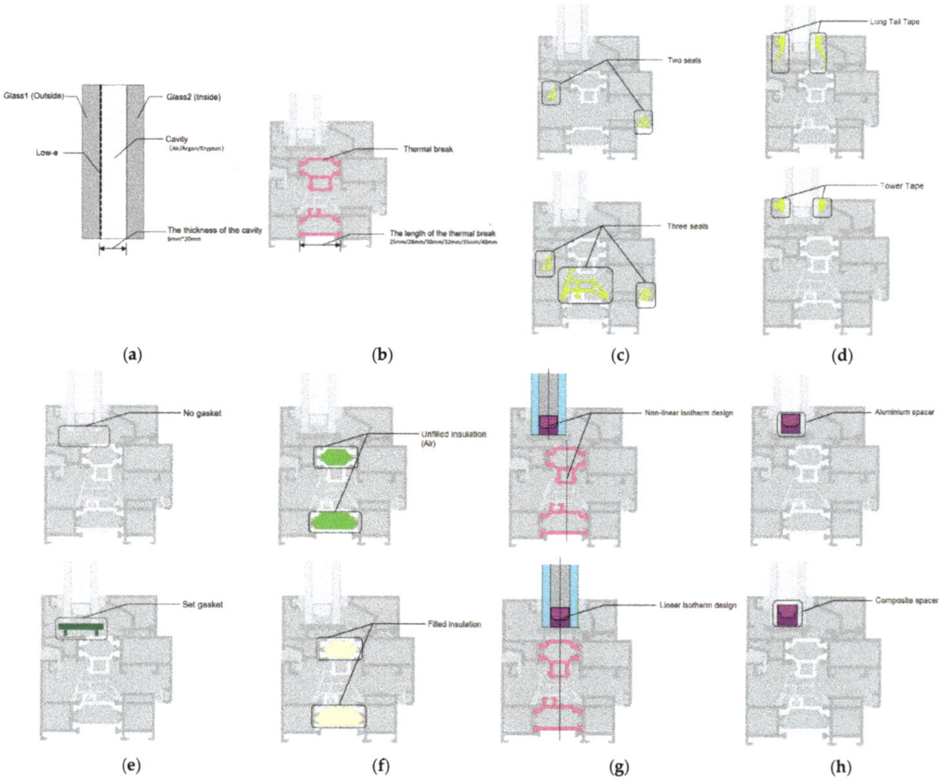

Figure 3. Graphical representation of the potential influencing factors on the thermal performance of broken-bridge aluminum exterior windows. (**a**) Double glazing; (**b**) thermal break; (**c**) number of sealing channels between the window frame and window sash; (**d**) the shape of the sealing tape; (**e**) whether or not to set the gasket under the glass; (**f**) whether the cavity is filled with insulation material; (**g**) isothermal design; (**h**) spacer types.

The material and length of the thermal break, as well as the composition of the broken-bridge aluminum alloy profile, are the primary factors that influence the outcome. Regarding the material, PA66 compounded with 25% glass fiber, which has been demonstrated to exhibit superior strength and performance, is selected as a fixed variable in the study. Concerning the variable of length, a total of six sizes are selected from the conventional range of 25 mm to 40 mm. These sizes, designated as 25 mm, 28 mm, 30 mm, 32 mm, 35 mm, and 40 mm, are employed as the quantitative variables in the subsequent analysis.

The sealing treatment between the window frame and the window sash incorporates two or three seals. The sealing tape between the glass and the profile is either tower shaped or long-tail shaped. The junction between the glass and the profile features either a set gasket or no gasket treatment. The broken-bridge cavity is filled with a heat preservation material and has not undergone a filling treatment. The isothermal design incorporates both linear and non-linear isothermal elements. The type of glazing spacer strips features an aluminum spacer and composite spacer. The aforementioned six factors are used as the qualitative variables according to their characteristics.

2.4. Methods of Analysis

2.4.1. Numerical Simulation

(1) Simulation Software Selection

At the present time, the widest range of applications and higher-maturity thermal performance calculation software have mainly been developed in the United States by the Lawrence Berkeley National Laboratory (LBNL) and their Window7, Therm7, and Optics7 series of software, and in Guangdong by the Guangdong Construction MQMC2010, with the title "building curtain wall, doors and windows thermal performance calculation software".

It is evident that Therm and MQMC both utilize two-dimensional finite element analysis methods to calculate the thermal performance of a frame [39,40]. The distinction between the two approaches lies in the methodology employed for the heat transfer calculation within the glass inlay area. Specifically, MQMC makes use of the line heat transfer coefficient calculation theory [41], while Therm employs the glass edge region calculation theory [42]. For the calculation of the frame heat transfer based on the glass edge region calculation theory, the glass edge (within 63.5 mm) region is designated as a special region, and the three-dimensional heat transfer influence of the glass inlay site is taken into account [43]. In the context of this study on the thermal performance of exterior windows constructed with glass as the primary material, Therm demonstrates enhanced applicability. Furthermore, the series of software developed by LBNL is characterized by its open-source nature, which facilitates wider utilization within academic research communities. Consequently, Window and Therm are selected as the primary simulation software for this study. As Therm primarily simulates steady-state two-dimensional heat transfer effects, the potential impact of the temperature gradients in corner regions and the three-dimensional geometry on the simulation outcomes is not addressed in this study.

(2) Calculation of environmental boundary conditions

JGJ/T151-2008 [44], entitled "Thermal Calculation Regulations for Building Doors, Windows and Glass Curtain Walls," stipulates that, in the design or evaluation of the thermal performance of stereotyped products within the categories of building doors, windows, and curtain walls, the standard calculation conditions must be utilized. In the execution of the actual engineering design, the boundary conditions for a project's calculation must be determined in accordance with the relevant design requirements. Moreover, the standard stipulates that the heat transfer coefficients of the windows, doors, and curtain wall products should be calculated using winter conditions, and that the shading coefficients of the windows, doors, and curtain wall products should be calculated using summer conditions [40,44]. The present study is principally concerned with the evaluation of exterior window products and a discussion of the influence of each factor on the heat transfer coefficient. For the purpose of this study, the following standard winter calculation conditions are selected:

Indoor air temperature (Tin) = 20 °C;
Outdoor air temperature (Tout) = −20 °C;
Indoor convective heat transfer coefficient (hc,in) = 3.6 W/(m^2·K);
Outdoor convective heat transfer coefficient (hc,out) = 16 W/(m^2·K);
Average indoor radiant temperature ($T_{rm,in}$) = T_{in};
Outdoor average radiation temperature ($T_{rm,out}$) = T_{out};
Solar irradiance (I_s) = 300 W/m^2;
Indoor and outdoor relative humidity (Relative Humidity) = 50%.

(3) Analog Material Selection

The material parameters associated with the glass, gas, and profiles used in the study are enumerated in Table 3.

Table 3. Parameter list for material selection.

Typology	Material Selection	Parameters		
Glass	Outer piece of glass (Glass1): CEV11-55T_5.CSG	Conductivity	Emissivity1	Emissivity2
		1.00 W/m·K	0.84	0.05
	Inner piece of glass (Glass2): Clear_5.CSG	Conductivity	Emissivity1	Emissivity2
		1.00 W/m·K	0.84	0.84
Gas	Air: 100% air	Conductivity		
		0.024069 W/m·K		
	Argon: 85% Ar mixed with 15% air	Conductivity		
		0.017425 W/m·K		
	Kryptonite: 95% krypton mixed with 5% air	Conductivity		
		0.009237 W/m·K		
Profile	Aluminum Aluminum alloys (anodized)	Conductivity	Emissivity	
		160.00 W/m·K	0.8	
	Sealing strip: Ethylene propylene diene monomer (EPDM)	Conductivity	Emissivity	
		0.25 W/m·K	0.9	
	Under-glass gasket: Polyvinylchloride (PVC) flexible, with 40% softer	Conductivity	Emissivity	
		0.14 W/m·K	0.9	
	Thermal break: Polyamide 6.6 with 25% glass fiber	Conductivity	Emissivity	
		0.30 W/m·K	0.9	
	Insulation in broken-bridge cavity: Polyurethane foam	Conductivity	Emissivity	
		0.05 W/m·K	0.9	

(4) Calculation of thermal performance of the whole window

The calculation of the thermal performance of the whole window is performed by utilizing the formula outlined within the NFRC standard system for determining the heat transfer coefficient for complete windows (Equation (1)) [45]:

$$U_t = \frac{\sum A_g U_g + \sum A_f U_f + \sum A_{eg} U_{eg}}{A_t} \tag{1}$$

where U_t is the heat transfer coefficient of the whole window [W//(m^2·K)]; A_g is the area of the central region of the window glass (m^2); U_g is the heat transfer coefficient in the middle region of the window glass [W/(m^2·K)]; A_f is the window frame area (m^2); U_f is the heat transfer coefficient of the window frame [W/(m^2·K)]; A_{eg} is the area of the window glass edge area (m^2); U_{eg} is the heat transfer coefficient in the edge region of the window glass [W/(m^2·K)]; and A_t is the area of the whole window (m^2).

Among these, the range of the central window glass area, the window glass edge area, and the window frame area are shown in Figure 4.

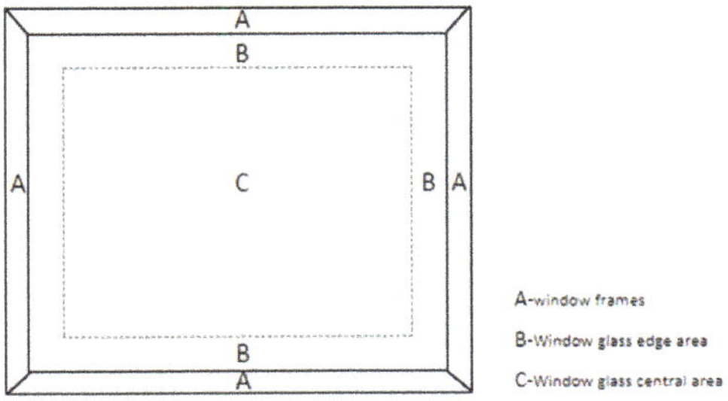

Figure 4. Computational model of the statistical method for glass edges.

(5) Validation of numerical simulation methods

The "Window" software is utilized to simulate the 5 mm (low-e) + 12 A + 5 mm insulating glass, with the glass manufacturer set as China South Glass. The simulation results indicate that the heat transfer coefficient of the glass is 1.6978 W/(m^2·K), which is 0.0522 W/(m^2·K) lower than the heat transfer coefficient of the glass shown in the atlas, at 1.75 W/(m^2·K), with an error rate of 2.98%.

According to the atlas, Therm software is utilized to establish the node models of typical objects (i.e., the 70 series of broken-bridge aluminum internal casement down-hung windows). The simulation is executed subsequent to the importation of the 5 mm (low-e) + 12 A + 5 mm insulating glass (Figure 5), and the simulation results are exhibited in Table 4.

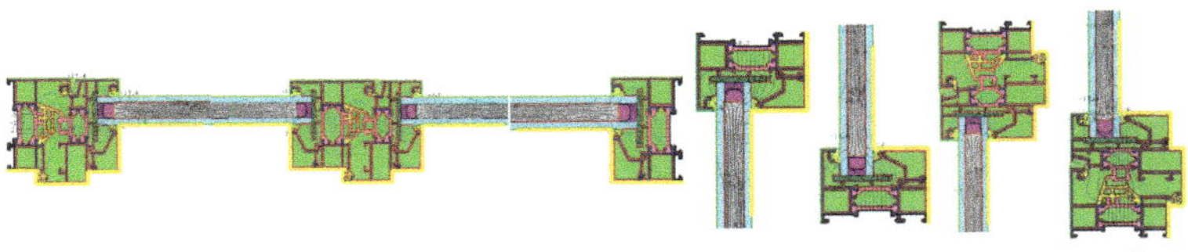

Node 1 Node 2 Node 3 Node 4 Node 5 Node 6 Node 7

Figure 5. Simulation of each node of a typical object.

The following example will illustrate this calculation: For a 2.0 m × 2.0 m window, the area of the window frame that is covered by the hole in the frame is approximately 20%, according to Equation (1). The heat transfer coefficient of the entire window is thus calculated to be 2.050 W/(m^2·K). A comparison of the aforementioned coefficient with the coefficient presented in the atlas, 1.96 W/(m^2·K), reveals a discrepancy of 0.09 W/(m^2·K), corresponding to an error rate of 4.5%.

In summary, the simulation results of the glass thermal performance and calculated whole window thermal performance, when compared with the atlas indices, exhibit an error rate of less than 5%. This is in alignment with the stipulation in the "Guidelines for Labeling the Energy-saving Performance of Doors and Windows in Buildings" that a maximum deviation of 5% is allowed between the test value and calculation value [46]. This demonstrates that the numerical simulation and calculation methods are both reasonable and feasible.

Table 4. Simulation of the thermal performance of each node of the 70 series broken-bridge aluminum tilt-and-turn window.

Region	Heat Transfer Coefficient of the Frame U_f	Heat Transfer Coefficient at the Edges U_{eg}	SHGC
Node 1	3.5050	2.1522	3.4916
Node 2	3.9583	2.1669	3.9967
Node 3	3.7038	2.1610	3.7907
Node 4	3.6671	2.1583	3.7519
Node 5	3.6664	2.1583	3.7511
Node 6	3.4722	2.1503	3.4553
Node 7	3.4739	2.1497	3.4560

2.4.2. Statistical Analysis

In order to achieve a comprehensive understanding of the factors influencing the thermal performance of broken-bridge aluminum windows, a rigorous investigative approach is undertaken. This investigation involves the meticulous examination of nine distinct factors, encompassing both the glass and profile properties. Each factor is studied in various modes, ensuring that the multifaceted nature of the factors is thoroughly explored.

(1) Single-variable approach

The univariate approach constitutes the examination of a single factor and the investigation of the impact of the independent variable on the dependent variable, whilst ensuring that only a single variable is altered at any given moment and that all other variables remain constant throughout the process; this enables the determination of the causal relationship between a single variable and the outcome [47].

The present investigation adopts the single-variable method to analyze the gas types and thicknesses in the cavity of the insulating glass. The investigation encompasses the impact of diverse thicknesses, ranging from 6 mm to 20 mm, on the performance of the glass when the cavity is filled with air, argon, and krypton, respectively. The optimal performance gas type and thickness are determined through longitudinal and transversal comparative analyses of the simulation results.

(2) Orthogonal test method

The orthogonal test method constitutes a design approach employed in the context of investigating multifactor and multilevel systems. Utilizing an orthogonal table to arrange the tests enables the systematic study of the effects that multiple factors and their interactions have on the experimental results. This can be achieved with a reduced number of tests [48]. A hallmark of this approach is the substitution of comprehensive tests with partial tests, a strategy that markedly reduces the number of tests required while concomitantly enhancing the efficiency of the process. The experimental results can be analyzed using variance and extreme variance analysis to clarify the trend of the influence of each factor on the indexes, the primary and secondary relationship between the factors, and the optimal combination [49].

It was hypothesized that there may be an interaction between the seven factors affecting the thermal performance of the profiles from the point of view of the nodal configuration. The length of the thermal break is known to affect the filling range of the insulation material in the bridging cavity. The other factors are weakly correlated, and the possibility of their interaction is small.

In the experimental investigation designed to ascertain the interaction between the length of the thermal break and the presence or absence of thermal insulation material within the bridging cavity, three sizes of cavity (25 mm, 32 mm, and 40 mm) were selected for the heat insulation strip; another variable was the presence or absence of thermal insulation material within the bridging cavity. The results of the experimental investigation are presented in Figures 6 and 7.

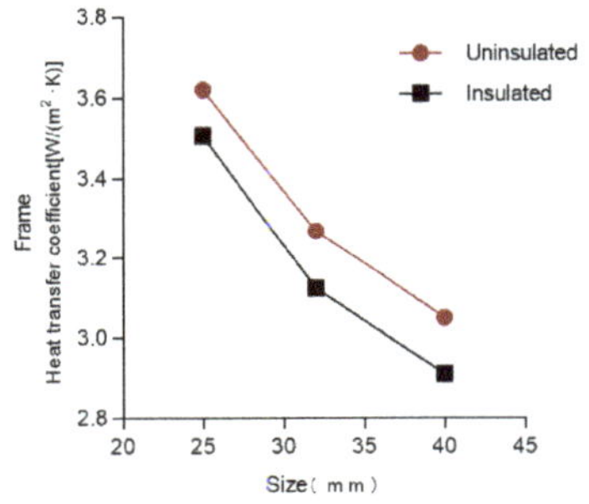

Figure 6. Simulation results of the thermal performance of frames with different lengths of thermal break (25 mm/32 mm/40 mm) and with or without the presence of filling insulation in the broken-bridge cavity.

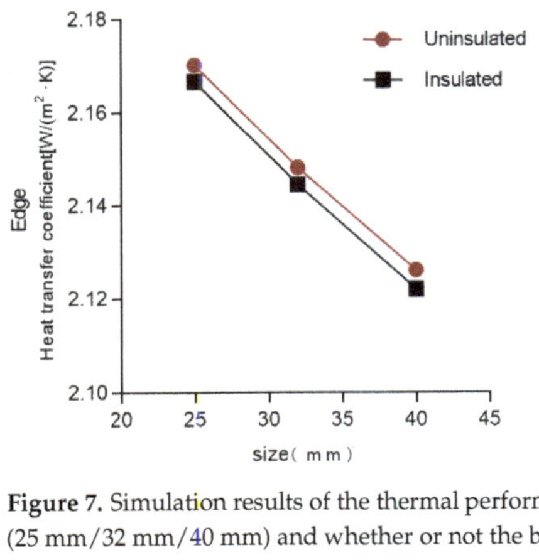

Figure 7. Simulation results of the thermal performance of edges with different insulating strip sizes (25 mm/32 mm/40 mm) and whether or not the bridging cavity was filled with insulation material.

The findings reveal a parallel linear relationship between the heat transfer coefficients of the frames and edges and the insulation material's presence or absence in the broken-bridge cavity. This observation suggests that there is no interaction between the two factors.

Given the low probability of an interaction between the length of the thermal break and the other factors, and given the large number of thermal break length variables, a single-variable method was used in the study. Within the range of 25–40 mm, six values were selected for investigation. The objective of this investigation was to examine the influence of varying lengths of the thermal break on the thermal performance of the profiles. Additionally, an attempt was made to establish a mathematical model of the heat transfer coefficient of the profiles through a data analysis.

To investigate the effect of the combined effect of multiple factors on the heat transfer coefficient of the profiles, the orthogonal test method was used to examine the other six factors affecting the thermal performance of the profiles. The objective was to determine the optimal combination scheme. The six aforementioned factors included the number of sealing channels between the window frame and the window sash, the shape of the sealing tape, whether or not to set gaskets under the glass, whether or not to fill the insulation material in the broken-bridge cavity, the isothermal design, and the type of glazing spacer used.

3. Results and Analysis

3.1. Characterization of the Impact of Each Variable

3.1.1. Characterization of the Impact of the Glass-Related Variables

Utilizing 5 mm (low-e) + 5 mm insulating glass as the constant entity, Window7.8 software was employed to simulate the filling of the cavity with various gas types and thicknesses, as illustrated in Figure 8. As is apparent from the figure, the different gas types and thicknesses had significant impacts on the thermal performance of the glass, which predominantly manifested in three aspects: Firstly, for the configuration of double-glazed glass composed of 5 mm (low-e) + 5 mm layers, there was no consistent decline in the heat transfer coefficient of the glass with an increase in the width of the interlayer between the gases, regardless of the gas type, either air, argon, or krypton. The figure indicates an initial reduction, a subsequent rise, and eventual stabilization. Secondly, for a specific hollow glass cavity, the thermal performance of the glass varied according to the different gas fillings used; in this case, the order of the heat transfer coefficient was air > argon > krypton. Thirdly, within the range of 6–20 mm, the three gases were observed to demonstrate optimal performance values. The optimal thickness of the air filling was determined to be 12 mm, with a heat transfer coefficient of 1.6978 W/(m^2·K); the optimal thickness of the argon filling was determined to be 12 mm, with a heat transfer coefficient of 1.446 W/(m^2·K); and the optimal thickness of the krypton filling was determined to be 8 mm, with a heat transfer coefficient of 1.2567 W/(m^2·K).

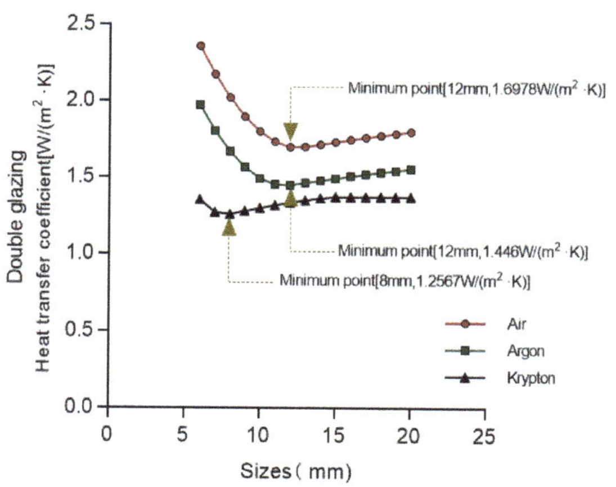

Figure 8. Thermal performance simulation results for insulating glass using different gas types and thicknesses.

In the scenarios involving filling with distinct gases, the variations in the thermal performance of the insulating glass could be predominantly attributed to the differing thermal inertia exhibited by these gases. Concerning the radiative heat transfer within a cavity, the superior thermal inertia of a given gas results in reduced heat transfer. The thermal inertia of a material is determined by its thermal conductivity; therefore, materials with low thermal conductivity exhibit superior thermal inertia. This ranking indicates that, following the gas filling, the thermal performance of the entire window is ranked as krypton > argon > air.

3.1.2. Characterization of the Impact of the Profile-Related Variables

(1) Influence of the length of the thermal break on the characteristics

The research object selected for analysis was the 70 series of broken-bridge aluminum interior casement down-hung windows, which have a total of seven nodes (see Figure 2). Node 7 was selected as the base node. Therm7 was employed to simulate the thermal performance of the nodes with thermal break lengths of 25 mm, 28 mm, 30 mm, 32 mm, 35 mm, and 40 mm, respectively; the simulation results are shown in Figure 9.

As illustrated in Figure 9, the length of the thermal break exerts a more substantial influence on the thermal performance of the profile nodes, particularly in the following two respects: Firstly, within the range of 25 mm to 40 mm, the impact of the thermal break length on the heat transfer coefficient of the frame surpassed that on the edge of the heat transfer coefficient. Secondly, with an increase in the length of the thermal break, both the heat transfer coefficients of the frame and the edge underwent a reduction, exhibiting a linear relationship. The primary reason for this is the superior thermal performance of the thermal break material in comparison to that of aluminum alloy. A composite strip composed of a 25% glass fiber polyamide 66 (PA66) blend exhibited a thermal conductivity of 0.30 W/(m·K). This value is equivalent to 1/530 of that observed in aluminum alloys, thereby significantly diminishing the rate of heat conduction.

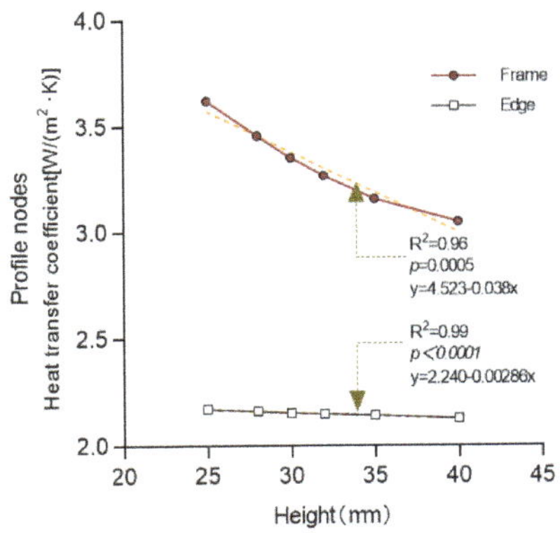

Figure 9. Thermal performance simulation results for different lengths of thermal break in broken-bridge aluminum alloy profiles.

The linear fitting of the data using GraphPad Prism10.0 resulted in model equations (Equations (2) and (3)) for the insulating strip length for the frame and edge heat transfer coefficients. The *p*-values of the two equations were 0.005 and <0.001, respectively. Furthermore, the R2 values were 0.96 and 0.99, indicating a strong correlation between the dependent and independent variables in both cases. The two models under consideration were well constructed, and the equations explained the relationship between the independent and dependent variables satisfactorily.

$$y = 4.523 - 0.038x \tag{2}$$

$$y' = 2.24 - 0.00286x \tag{3}$$

In the equations, y is the heat transfer coefficient of the frame, y′ is the heat transfer coefficient in the edge region, and x is the length of the thermal break.

In practice, the aforementioned two formulas can be used to swiftly ascertain the thermal performance of various lengths of thermal breaks of the 70 series of broken-bridge aluminum exterior window profiles under conventional construction.

(2) Characterization of the impact of the other variables

The six factors other than the length of the thermal break, each of which contained two levels, were selected and appropriately altered in an orthogonal table of type $L_{12}(2^{11})$, resulting in an orthogonal table of type $L_{12}(2^{6})$, whose orthogonality was unaffected. Given that the interaction between the factors was not considered, the factors were only entered into the table in the position corresponding to the column number at the top of the chosen orthogonal table, and one factor occupies one column. The experiment thus yielded 12 groups of design schemes. The results of this experiment are displayed in Table 5, alongside the design scheme for each group and the simulation.

The frame and edge heat transfer coefficients are pivotal parameters in the calculation of the heat transfer coefficient of the entire window. This study proposes an integrated heat transfer coefficient (U_z) of the profile based on the logic of calculating the heat transfer coefficient of the whole window. This process determines the significance of the influence of each factor on the heat transfer coefficient of the profile. The integrated heat transfer coefficient (U_z) is defined as the heat transferred per unit of time through a specific profile or edge area, given a unit area. In steady-state conditions, when the air on both sides of the profile or edge area is at a constant temperature difference, the coefficient can be calculated using Equation (4), as follows:

$$U_z = \frac{\sum A_f U_f + \sum A_{eg} U_{eg}}{A_f + A_{eg}} \tag{4}$$

where U_z is the integrated heat transfer coefficient of the profile [W/(m^2·K)]; A_f is the window frame area (m^2); U_f is the heat transfer coefficient of the window frame [W/(m^2·K)]; A_{eg} is the area of the window glass edge area (m^2); and U_{eg} is the heat transfer coefficient in the edge region of the window glass [W/(m^2·K)].

In instances where U_f and U_{eg} can be derived from a Therm simulation, the calculation of U_z necessitates the solution of A_f and A_{eg}. The dimensions of the node window frame are 87 mm, and the window glass edge area is 63.5 mm. In the absence of a specified window type, the lengths of the window frame and the window glass edge (h) are considered to be in approximate agreement. This results in the approximate Equation (5) for the integrated heat transfer coefficient of the node profile, as follows:

$$U_z = \frac{\sum A_f U_f + \sum A_{eg} U_{eg}}{A_f + A_{eg}} = \frac{0.087hU_f + 0.0635hU_{eg}}{0.087h + 0.0635h} = \frac{0.087U_f + 0.0635U_{eg}}{0.1505} \tag{5}$$

Utilizing the aforementioned equation, the integrated heat transfer coefficients of the profiles for the twelve groups of schemes of orthogonal tests can be calculated; the results are displayed in Table 5.

Table 5. Orthogonal test data statistics.

Serial Number	Considerations						Heat Transfer Coefficient of the Frame	Edge Heat Transfer Coefficient	Combined Heat Transfer Coefficient of Profiles
	Spacer Type	Whether the Bridge is Filled with Insulation Material or Not	Number of Sealing Channels	Under-Glass Gasket	Adhesive Strip Shape	Isothermal Design			
1	Aluminum spacer	Insulated	2 channels	Gasketless	Tower shaped	Straight isotherm	4.087	2.1324	3.2623
2	Aluminum spacer	Insulated	3 channels	Gasketed	Tower shaped	Non-linear isotherm	3.3476	2.1488	2.8418
3	Aluminum spacer	Insulated	3 channels	Gasketed	Long tail	Non-linear isotherm	3.3535	2.149	2.8453
4	Aluminum spacer	Uninsulated	2 channels	Gasketed	Long tail	Straight isotherm	4.0234	2.1453	3.231
5	Aluminum spacer	Uninsulated	2 channels	Gasketless	Long tail	Non-linear isotherm	4.0199	2.1673	3.2382
6	Aluminum spacer	Uninsulated	3 channels	Gasketless	Tower shaped	Straight isotherm	3.739	2.1108	3.052
7	Composite spacer	Insulated	2 channels	Gasketed	Long tail	Straight isotherm	3.6025	1.8995	2.884
8	Composite spacer	Insulated	2 channels	Gasketless	Tower shaped	Non-linear isotherm	3.7036	1.9515	2.9643
9	Composite spacer	Insulated	3 channels	Gasketless	Long tail	Straight isotherm	2.9984	1.8667	2.5209
10	Composite spacer	Uninsulated	2 channels	Gasketed	Tower shaped	Non-linear isotherm	3.7066	1.949	2.965
11	Composite spacer	Uninsulated	3 channels	Gasketed	Tower shaped	Straight isotherm	3.1158	1.8707	2.5905
12	Composite spacer	Uninsulated	3 channels	Gasketless	Long tail	Non-linear isotherm	3.1704	1.9119	2.6394

The level of each factor was characterized by the numbers "1" and "2" (Table 6). The orthogonal test data were subsequently subjected to an analysis of variance (ANOVA) using SPSS29.0 (Table 7). The results of the above analysis were used to determine the significance of each factor through the p-value. Furthermore, the F-value was used to provide additional insight into the relative importance of each factor. Table 7 demonstrates that, for the comprehensive heat transfer coefficient of the profiles, the type of spacer and the number of sealing channels are the significant factors; the presence or absence of insulation material in the cavity is a more significant factor; and the presence or absence of a gasket under the glass, the shape of the sealing tape, and the isothermal line design are the non-significant factors. The sequence of significance is as follows: the number of sealing channels > the type of spacer > the presence or absence of insulation material in the cavity.

Table 6. Table of factor levels.

Level	Spacer Type	Whether or Not There Is Insulation in the Broken-Bridge Cavity	Number of Sealing Channels	Under-Glass Gasket	Shape of Sealing Tape	Isothermal Design
1	Aluminum spacer	Insulated	2 channels	Gasketed	Tower shaped	Straight isotherm
2	Composite spacer	Uninsulated	3 channels	Gasketless	Long tail	Non-linear isotherms

Table 7. Analysis of variance.

Term (in a Mathematical Formula)	Square Sum	Degrees of Freedom	Mean Square	F	p	R^2	Adjustment of R^2
Intercept	102.286	1	102.286	46,013.532	0.000 ***	0.984	0.965
Spacer type	0.303	1	0.303	136.258	0.000 ***		
Whether or not there is insulation in the broken-bridge cavity	0.013	1	0.013	5.923	0.059 *		
Number of sealing channels	0.352	1	0.352	158.296	0.000 ***		
Under-glass gasket	0.009	1	0.009	3.827	0.108		
Shape of sealing tape	0.008	1	0.008	3.769	0.11		
Isothermal design	0	1	0	0.082	0.786		

Note: "***" indicates that the factor has a significant impact on the result; "*" indicates that the factor has a relatively significant impact on the result; no mark indicates that the factor has no significant impact on the result.

3.2. *Combination of Variables with the Goal of Thermal Performance Improvement*

3.2.1. Optimal Combination Based on Mathematical Analysis

It is important to note that, although some of the factors were not significant according to the ANOVA, this does not necessarily indicate that they do not affect any of the combined heat transfer coefficients of the profiles; they may simply not reach the level of statistical significance. When determining the optimal combination, it is imperative to consider all factors within the research scope. Deriving the order of influence of each factor on the integrated heat transfer coefficient of the profiles and the optimal combination scheme is facilitated by a polar analysis, and this analysis is supported by the orthogonal test data, the results of which are shown in Table 8 below.

The results of the extreme difference analysis indicate that the optimal level combination is a composite spacer strip + broken-bridge cavity with thermal insulation + three seals + gaskets under the glass + long-tailed tape + non-linear isotherms. This conclusion is supported by a comparison of the K1 avg value with the K2 avg value in the longitudinal direction. It is noteworthy that this combination does not appear in the orthogonal test program. According to the program developed for the establishment of the model and simulation, the node frame heat transfer coefficient is 2.9711 W/m^2·K, the edge area heat transfer coefficient is 1.9035 W/m^2·K, and the calculation of the profile comprehensive heat transfer coefficient is 2.5206 W/m^2·K. A comparison of this combination with the smallest value for the orthogonal test on the nineth group of data (with a node frame heat transfer coefficient of 2.9984 W/m^2·K, an edge area heat transfer coefficient of 1.8635 W/m^2·K, and a profile comprehensive heat transfer coefficient of 2.5209 W/m^2·K) reveals that the latter is almost identical, albeit smaller. This suggests that the combination is optimal.

Table 8. Extreme variance analysis.

Category	Spacer Type	Whether or Not There Is Insulation in the Broken-Bridge Cavity	Number of Sealing Channels	Under-Glass Gasket	Adhesive Strip Shape	Isothermal Design
K1	18.471	17.319	18.545	17.358	17.676	17.541
K2	16.564	17.716	16.49	17.677	17.359	17.494
K1 avg	3.078	2.886	3.091	2.893	2.946	2.923
K2 avg	2.761	2.953	2.748	2.946	2.893	2.916
R	0.318	0.066	0.342	0.053	0.053	0.008
Optimal level of factors	Composite spacer	Insulated	3 seals	Gasketed	Long-tail tape	Non-linear isotherms
Order of priority of factors	Number of sealing channels > spacer types > whether or not the broken-bridge cavity is filled with insulation > under-glass gasket > shape of sealing tape > isothermal design					

In addition, a horizontal comparison of the R-values indicates that the order of influence of each factor on the integrated heat transfer coefficient of the profile is as follows: number of sealing channels > spacer types > whether or not there is insulation in the broken-bridge cavities > under-glass gaskets > shape of sealing tape > isothermal design. This ordering result is consistent with the ordering of the significant factors in the ANOVA, and it is evident from these analyses that the latter three factors, which are non-significant, exert a lesser influence on the comprehensive heat transfer coefficient of profiles.

3.2.2. Performance Simulation of the Optimal Combination of Whole Windows

In order to achieve the best possible performance from a window, it is necessary to consider the factors that influence the performance of the glass and the profile in conjunction. In order to explore the extreme value of the optimal combination on the performance of the whole window within the scope of this study, 5 mm (low-e) + 8 Kr + 5 mm insulating glass, as well as a 40 mm long thermal break, were selected in combination with the six factors mentioned above, specifically 5 mm (low-e) + 8 Kr + 5 mm insulating glass + 40 mm long thermal break + composite spacer strips + broken-bridge cavities with insulation + 3 seals + gaskets + long-tailed tape + non-linear isothermal design.

The nodes of the object under study were optimized according to the optimal combination scheme of the whole window performance (Figure 10). The optimized simulation results were then compared with the simulation results of a typical object (Tables 9 and 10). The findings indicate a 25.98% reduction in the heat transfer coefficient of the glass, with the average reduction rate for the heat transfer coefficient of the frame and the edge of each node of the profile ranging from 29% to 35%.

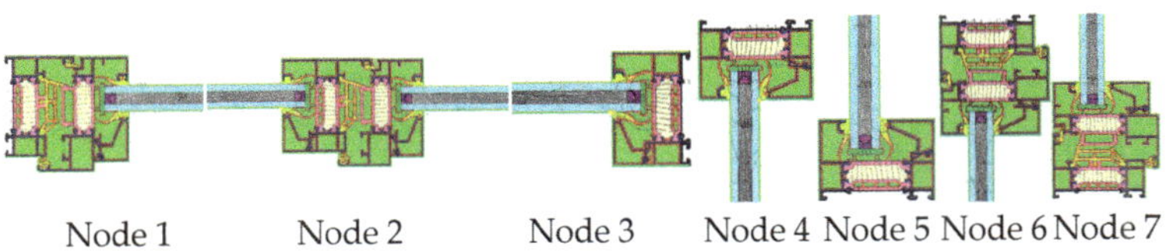

Figure 10. Simulation of each node after optimizing a typical object by the optimal combination.

Table 9. Comparison of glass heat transfer coefficients.

Heat Transfer Coefficient of Glass (5 mm (low-e) + 12 A + 5 mm)	Glass Selection (5 mm (low-e) + 8 Kr + 5 mm) Heat Transfer Coefficient	Reduction Rate
1.6978	1.2567	25.98%

Table 10. Comparison of heat transfer coefficients at each node of profile.

Node	Heat Transfer Coefficients at Each Node of a Typical Object		Heat Transfer Coefficients at Each Node for the Optimal Combination of Whole Window Performance		Peak Reduction Rate (Frame/Edge)
	Heat Transfer Coefficient of the Frame, U_f	Heat Transfer Coefficient at the Edge, U_{eg}	Heat Transfer Coefficient of the Frame, U_f	Heat Transfer Coefficient at the Edge, U_{eg}	
Node 1	3.5050	2.1522	2.4175	1.4992	31.03%/30.34%
Node 2	3.9583	2.1669	2.6076	1.5089	34.12%/30.37%
Node 3	3.7038	2.1610	2.5283	1.5123	31.74%/30.02%
Node 4	3.6671	2.1583	2.5217	1.5141	31.23%/29.85%
Node 5	3.6664	2.1583	2.5218	1.5141	31.22%/29.85%
Node 6	3.4722	2.1503	2.4240	1.4951	30.19%/30.47%
Node 7	3.4739	2.1497	2.4225	1.4945	30.27%/30.48%

To illustrate this point, consider a whole window measuring 2.0 m × 2.0 m. Utilizing Equation (1), the optimal combination of the whole window heat transfer coefficient is calculated to be 1.469 W/(m^2·K). Compare this to the typical simulation of the object, which has a calculated value of 2.05 W/(m^2·K). The reduction rate is therefore 0.581 W/(m^2·K), which is equivalent to a reduction of 28.34%.

The enhancement of the performance of the traditional 70 series aluminum alloy windows not only ensures adherence to elevated standards of energy-saving requirements but also extends the range of its applications. Concurrently, in comparison with the typical target construction node, the optimal combination of the node inside the window frame, the window, and the glass edges does not appear to result in condensation. Furthermore, the sound insulation performance is enhanced by the bridging cavity, which is filled with thermal insulation materials.

3.2.3. Cost-Effectiveness of the Optimal Combination of Whole Windows

Cost-effectiveness is frequently analyzed from the perspective of the whole life cycle, as illustrated by the following examples. The initial cost, payback time, energy cost savings, and total life cycle savings are the primary factors to be considered [50,51]. The present study focused on the improvement of the thermal performance of aluminum with broken-bridge windows. A cost–benefit analysis was carried out from the perspectives of the incremental cost and a price comparison of products with the same performance.

The enhancement of the optimal combination scheme of the entire window over the typical object was achieved through the following adjustments:

- The glass selection was adjusted from 5 mm (low-e) + 12 A + 5 mm to 5 mm (low-e) + 8 Kr + 5.
- The length of the thermal break was adjusted from 25 mm to 40 mm.

- The spacer tape between the glass was adjusted from an aluminum spacer to an composite spacer.
- The broken-bridge cavity was filled with insulation material.
- The shape of the sealing tape between the glass and the profile was adjusted from tower-shaped tape to a long-tailed tape shape.

The financial implications of each factor are delineated in Table 11 below.

Table 11. Typical objects' difference factor market costs from optimal portfolio.

Serial Number	Type of Factor	Specific Factors	Costs (Manufacturing, Production, etc.)
1	Spacer type	Aluminum spacer	About 1.5 RMB/m
		Composite spacer	About 8.5 RMB/m
2	Whether or not there is insulation in the broken-bridge cavity	/	/
		Polyurethane foam	About 1.0 RMB/meter
3	Gas type	Atmosphere	/
		Krypton	About 450 RMB/m^3
4	Length of the thermal break	25 mm	About 2.5 RMB/m
		40 mm	About 7.0 RMB/m
5	Shape of the sealing tape	Tower-shaped tape	About 1.0 RMB/m
		Long-tailed tape	About 1.5 RMB/m

Note: Cost information for each factor was obtained from www.1688.com (accessed on 12 February 2025).

The calculation of the incremental cost of the optimal combination scheme for the more typical object is determined by the cost–price difference of each factor. The result indicates that the incremental cost is approximately 80 RMB/m^2. In this estimation scheme, the relevant cost is ignored due to the uncertainty surrounding labor and production line upgrading. According to the quotes from the relevant manufacturers in China, the market price of the typical object is approximately 500 RMB/m^2. This can serve as the basis for estimating the market price of the optimal combination scheme to be approximately 580 RMB/m^2.

The prevailing value of the common heat transfer coefficient is 1.4 W/(m^2·K) for external window products in the 100 series of aluminum broken-bridge windows with three-glass and two-cavity hollow low-E glass. The market price of these products is approximately 850 RMB/m^2. The optimal combination proposed in this study exhibits a price difference of approximately 270 RMB/m^2. This finding indicates that the study demonstrates superior economic benefits and market application value.

4. Summary and Discussion

The present study investigated the factors influencing the thermal performance of aluminum windows, specifically focusing on their glass and profiles. With regard to the glass, the primary investigation focused on the impact of the insulating glass cavity gas type and the thickness of the glass on the performance, with krypton gas demonstrating superior performance, followed by argon, and air proving to be the least effective. Additionally, the optimal thickness of the different gases was identified to be 12 mm for air and argon, and 8 mm for krypton. The study established that regardless of the type of gas used, the heat transfer coefficient of the glass exhibited an initial decrease, followed by an increase, and subsequently stabilized with an increasing thickness of the cavity.

In terms of the profiles, first, the effect of heat-insulating strips on the performance of the profiles was explored. The study demonstrated that the heat transfer coefficients of both the frames and edges decreased linearly with an increase in the length of the thermal break. Concurrently, a linear relationship was identified and a linear equation was constructed, which enabled the rapid deduction of the thermal performance of conventionally constructed broken-bridge aluminum under different thermal break lengths. Second, the other factors affecting the thermal performance of the profiles were studied using the orthogonal test method. Through an analysis of variance (ANOVA), the type of spacer strip, whether the broken-bridge cavity was filled with insulation material, and the number of sealing channels were found to be the most significant factors. The significance order of these factors was determined to be the number of sealing channels > the type of spacer strip > whether the broken-bridge cavity was filled with insulation material. Whether or not to set the gasket under the glass, the shape of the sealing tape, and the isothermal design were found to be non-significant factors. Finally, the optimal combination of factors was found to be, according to a polar analysis, a composite spacer + broken-bridge cavity with thermal insulation + three seals + under-glass gaskets + long-tailed tape + a non-linear isothermal line.

The optimal combination of factors for the entire window, including the glass and profile, is hereby proposed as follows: 5 mm (low-e) + 8 Kr + 5 mm insulating glass + 40 mm thermal break + composite spacer + broken-bridge cavity with thermal insulation + three seals + under-glass gaskets + long-tailed tape + a non-linear isothermal line. The heat transfer coefficient of the entire window is calculated to be up to 1.469 $W/m^2{\cdot}K$, representing a 28.34% reduction compared to that of a typical object. Concurrently, the cost–benefit analysis indicates that this window exhibits superior performance enhancement with a minimal incremental cost, thereby expanding its application scope. Regarding China specifically, the GB 55015-2021 "General Specification for Energy Efficiency and Renewable Energy Utilization in Buildings" stipulates that these standards apply to all buildings, except those with a high window-to-wall ratio in cold regions. This is particularly evident in the context of numerous low-cost projects, where this product's popularity and application value are amplified; thus, it will play a pivotal role in reducing building energy consumption.

While the present study focused on the 70 series, its findings are also applicable to other models due to the shared structural characteristics of broken-bridge aluminum exterior windows. To illustrate this point, consider the 60 series of aluminum windows. According to the optimal structure, the heat transfer coefficient of the entire window is reduced from 2.12 $W/m^2{\cdot}K$ to 1.568 $W/m^2{\cdot}K$, representing a 26.0% decrease. Moreover, despite the study's proposal of the optimal combination of parameters, the following points must be considered during actual engineering applications: Firstly, the process for extracting krypton gas is intricate and technically challenging. The requirements for the production and operation procedures are exceedingly stringent, and the cost is three times that of argon. However, the application of argon can be reasonably selected according to the budget [52,53]. Secondly, if the sealing performance of the hollow glass cavity is inadequate, krypton and argon gases may escape too quickly, thereby affecting the thermal performance of the entire window [54]. The sealing performance of the cavity is a critical index in the selection process. Thirdly, it is essential to ensure that the length of the thermal break is not increased, particularly for low-series profiles. Additionally, the thermal performance and material strength should be evaluated comprehensively. It has been demonstrated that, particularly at the stress concentration, these elements have the potential to become the weakest link in the structural loading, affecting the overall stability and integrity [53]. Fourthly, it is important to note that the simulation results are subject to variation due to shifts in the environmental boundary conditions or window orientation [55,56]. The results of this study are based on the standard calculation conditions specific to Chinese winters. However, if the simulation is based on the ISO (EN) standard system or the NFRC standard system (NFRC100/NFRC200), the results may vary. Conversely, the conclusions may vary if the window orientation is changed.

In subsequent research, we will collaborate with manufacturers to sample and experimentally test the identified optimal combination. We will then compare and analyze the test data with the simulation results. This will allow us to verify the practical application effect and manufacturing feasibility of the identified optimal combination.

Author Contributions: Conceptualization, F.Y. and L.L. (Lin Liu); methodology, Y.S. and D.W.; validation, F.Y.; formal analysis, L.L. (Lei Li); investigation, Y.Z. and L.S.; resources, F.S. and Y.S.; data curation, Y.Z.; writing—original draft preparation, L.L. (Lei Li); writing—review and editing, F.Y. and Y.S.; visualization, L.L. (Lei Li) and L.S.; supervision, L.L. (Lin Liu); project administration, F.S.; funding acquisition, D.W. All authors have read and agreed to the published version of the manuscript.

Funding: China MCC5 Group Corp. Ltd., China: WY2023B010.

Data Availability Statement: The data presented in this study are available in [Research on the Factors Influencing Broken-Bridge Aluminum Exterior Windows with the Goal of Thermal Performance Improvement]. Your prompt attention to this matter is greatly appreciated, as it helps us maintain the highest standards of research integrity and ensures the reproducibility of your work. If you have any questions or need further clarification, please do not hesitate to contact us.

Conflicts of Interest: Author Feining Yang, Yu Sun, Lei Li, Yu Zhang, Lingyun Sun, Dong Wang and Fengjun Sun are employed by the China MCC5 Group Corp. Ltd. The remaining author declare that the research was conducted in the absence of any commercial or financial relationships that could be construed as a potential conflict of interest.

References

1. U.S. Energy Information Admnistration. *Internaltional Energy Outlook 2019*; U.S. Energy Information Admnistration: Washington, DC, USA, 2019; pp. 25–150.
2. IEA. *Roadmap for Energy-Efficient Buildings and Construction in ASEAN*; IEA: Paris, France, 2022; Available online: https://www.iea.org/reports/roadmap-for-energy-efficient-buildings-and-construction-in-the-association-of-southeast-asian-nations (accessed on 15 May 2025).
3. International Energy Agency. *Net Zero by 2050: A Roadmap for the Global Energy Sector*; International Energy Agency: Paris, France, 2021; p. 224. Available online: https://www.iea.org/reports/net-zero-by-2050 (accessed on 15 May 2025).
4. Qiu, C.; Yang, H.; Zhang, W. Investigation on the energy performance of a novel semi-transparent BIPV system integrated with vacuum glazing. *Build. Simul.* **2019**, *12*, 29–39. [CrossRef]
5. Chel, A.; Kaushik, G. Renewable energy technologies for sustainable development of energy efficient building. *Alex. Eng. J.* **2018**, *57*, 655–669. [CrossRef]
6. Kim, H.; Yeo, M. Thermal Bridge Modeling and a Dynamic Analysis Method Using the Analogy of a Steady-State Thermal Bridge Analysis and System Identification Process for Building Energy Simulation: Methodology and Validation. *Energies* **2020**, *13*, 4422. [CrossRef]
7. Grynning, S.; Gustavsen, A.; Time, B.; Jelle, B.P. Windows in the buildings of tomorrow: Energy losers or energy gainers? *Energy Build.* **2013**, *61*, 185–192. [CrossRef]
8. Cuce, E.; Cuce, P.M. Vacuum glazing for highly insulating windows: Recent developments and future prospects. *Renew. Sustain. Energy Rev.* **2016**, *54*, 1345–1357. [CrossRef]
9. Uddin, M.M.; Jie, J.; Wang, C.; Zhang, C.; Ke, W. A review on photovoltaic combined vacuum glazing: Recent advancement and prospects. *Energy Build.* **2023**, *286*, 112939. [CrossRef]

10. Huang, J.; Chen, X.; Peng, J.; Yang, H. Modelling analyses of the thermal property and heat transfer performance of a novel compositive PV vacuum glazing. *Renew. Energy* **2021**, *163*, 1238–1252. [CrossRef]
11. Pont, U.; Wölzl, M.; Schober, P.; Khosravi, S.N.; Schuss, M.; Mahdavi, A. Recent progress in the development of windows with vacuum glass. *MATEC Web Conf.* **2019**, *282*, 02020. [CrossRef]
12. Lyu, Y.; Zheng, S.; Zhuo, K.; Su, H.; Li, J.; Liu, C. Experimental and numerical performance investigation of a water-flow window with vacuum-glazing insulation. *Sol. Energy* **2022**, *233*, 259–270. [CrossRef]
13. Li, D.; Zhang, C.; Li, Q.; Liu, C.; Arıcı, M.; Wu, Y. Thermal performance evaluation of glass window combining silica aerogels and phase change materials for cold climate of China. *Appl. Therm. Eng.* **2020**, *165*, 114547. [CrossRef]
14. Schultz, J.M.; Jensen, K.I.; Kristiansen, F.H. Super insulating aerogel glazing. *Sol. Energy Mater. Sol. Cells* **2005**, *89*, 275–285. [CrossRef]
15. Wu, S.; Sun, H.; Duan, M.; Mao, H.; Wu, Y.; Zhao, H.; Lin, B. Applications of thermochromic and electrochromic smart windows: Materials to buildings. *Cell Rep. Phys. Sci.* **2023**, *4*, 101370. [CrossRef]
16. Liang, R.; Sun, Y.; Aburas, M.; Wilson, R.; Wu, Y. Evaluation of the thermal and optical performance of thermochromic windows for office buildings in China. *Energy Build.* **2018**, *176*, 216–231. [CrossRef]
17. Piccolo, A. Thermal performance of an electrochromic smart window tested in an environmental test cell. *Energy Build.* **2010**, *42*, 1409–1417. [CrossRef]
18. Bai, T.; Li, W.; Fu, G.; Zhang, Q.; Zhou, K.; Wang, H. Dual-band electrochromic smart windows towards building energy conservation. *Sol. Energy Mater. Sol. Cells* **2023**, *256*, 112320. [CrossRef]
19. Gustavsen, A.; Grynning, S.; Arasteh, D.; Jelle, B.P.; Goudey, H. Key elements of and material performance targets for highly insulating window frames. *Energy Build.* **2011**, *43*, 2583–2594. [CrossRef]
20. Van Den Bergh, S.; Hart, R.; Jelle, B.P.; Gustavsen, A. Window spacers and edge seals in insulating glass units: A state-of-the-art review and future perspectives. *Energy Build.* **2013**, *58*, 263–280. [CrossRef]
21. Paulos, J.; Berardi, U. Optimizing the thermal performance of window frames through aerogel-enhancements. *Appl. Energy* **2020**, *266*, 114776. [CrossRef]
22. Xiaoyan, Y. Analysis of the influence of building exterior window profiles on heat transfer coefficient. *Green Build.* **2022**, *14*, 72–76.
23. Appelfeld, D.; Hansen, C.S.; Svendsen, S. Development of a slim window frame made of glass fibre reinforced polyester. *Energy Build.* **2010**, *42*, 1918–1925. [CrossRef]
24. Pittie, T.; Ahmad, S.; Singh, H.; Rashid, M.M.; Adhikari, S.; Shetty, S.; Krishnan, N.M.A. A combined experimental and numerical analysis framework for the design of aluminum windows. *J. Build. Eng.* **2024**, *82*, 108259. [CrossRef]
25. Simko, T.; Moore, T. Optimal window designs for Australian houses. *Energy Build.* **2021**, *250*, 111300. [CrossRef]
26. Van Den Bossche, N.; Buffel, L.; Janssens, A. Thermal Optimization of Window Frames. *Energy Procedia* **2015**, *78*, 2500–2505. [CrossRef]
27. Chen, H.; Liu, Y.; Zhang, L.; Li, W.; Li, C.; Huang, L.; Jiang, S.; Luo, T. Research on new heat-insulating aluminum alloy profiles for building exterior windows. *Build. Sci.* **2021**, *37*, 170–174+210. [CrossRef]
28. Wan, C.; Wang, H.; Sun, S.; Wang, L. Simulation study on factors affecting thermal performance of aluminum alloy windows. *Constr. Sci. Technol.* **2014**, *1*, 62–65. [CrossRef]
29. Jin, J.; Yu, H. Analysis of key technology of thermal performance of heat-insulating aluminum alloy profiles. *Sichuan Build. Sci. Res.* **2017**, *43*, 132–135.
30. Xi, Y.; Liu, W. Research on heat transfer coefficient of building exterior windows based on different measures. *New Build. Mater.* **2020**, *47*, 20–24+34.
31. Chen, J.; Shen, R. Research on isothermal distribution and heat transfer coefficient of thermal insulation profiles. *Doors Windows* **2011**, *5*, 46–51.
32. *22J603-1*; Aluminum Alloy Windows and Doors. China Institute of Building Standard Design & Research: Beijing, China, 2022.
33. *GB55015-2021*; General Specification for Building Energy Efficiency and Renewable Energy Utilization. International Energy Agency: Paris, France, 2021.
34. Amaral, A.R.; Rodrigues, E.; Gaspar, A.R.; Gomes, Á. A thermal performance parametric study of window type, orientation, size and shadowing effect. *Sustain. Cities Soc.* **2016**, *26*, 456–465. [CrossRef]
35. Troup, L.; Phillips, R.; Eckelman, M.J.; Fannon, D. Effect of window-to-wall ratio on measured energy consumption in US office buildings. *Energy Build.* **2019**, *203*, 109434. [CrossRef]
36. Marino, C.; Nucara, A.; Pietrafesa, M. Does window-to-wall ratio have a significant effect on the energy consumption of buildings? a parametric analysis in Italian climate conditions. *J. Build. Eng.* **2017**, *13*, 169–183. [CrossRef]
37. Teich, M.; Scherer, C.; Schuster, M.; Brandenstein, M.; Elstner, M. Reuse and remanufacturing of insulated glass units. *Glass Struct. Eng.* **2024**, *9*, 339–356. [CrossRef]

38. Mert Cüce, A.P.; Cüce, E.; Şen, H. Numerical analysis of thermal insulation performance of double glazing products using air, argon and krypton. In Proceedings of the II-International Conference on Global Practice of Multidisciplinary Scientific Studies, Batumi, Georgia, 26–28 July 2022; pp. 1480–1486.
39. Mo, Y.; Wang, C.; Kassem, M.A.; Wang, D.; Chen, Z. Optimizing Window Configurations for Energy-Efficient Buildings with Aluminum Alloy Frames and Helium-Filled Insulating Glazing. *Sustainability* **2024**, *16*, 6522. [CrossRef]
40. Chen, Z.; Li, Q.; Zhang, L.; Meng, Q. Thermal properties of Aluminum & PVC co-extrusion sliding window. In Proceedings of the 2011 International Conference on Multimedia Technology, Hangzhou, China, 26–28 July 2011; pp. 5660–5663. [CrossRef]
41. *2006BAJ02B06*; Research on Thermal Calculation Standard System of Chinese and Foreign Building Doors, Windows and Curtain Walls: One of the Special Research Contents of National Eleventh Five-Year Plan "Research and Development of Typical Building Window Systems for Regional Use". China Building Metal Structure: Beijing, China, 2011; Volume 2, pp. 33–38.
42. Curcija, D.C.; Hart, R.G. *Modeling Thermal Performance of VIG: WINDOW and THERM Software Tools*; Proc. Presented at the 1st China (Qingdao) International Glass & Equipment Technology Seminar; Lawrence Berkeley National Laboratory: Alameda County, CA, USA, 2018.
43. Mitchell, R.; Kohler, C.; Curcija, D.; Zhu, L.; Vidanovic, S.; Czarnecki, S.; Arasteh, D.; Carmody, J.; Huizenga, C. *THERM 7/WINDOW 7 NFRC Simulation Manual*; Lawrence Berkeley National Lab. (LBNL): Berkeley, CA, USA, 2017.
44. *JGJ/T151-2008*; Thermal Calculation Regulations for Building Doors, Windows and Glass Curtain Walls. China Construction Industry Press: Beijing, China, 2009.
45. *NFRC 100:2004*; Procedure for Determining Fenestration Product U-Factors. NFRC: Greenbelt, MD, USA, 2004.
46. *RISN-TG013-2012*; Guidelines for Labeling Energy-Saving Performance of Building Windows and Doors. NFRC: Greenbelt, MD, USA, 2012.
47. Mead, R. *The Design of Experiments*; Wikimedia: San Francisco, CA, USA, 1990.
48. Zhu, J.; Chew, D.A.S.; Lv, S.; Wu, W. Optimization method for building envelope design to minimize carbon emissions of building operational energy consumption using orthogonal experimental design (OED). *Habitat Int.* **2013**, *37*, 148–154. [CrossRef]
49. Li, Q.; Hu, H.; Ma, L.; Wang, Z.; Arıcı, M.; Li, D.; Luo, D.; Jia, J.; Jiang, W.; Qi, H. Evaluation of energy-saving retrofits for sunspace of rural residential buildings based on orthogonal experiment and entropy weight method. *Energy Sustain. Dev.* **2022**, *70*, 569–580. [CrossRef]
50. Omle, I.; Kovács, E.; Bolló, B. Applying recent efficient numerical methods for long-term simulations of heat transfer in walls to optimize thermal insulation. *Results Eng.* **2023**, *20*, 101476. [CrossRef]
51. Marwan, M.; Jamal, J.; Hamid, A.; Nasir, N.; La Nafie, N.A.; Gunawan, A.; Syamsuddin, S.; Razak, B.A.; Musaruddin, M. Optimize electrical energy cost of air conditioning considering to different wall characteristics. *Results Eng.* **2023**, *17*, 100990. [CrossRef]
52. Jelle, B.P.; Hynd, A.; Gustavsen, A.; Arasteh, D.; Goudey, H.; Hart, R. Fenestration of today and tomorrow: A state-of-the-art review and future research opportunities. *Sol. Energy Mater. Sol. Cells* **2012**, *96*, 1–28. [CrossRef]
53. Wakili, K.G.; Raedle, W.; Krammer, A.; Uehlinger, A.; Schüler, A.; Stöckli, T. Ug-value and edge heat loss of triple glazed insulating glass units: A comparison between measured and declared values. *J. Build. Eng.* **2021**, *44*, 103031. [CrossRef]
54. Wolf, A.T. Studies into the life-expectancy of insulating glass units. *Build. Environ.* **1992**, *27*, 305–319. [CrossRef]
55. Altun, A.F. Determination of Optimum Building Envelope Parameters of a Room concerning Window-to-Wall Ratio, Orientation, Insulation Thickness. *Buildings* **2022**, *12*, 383. [CrossRef]
56. Alghoul, S.K.; Rijabo, H.G.; Mashena, M.E. Energy consumption in buildings: A correlation for the influence of window to wall ratio and window orientation in Tripoli, Libya. *J. Build. Eng.* **2017**, *11*, 82–86. [CrossRef]

buildings

MDPI

Article

Investigation on Thermal Environment of Urban Slow Lane Based on Mobile Measurement Method—A Case Study of Swan Lake Area in Hefei, China

Mengyuan Li [1], Taotao Shui [2,3,4,*], Linpo Shi [5] and Ruxue Cao [4]

1 School of Architecture and Planning, Anhui Jianzhu University, Hefei 230601, China
2 Anhui Institute of Urban-Rural Green Development and Urban Renewal, Hefei 230601, China
3 BIM Engineering Centre of Anhui Province, Hefei 230601, China
4 School of Environment and Energy Engineering, Anhui Jianzhu University, Hefei 230601, China; 13477377072@163.com
5 School of Art, Anhui Jianzhu University, Hefei 230601, China
* Correspondence: shuitaotao@ahjzu.edu.cn

Abstract: Numerous issues with the urban thermal environment have been brought on by the rapid development of urbanization. The thermal climate of the slow lane, a major urban activity area, is directly tied to the well-being and comfort of city dwellers. The Swan Lake area in Hefei was chosen as the research site for this paper. The mobile measurement method was used to determine the heat island intensity distribution of the slow lane in each season of the year. The effects of building density, the percentage of permeable underlying surface, and shading on the slow lane's thermal environment were then thoroughly examined. According to the study, the distribution of heat island intensities along the mobile measurement route varies significantly depending on season, as well as time of year. Summer and winter have the most notable variations in the distribution of heat island intensities along the mobile measurement route; the summer values range from 0.1 to 4, while the winter values range from −0.3 to 3. The results showed a maximum difference of 30.2 °C in surface temperature (Ts) readings and 11.9 °C in air temperature (Ta) readings between the identical sites with and without shading, according to tests conducted at four typical mobile measurement locations along the mobile measuring route. The shading factor has a greater effect on the slow lane's thermal environment than permeable underlying surface and building density, as seen by the standardized coefficient of shading being significantly higher than both of these factors. With a standardized coefficient of shading of −0.493 in the winter and a standardized coefficient of shading of −0.517 in the summer, the effect of the shading factor on the thermal environment is more noticeable in the summer.

Keywords: mobile measurements; urban heat islands; underlying surface; slow lane; shading factor

1. Introduction

The rapid growth of urbanization in the current atmosphere of global warming has resulted in an increasing amount of issues with the urban outdoor thermal environment [1–3]. Specifically, 38–40% of urban land cover comprises slow lanes, which are among the most crucial routes for short-distance transit [4]. Slow lanes are a vital type of urban space, closely tied to residents' daily lives. Particularly, slow lanes within cities, as key components of the slow traffic system, connect communities and surrounding areas, offering higher usage frequency and a more direct experience [5]. The thermal environment of the slow lane exerts a more pronounced influence on pedestrians, with the study indicating a temperature rise of 1 °C in the hot summer conditions. Heat-related morbidity and mortality will rise by 1% to 3% [6], thus impacting public health [7–9]. Consequently, a comprehensive analysis of the underlying surface influence on the thermal conditions of the slow lane is crucial for alleviating the urban heat island effect and establishing a comfortable thermal environment through judicious urban planning.

The urban thermal environment is influenced by various elements, including the quantity of solar radiation received at the surface of the urban area and its conversion into sensible heat [10], urban physical characteristics (e.g., urban layout, type and structure of the underlying surface) [11–13], socio-economic aspects (e.g., land use) [14], and demographic attributes (e.g., age, income, and population density) [15,16]. The urban underlying surface is a factor determining alterations in the thermal environment [17–19]. The primary focus of the study is to analyze the effect of the underlying surface temperature on the thermal environment. Specifically, by examining the impact of subsoil type and pavement material on the slow lane temperature [20], it was determined that augmenting the natural subsoil cover within urban areas can mitigate the intensity of the urban heat island (UHI) effect and the likelihood of warming [21]. Furthermore, the detrimental influence of bare soil on the thermal environment ranks just below that of impervious slow lane. Additionally, it was observed that shading can markedly decrease the average radiant temperature [22], as shading facilitates the reduction of reflected longwave radiation from the ground. Furthermore, shading diminishes the reflection of longwave radiation from the ground [23]. Additionally, the cooling capacity of urban tree shading significantly influences the urban thermal environment [24,25]. However, numerous factors affect the thermal environment, and the mechanisms involved are intricate, with synergistic interactions being among these factors, complicating the analysis. The current analysis of the underlying surface's impact on the thermal environment is limited, focusing solely on a single factor. Therefore, further investigation is required to examine the combined effects of the underlying surface, building density, and shading distribution on the urban thermal environment; thus, optimizing the urban heat island effect has become urgent.

Different underlying surface parameters play a key role in facilitating thermal mitigation strategies and human adaptation measures. Changes in urban land use patterns and morphological characteristics have significantly increased urban temperatures, resulting in higher temperatures in the city than in the suburbs, creating the urban heat island (UHI) [26]. The UHI directly or indirectly affects the thermal comfort and well-being of urban residents [27]. Consequently, an increasing number of researchers are focusing on the UHI and using it to assess and improve the urban thermal environment [28]. However, current research on the thermal environment of slow lanes has mainly focused on the effects of the geometry of the slow lanes on the microclimate generated [29], while there have been few comprehensive analyses on the effects of different underlying surface parameters on the intensity distribution of heat islands in slow lanes.

To address these limitations, researchers have proposed several methods. The current methodology for studying the impact of thermal environments relies on field observations and numerical simulations. Given the multitude of factors influencing the actual thermal environment, numerical simulations struggle to adequately account for the intricate effects of these variables. Consequently, contemporary research on urban thermal environments predominantly utilizes field testing [30], which encompasses fixed observations, mobile measurements [31], and remote sensing technologies [32]. The thermal environment data acquired through fixed observation methods are precise and reveal the dynamic fluctuation patterns of long-term testing [33]. However, the local spatial diversity and complexity necessitate that the study of the local thermal environment spatially encompasses as many representative locations as possible [34]. This approach would require substantial resources and a workforce for real-time monitoring via fixed meteorological stations. Furthermore, remote sensing technology is constrained by the complexity of the field surroundings and cannot adequately characterize the thermal environment of medium urban areas [35]. Remote sensing technology, constrained by the complexities of the field environment, fails to adequately characterize the thermal environment of urban small-scale areas. It offers only an instantaneous surface temperature distribution image, necessitating inversion of measurement data through the quantitative relationship with air temperature, which significantly impacts the accuracy of spatial thermal environment calculations at pedestrian height. In contrast, mobile measurement technology can acquire thermal environment parameters (e.g., air temperature and humidity) from several representative locations using a limited number of instruments, facilitating the analysis of the spatial and temporal distribution and characteristics of urban thermal environment parameters [36,37]. Mobile measurements can ascertain the temperature distribution in urban areas that are inaccessible to vehicles, as well as the temperature profiles of meteorological stations at monitoring points [38]. Additionally, mobile tests are more cost-effective for measuring and capturing gradual changes in thermal conditions that vary with the landscape [39]. Sundborg [40] (1951) was the first to employ mobile measurement in Sweden to investigate the urban heat island (UHI) effect; Mizuno M [41] and Leconte, F [42] (2015) examined the mechanism by which urban structure affects a city's thermal climate by mobile measurement. Liu Lin [43] (2017) employed mobile measurement to investigate local-scale urban heat island (UHI) characteristics across various urban patterns; Cao C [44] (2020) utilized mobile measurement to examine the correlation between urban vegetation distribution and the mitigation of high humidity and temperature issues in Nanjing; and Emery, J [45] (2021) analyzed mobile measurement of air temperature in eastern France to assess the influence of urban form on micro-scale temperature variations. Chafer, M [46] (2022) examined the effects of various urban forms (e.g., high-rise and low-rise structures, green spaces) on the urban microclimate in Singapore, as well as the intensity of the urban heat island (UHI) effect associated with these forms, utilizing mobile measurements. In recent years, mobile measurements have been extensively utilized in thermal environment research.

This paper focuses on the Swan Lake area in Hefei City, examining the thermal environment of the slow lane using the mobile measurement method, primarily analyzing the effects of different underlying surface parameters on the pedestrian-level thermal conditions of the slow lane. Section 2 of this paper primarily delineates the study area and the measurement method. Section 3 initially evaluates the accuracy of four temporal correction models pertaining to the mobile measurement method. Subsequently, it examines the distribution of heat island intensity along the mobile measurement route across various seasons and times, as well as the shading distribution at typical locations on the mobile test route during different time intervals. It further investigates the resultant variations in surface and air temperatures, culminating in a comprehensive analysis of the thermal environment at varying pedestrian heights on the slow lane. The impact of varying building densities, proportions of the permeable underlying surface, and shading on the thermal conditions of the slow lane were also examined.

2. Materials and Methods

2.1. Study Area

Hefei is situated in eastern China, within the Jianghuai Plain, characterized by an undulating landscape and minimal geographical variation. The subtropical humid monsoon climate is characterized by chilly winters and hot summers. In recent years, Hefei has experienced significant social and economic development, transitioning from a "ring city" to a "lake" configuration, resulting in an expansion of the urban built-up area. The city has been constructed on an increasingly expansive scale, transitioning from "lakefront" to "embrace the lake". The rise in population density distribution, along with the nature of urban surface coverage and building design, has led to alterations that contribute to the complexity of spatial diversity [47]. The vicinity of Swan Lake in Hefei, a representative region of the city, was chosen as the study area to encompass a diverse array of built environment types, as illustrated in Figure 1. The study area encompasses multifunctional zones, including residential districts, commercial centers, entertainment venues, and urban parks, primarily featuring a mix of medium-to-high-rise and low-rise structures with diverse architectural designs, thereby facilitating a comprehensive sample of local climate zones (LCZs). The LCZ scheme has been utilized to assess the spatial morphology of the case area [48]. This paper employs a neighborhood-based LCZ classification method [49] to categorize the urban neighborhood units within the case area into various LCZ types, thereby illustrating the heterogeneity of urban morphology, as depicted in Figure 1.

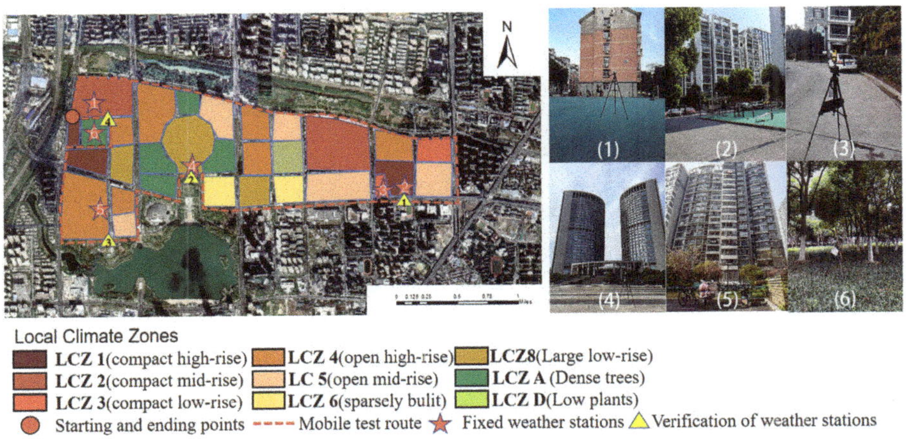

Figure 1. Mobile measurement route and local climate zone (LCZ) distribution, along with the mobile measurement program (1 to 6, depicting the vicinity of permanent weather stations FS01-FS06).

2.2. Research Methods

2.2.1. Mobile Measurement Program

This study involved the creation of a closed mobile route encompassing forty-one mobile test locations surrounding Swan Lake in Hefei, utilizing six concurrently running fixed weather stations and four validating weather stations, as seen in Figure 1. The research location is situated near the water, with negligible traffic congestion. The mobile test route included a total length of 10 km, encompassing 41 mobile measurement locations with varying surroundings and configurations, from dispersed low-rise building zones to densely crowded urban centers. Six fixed weather stations (HOBO-U23-002), designated as FS01 through FS06, served as time-adjusted reference stations, while four validation weather stations, labeled VSO1 through VS04, were employed to assess the errors of the four temporal correction models. To address the various underlying surface conditions of the reference stations, the six station LCZ types—LCZ1, LCZ2, LCZ3, LCZ4, and LCZ8—along with the site environments of these four weather stations, are presented.

A mobile measurement was conducted using an e-bike equipped with a portable, radiation-shielded external temperature and humidity recorder (HOBO U23-002) and a GPS to identify the 41 test sites. The apparatus was fixed to the e-bike at a height of 1.5 m above the ground, as illustrated in Figure 2. Data from the air temperature and relative humidity logger, together with GPS, were recorded at one-minute intervals to assure the precision of each test site. An illuminance meter (DLX-1082) was employed to document the shading along the moving route, and it was calibrated using data from ArcGIS. Furthermore, the e-bikes maintained an average velocity of 20–30 km/h and primarily traversed slow lanes to approach the local thermal environment while minimizing exposure to thermal emissions from vehicular traffic.

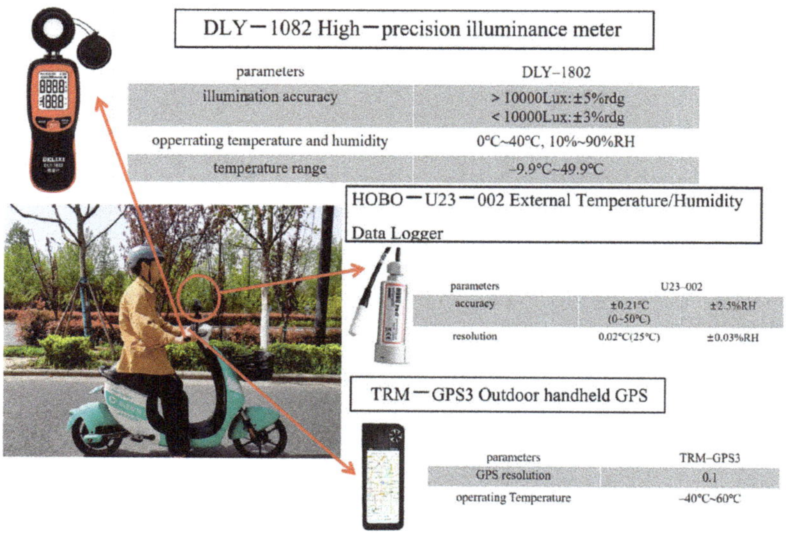

Figure 2. Mobile measurement tools and equipment.

The mobile polls were conducted on 18 August 2023; 18 November 2023; 23 January 2024; and 12 April 2024, with testing occurring at 9:00, 11:00, and 15:00, respectively. Initially, to guarantee the safety and comfort of the measures, it is essential to consider that rain may disrupt the efficacy of unsheltered mobile surveys conducted on an e-bike. Moreover, intense precipitation and damp circumstances can impair the GPS signal transmission and diminish the positional accuracy of mobile surveys, while the spatial attributes of the local thermal environment on rainy days may significantly differ from those on dry days. Thermal environment parameters were concurrently documented at stationary weather stations at one-minute intervals. Although mobile measurement has been recognized by many scholars for the study of thermal environments and the heat island effect in urban spaces, there is a very important problem in the application of mobile measurement, which is the heterogeneity of thermal environment parameters obtained by mobile testing methods in time and space. First, these mobile measurement data are recorded by the mobile equipment moment by moment during the observation along the route, and with the change in background meteorological parameters, the non-simultaneous data cannot be directly compared. Second, it is known from the literature that the formation of thermal environment parameters at different location points is usually affected by the urban spatial morphology and different types of underlying surface configurations, and the amount of changes in thermal environment parameters at different location points will be similarly changed by the thermal effects of the environment in which they are located. The routes of mobile tests usually pass through complex urban neighborhoods, which are greatly influenced by the surrounding dense buildings and traffic flow, etc. Therefore, with the changes of the underlying surface characteristics during the mobile observation process, the observation data obtained by the mobile tests are spatially heterogeneous, and the same cannot be directly compared and analyzed. From the above, it can be seen that although the mobile test method has the advantage of covering a large local area and diversified geographical conditions, the spatial and temporal heterogeneity of the test data makes the mobile test method itself have limitations in its application. Therefore, the problem of temporal heterogeneity of mobile measurements with temporal corrections is addressed in subsequent papers.

2.2.2. Introduction of the Classical Temporal Correction Model

The article contrasts four typical spatial and temporal correction models, which are as follows: the "Single" temporal correction model, the "Multiple" temporal correction model, the "Multiple-distance" temporal correction model, and the "Multiple-distance-underlying surface" temporal correction model. These models exhibit distinct attributes in temporal correction. Table 1 presents the specific spatial and temporal correction models.

Table 1. Exposition of four distinct temporal correction models.

Models	Station Number	Calculation Formula
"Single" temporal correction model [50]	N = 1	$y_{j,tr} - y_{j,tj} = x_{j,tr} - x_{j,tj}$
"Multiple" temporal correction model [41]	N > 1	$y_{j,tr} - y_{j,tj} = \sum_{i=1}^{N} (x_{i,tr} - x_{i,tj}) / N$
"Multiple-distance" temporal correction model [32]	N > 1	$y_{j,tr} - y_{j,tj} = \sum_{i=1}^{N} (x_{i,tr} - x_{x,tj}) * l_{ij}^{-2} / \sum_{i=1}^{N} l_{ij}^{-2}$
"Multiple-distance-underlying surface" temporal correction model [43]	N > 1	$y_{j,tr} - y_{j,tj} = \sum_{i=1}^{N} [k_{ij}(x_{i,tr} - x_{i,tj})] / N$ $k_{ij} = A * \frac{p_j}{p_i} + B * \frac{l_{ij}}{l} + c$

Note. N: quantity of meteorological stations; $y_{j,tr}$: the thermal environment parameter at mobile test location j during the uniform reference moment tr; $y_{j,tj}$: the thermal environment parameter at mobile test location j at test time t_j; $x_{i,tr}$: the thermal environment parameter at fixed weather station i during the uniform reference moment tr; $x_{i,tj}$: the thermal environment parameter at fixed weather station i at test time t_j. l_{ij}: the precise distance between fixed weather station i and mobile location j; l denotes the overall length of the mobile route. k_{ij}: the influence of underlying surface conditions and the distance between a fixed weather station and mobile site j on the correlation coefficient; l: the total length of the mobile route; Pi: the underlying surface characteristic parameter of station i; P_j: the underlying surface characteristic parameter of mobile location j; A: the coefficient of the subsurface characteristic term; B: the coefficient of the spatial distance scale term; C: the coefficient of the constant term.

This paper also compares the accuracy of different temporal correction models by means of mean absolute error (MAE) and root mean square error (RMSE), which are calculated using Equations (1) and (2), respectively.

$$MAE = \sum_{y=1}^{4} \left| v_{ya} - v_{yc} \right| / 4 \quad (1)$$

$$RMSE = \sqrt{\frac{\sum_{y=1}^{4}\left(v_{ya} - v_{yc}\right)^2}{4}} \quad (2)$$

2.2.3. Local Heat Island Intensity

Mobile measurements are implemented inside the study region to acquire air temperature data at several test points along the mobile route during different seasons, which are crucial for quantitatively analyzing the thermal environment disparities across the mobile test stations. This paper delineates the disparities in the thermal environments of mobile test sites through localized heat island intensity (LUHI) values [51,52], and it compares the adjusted air temperatures at these test sites along mobile routes to those recorded at reference meteorological stations in suburban areas, thereby facilitating the precise calculation of LUHI values at multiple temporal points for the mobile test sites; furthermore, the suburb location is chosen to be Hefei Xinqiao Airport. The precise computation of the local heat island intensity value (LUHI) is shown in Equation (3). LUHII is the local heat island intensity at location point *i* at the observation time *t;* Tit represents the temperature at location point *i* at the observation time t; and T_{st} indicates the temperature of the suburb at the observation time *t*.

$$LUHI = T_{it} - T_{st} \quad (3)$$

2.2.4. ArcGIS Spatial Interpolation Method

This work employed the spatial interpolation approach to derive a continuous spatial field of the measured parameters [53], utilizing recorded measurement point data to anticipate data for unknown regions. The Kriging method is utilized in spatial interpolation as a regression procedure for spatial modeling and the prediction of stochastic processes, relying on covariance function and variational function models. In contrast to other spatial interpolation techniques, the Kriging method emphasizes the impact of spatially extensive autocorrelation of regional variables on the weights. Therefore, we used the "Geostatistical Analyst" module of ArcGIS 10.8 software for spatial interpolation of LUHI, so as to obtain the temporal distribution of LUHI along the whole moving route.

3. Research Results

3.1. Comparison of Error Results of Four Temporal Correction Models

The validation outcomes of Ta were articulated through the mean absolute error (MAE) and root mean square error (RMSE) by comparing the reference and adjusted temperatures (Ta) to each spatiotemporally corrected model, as illustrated in Figure 3. The formulas for computing the MAE and RMSE are delineated as Equation (2) and Equation (3), respectively.

In this context, v_{ya} and v_{yc} represent the temperature recorded by the validated weather station at the specific moment and the temperature at that moment following the adjustment of location points along the mobile measurement route, respectively; y signifies the identifier for the validated weather stations; and the adjusted air temperature values derived from various temporal correction models are illustrated in Figure 3. Despite the small error of the "Multiple-distance-underlying surface" temporal correction model, the large difference in the temperature between validation point two and validation three and the corresponding measurement points on the moving route is due to the fact that the area where validation point two is located is a plaza, which is not surrounded by any building or environment for shade, while validation point three is surrounded by a high-rise building, which is able to provide shade, hence the large difference in the temperature between the two validation points.

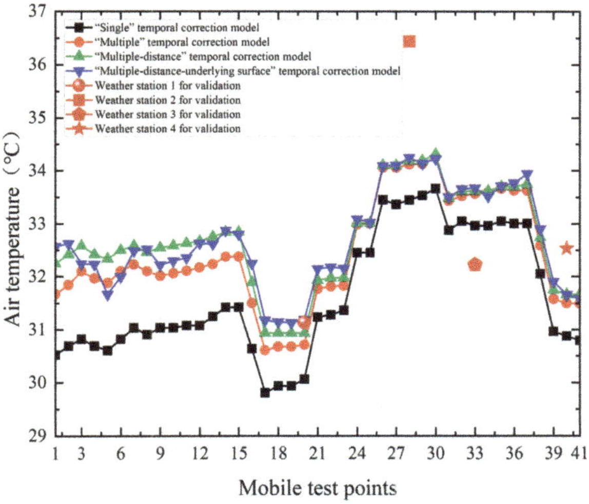

Figure 3. Comparison of correction outcomes from various temporal correction models.

The accuracies of several temporal correction models are evaluated based on the MAE and RMSE error values, as presented in Table 2. The "Multiple-distance-underlying surface" temporal correction model exhibits the lowest MAE value for Ta, succeeded by the "Single" temporal correction model, the "Multiple" temporal correction model, and the "Multiple-distance" temporal correction model. The MAE value of the "Multiple-distance-underlying surface" temporal correction model is approximately 1.144 °C, but the MAE values of the other temporal correction models exceed 1.2 °C. Additionally, the RMSE value of the "Multiple-distance-underlying surface" model is 1.387 °C. Furthermore, the RMSE value of Ta for the remaining three spatial and temporal correction models exceeds that of the "Multiple-distance-underlying surface" temporal correction model. Consequently, the "Multiple-distance-underlying surface" temporal correction model exhibits superior accuracy compared to the alternative models. The error results imply a significant fluctuation in the MAE and RMSE values of the "Single" temporal correction model compared to the other three correction models. The analysis indicates that the underlying surface properties exert the most substantial influence on the correlations among the various places, as informed by the four models' influences.

Table 2. Temperature correction error results based on four temporal correction models.

Models	**Air Temperature**	
	MAE (°C)	**RMSE (°C)**
"Single" temporal correction model	1.615	1.829
"Multiple" temporal correction model	1.282	1.454
"Multiple-distance" temporal correction model	1.236	1.410
"Multiple-distance-underlying surface" temporal correction model	1.144	1.387

3.2. Distribution of Heat Island Intensity Along the Movement Route

The precise distribution of heat island intensity along the moving routes was examined in relation to the four zones depicted in Figure 4. In the comprehensive analysis of Figure 5, the LUHI values in the Hefei Swan Lake research region throughout various moments of the spring testing period predominantly ranged from −3 °C to 2.5 °C along the mobile measurement routes. The spatial distribution of LUHI values across various moments on mobile routes exhibits considerable spatial disparities; nevertheless, the variability of the heat island intensity distribution over the entire routes in spring is not substantial. Zone 2 primarily consists of a dense high-rise building complex, with LUHI values in this area progressively increasing over time, while the heat island intensity at 9:00 is minimal. The maximum LUHI value in this region attains 1.95 °C post 11:00. Figure 6 illustrates the spread of local heat island intensity along the summer mobile measuring route. The disparity in heat island intensity distribution along the summer mobile route is more pronounced than in spring; however, Zone 3 maintains a higher LUHI value. Despite the presence of green space in this area, it is significantly influenced by solar radiation due to its open nature, resulting in an elevated LUHI value. Zone 1 comprises the People's Park, where the abundant greenery significantly mitigates heat island intensity, particularly during the summer months. Zone 2 and Zone 4 are both high-rise residential districts; however, a notable disparity exists in the distribution of heat island intensity. This variation arises because the mobile route in Zone 2 is obscured by a viaduct, resulting in a lower LUHI value for this route due to the combined shading effects of the building and the viaduct, while the LUHI value for the surrounding area remains elevated. The LUHI values are diminished, attaining 0.1 °C, approaching the LUHI values of the park adjacent to Zone 1. In contrast, the shading along the mobility route in Zone 3 is inconsistent, resulting in elevated LUHI values, indicative of greater heat island intensity in that region. Figure 7 depicts the distribution of local heat island intensities along the mobile measurement route in autumn, revealing a distribution that closely resembles that of spring and exhibits minimal variability in heat island intensities. Figure 8 illustrates the spread of heat islands during the winter measuring period, revealing significant variations in heat island strength at different times. The LUHI value of Zone 1 remains low, exhibiting the "cold island effect". Zones 3 and 4 demonstrate a more pronounced heat island effect, with the maximum LUHI value reaching 3 °C. Zone 2's LUHI value, influenced by shading, also attains 3 °C. Zone 2 maintains a reduced LUHI, which is attributable to its shade impact.

Figure 4. Four zones on the mobile measurement route.

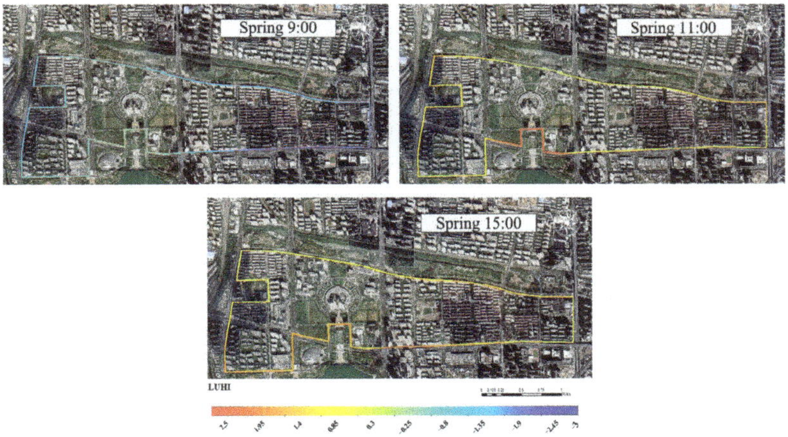

Figure 5. Distribution of LUHI values of mobile measurement route in the spring.

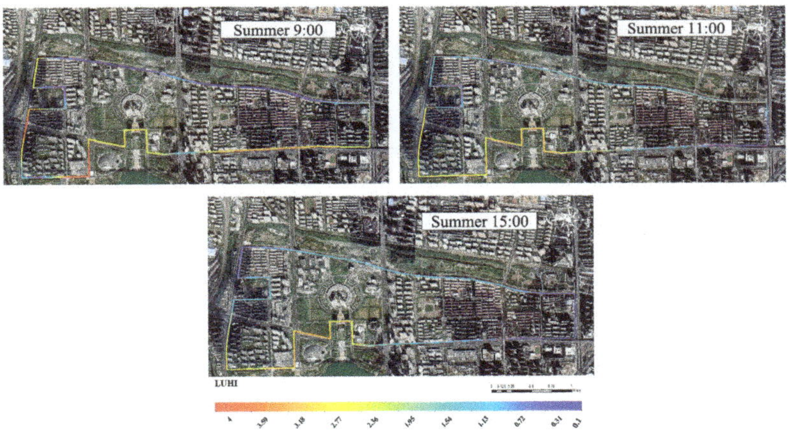

Figure 6. Distribution of LUHI values of mobile measurement route in the summer.

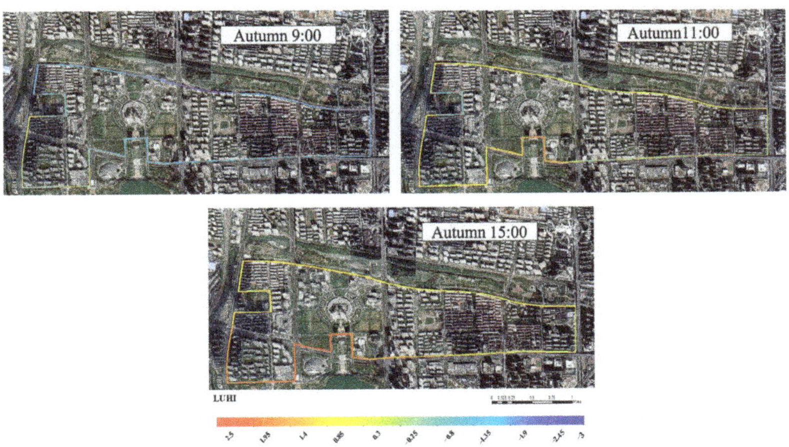

Figure 7. Distribution of LUHI values of mobile measurement route in the autumn.

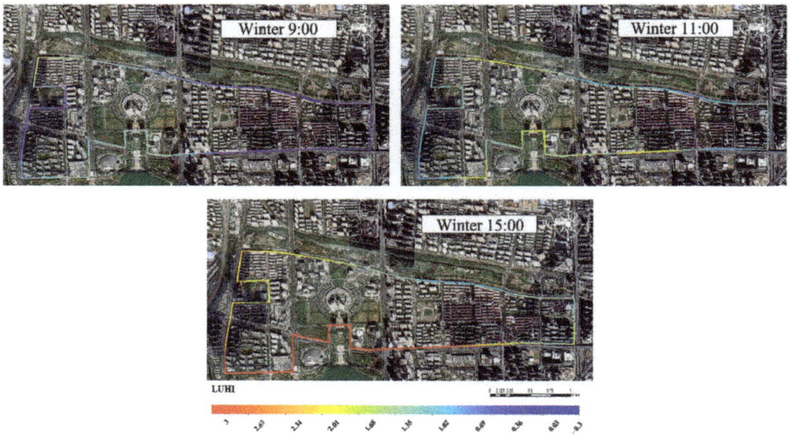

Figure 8. Distribution of LUHI values of mobile measurement route in the winter.

Analysis of the annual distribution of heat island intensity reveals significant variability during spring and autumn movement routes, with LUHI values spanning from −3 to 2.5 °C. In contrast, the variability during summer movement routes is more pronounced, exhibiting LUHI values from 0.1 to 4 °C. Despite Zone 3 having a greater proportion of green space, its LUHI value is elevated, contradicting the common belief that green space mitigates the heat island effect [54]. Additionally, the LUHI value of Zone 2 is markedly higher than that of the other regions at 15:00. Throughout the year, the same high-rise buildings encircle Zone 3; however, the LUHI value is elevated in comparison to Zone 4. Since Block 3 is surrounded by a square area, the temperature in the square area is very high and the hot air from the square flows into the area of the moving route, causing the moving route to show higher LUHI values. Consequently, we hypothesized that the shading effect influences the distribution of heat island intensity along the mobile routes. Four representative measurement points were subsequently selected for a more in-depth analysis of the underlying causes of this phenomenon. Refer to Section 3.3.

3.3. *Impact of Shading on UHI on Mobile Route*

This section analyzes the impact of shade on the thermal environment of the slow lane by selecting representative measurement stations throughout the route, with field observations conducted on 28 September 2024 from 6 a.m. to 6 p.m. The positions of the four designated measuring spots are illustrated in Figure 9. Fixed weather stations (HOBO-U23-002) were installed at an elevation of approximately 1.5 m above ground level, designated F1 to F4, and surface temperatures were taken and documented at 30 min intervals. The Ts was measured and documented every 30 min using an infrared thermal imager (UTi160s) to capture images of the area surrounding the weather stations, with infrared images taken bi-hourly. Additionally, a black sphere recorder (Taiwan Hengxin AZ8758) was positioned at measurement points two and three to record the temperature of the black sphere every 5 min. The instrumentation locations are illustrated in Figure 9, with the black spheres having resolutions of 0.1 °C, 1.5 °F, and 0.5 °C. The black spheres were positioned roughly 1 m above the ground. The black sphere recorder has a resolution of 0.1 °C and an accuracy of ±3 °C under outdoor conditions, while the infrared thermal imager (UTi160s) has a resolution of 0.1 °C and an accuracy of ±2 °C. The black sphere recorder (Taiwan Hengxin AZ8758) was positioned at Station 2 and Station 3 to monitor the black sphere temperature at 5 min intervals.

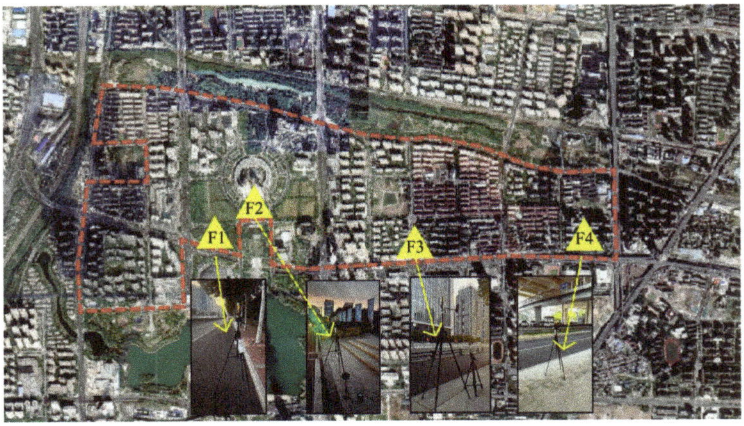

Figure 9. Location of the four fixed weather stations.

The influence of shading variations on the thermal conditions of the mobile route was meticulously examined by collecting data on shading alterations, surface temperatures, meteorological parameters, and thermograms at the sites of the four stationary meteorological measurement stations. Figure 10 illustrated the surface temperature and shadow variation data collected from the four designated sites. The figure illustrated the variations in air temperature (Ta) at four measurement locations at various times during the day. According to the trend of surface temperature variations, the observation period was segmented into the following three intervals: period 1 (06:00–12:00), period 2 (11:00–16:00), and period 3 (16:00–18:00). The air temperature (Ta) trend at several measurement places exhibited similarity throughout the observation period, consistently rising throughout time periods 1 and 2, peaking at approximately 14:00, and subsequently declining, as illustrated in Figure 11.

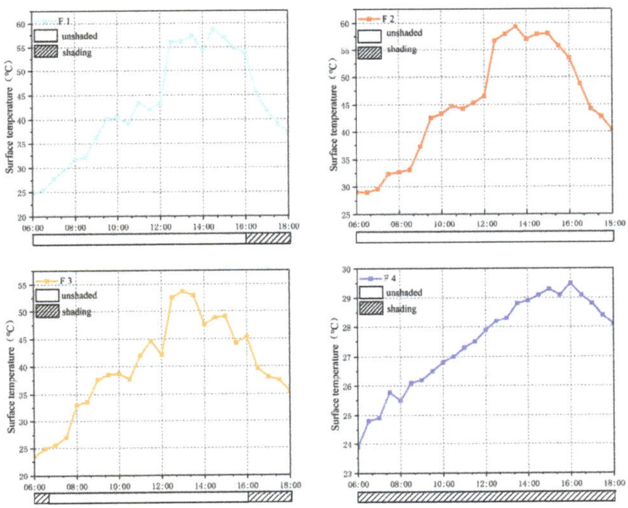

Figure 10. Distribution of surface temperature and shade at the four fixed sites.

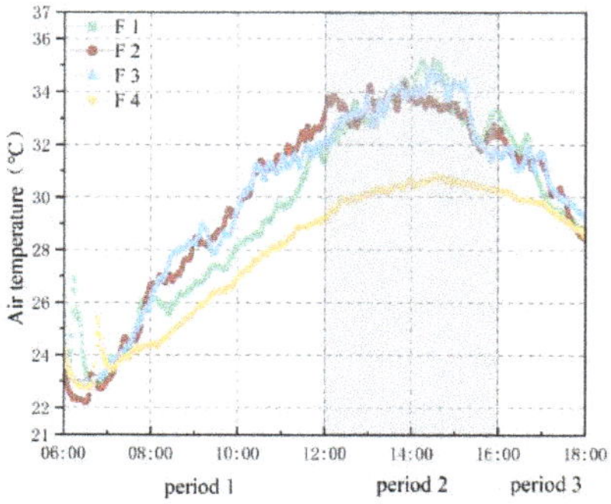

Legend	Average temperature (°C)			
	Period 1	Period 2	Period 3	Total
F1	26.91	33.42	30.63	30.32
F2	27.68	33.30	30.66	30.55
F3	27.73	33.39	30.77	30.63
F4	25.85	30.31	29.57	28.57

Figure 11. Average air temperature at four fixed measurement points.

The disparities in Ta among F1, F2, and F3 are minimal; however, the discrepancies in Ta between F4 and the other three sites are significant, with a maximum average difference in Ta ranging from 3 to 5 °C at various periods during the day, as shown in Figure 11. In comparing the average hourly Ta, only F4 exhibits a substantial difference from F1, F2, and F3, with F4 regularly recording the lowest Ta, averaging 1.7–2.1 °C lower than F1, F2, and F3. Among all measurement stations, F3 recorded the highest Ta during time period 1; however, this diminished markedly after 14:00. Conversely, F2 exhibited the highest Ta in time period 2, while F3 again had the highest Ta in time period 3. Moreover, F4 exhibited the shade structure of a viaduct and reported a somewhat higher Ta compared to the other three measurement points over the three period intervals. The maximum disparity in Ta readings between shaded and unshaded Ta at measurement points FI and F3 was approximately 11.9 °C, while F2 emerged as the hottest location at the onset of time period 2 due to the escalation of solar radiation. The mean surface temperature variations are significantly diverse throughout the observation period, arranged in increasing order, as follows: F4 < F3 < F1 < F2, with F2's mean surface temperature being 17.4 °C greater than that of F4.

Figure 12 illustrates that the shading circumstances influenced both the magnitude and distribution of the average Ts and air temperature at various measurement locations. At 8:00 am, a distinct contrast in the coloration of the images adjacent to F1 and F3 is evident, alongside a notable disparity in the Ts of the slow lane, which is attributable to the shading effects of the trees. The Ts of F1 registers at 24.1 °C under shade and 28.3 °C in unshaded conditions, while F3 exhibits a Ts of 27.5 °C when shaded and 29.4 °C when unshaded. Currently, F4 was influenced by the viaduct's shade, resulting in a Ts of 20.8 °C, but F2, situated in an unobstructed area, experienced a Ts of 29.8 °C during the testing period. At 14:00, when the sun's elevation angle rises, the canopy gap widens, allowing more direct sunlight to reach the ground, resulting in an increased distribution of high-temperature zones. The thermal distributions of F1, F2, and F3 illustrate the influence of shading on the surface temperature. Specifically, beneath the complete shading of the viaduct, the Ts of the road at F4 is lower than that at the other three measurement points, registering at 23.7 °C. Conversely, F2 emerges as a high-temperature zone, exhibiting a Ts of 45.6 °C, followed by F1 and F3. F1 experiences a surface temperature of 49.6 °C without shading and 45.4 °C with shading, while F3 records a Ts of 40.4 °C without shading and 37.5 °C with shading. The temperature changes depicted on the maps indicate that shading substantially influences temperature.

Further analysis of the effects of summer and winter shading strategies on the thermal environment revealed that shading variability on the summer and winter movement routes was reflected in the type of trees. Deciduous broadleaf trees are planted along the street on the slow lane in Hefei, and the trees planted along the moving lines in the study area were deciduous broadleaf trees alternating with evergreen trees, sycamore, and magnolia respectively. Therefore, in the summer months, greater shading was provided to the slow lane due to the alternation of the two types of trees. In winter, the sycamore trees, although deciduous, still provided shading due to their large crowns, alternating with magnolias.

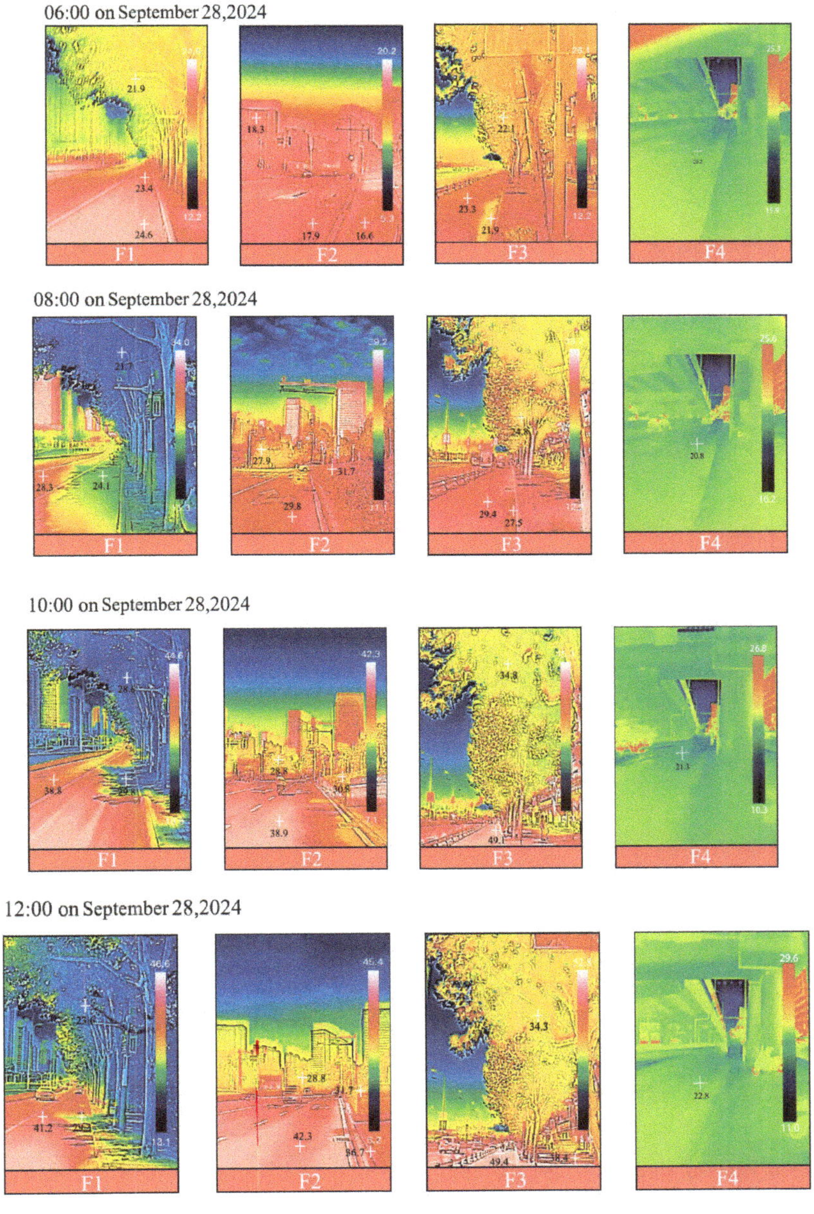

Figure 12. *Cont.*

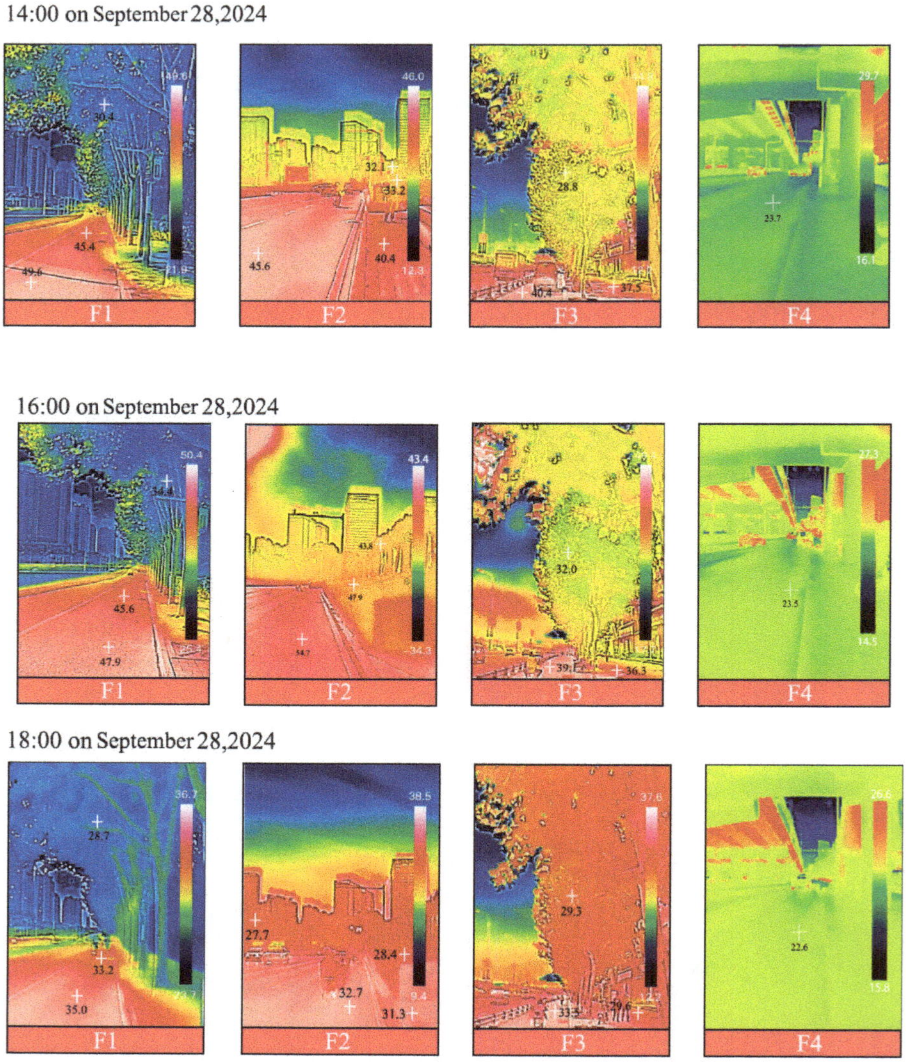

Figure 12. Thermal images of four measurement points taken at different times of the year.

3.4. Analysis of the Factors Influencing the Hot Loop of a Slow-Moving Road

3.4.1. Permeable Underlying Surface (PSF)

Initially, the impact of the permeable underlying surface on the air temperature along the mobile route is examined, with the results of correlating air temperature to the percentage of permeable underlying surface illustrated in Figure 13. This indicates a positive correlation between the permeable underlying surface and air temperature, suggesting that the permeable underlying surface does not substantially enhance the air temperature. Furthermore, the permeable underlying surfaces at the mobile measuring points consist entirely of trees interspersed with shrubs, and the varying percentages of underlay surrounding the measuring point are noted. The permeable underlying surface primarily consists of grass and plants, and it does not effectively facilitate cooling in areas F1, F2, and F3, surrounding the underlay.

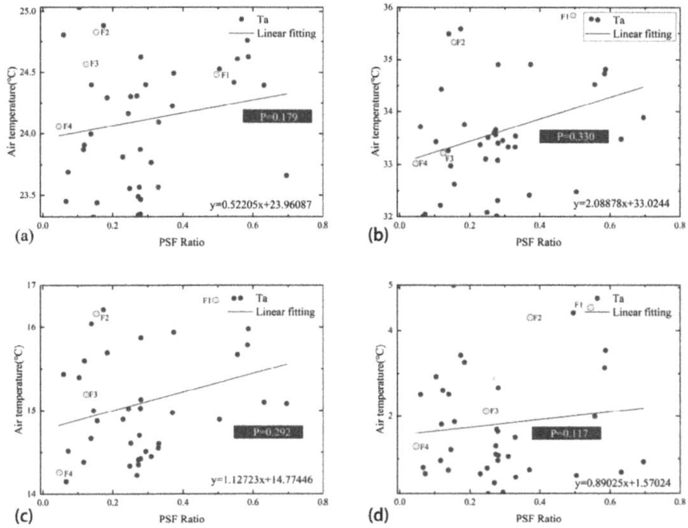

Figure 13. Correlation analysis between PSF and Ta ((**a**) for spring, (**b**) for summer, (**c**) for autumn, (**d**) for winter).

3.4.2. Building Density (BD)

Concerning the influence of building density on air temperature along the transit route, as illustrated in Figure 14, there exists a negative correlation between building density and air temperature. This suggests that increased building density may enhance the thermal environment of the slow-moving road to a certain degree. Furthermore, Figure 14b indicates that, during summer, measurement points F3 and F4 fall below the fitted line, while measurement points F1 and F2 are positioned above it. Notably, the locations of measurement points F3 and F4 correspond to an area characterized by high-density, high-rise buildings. The resultant building shade can enhance the thermal conditions of the slow lane.

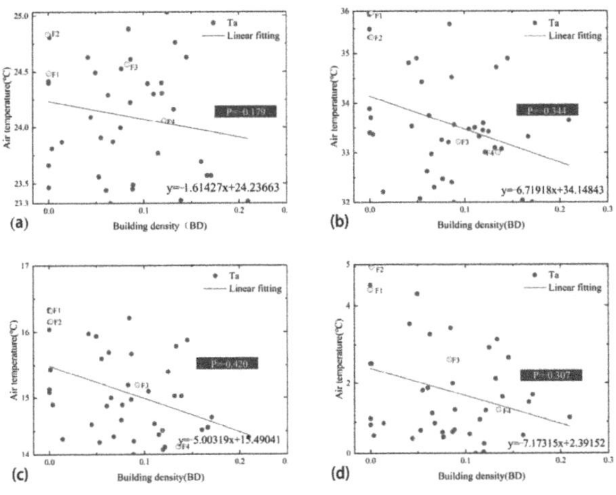

Figure 14. Correlation analysis between BD and Ta ((**a**) for spring, (**b**) for summer, (**c**) for autumn, (**d**) for winter).

3.4.3. Shading Ratio (SR)

Figure 15 illustrates the correlation analysis between air temperature variation and shading rate at pedestrian height throughout the year. The shading rate is derived from satellite imagery processed in ArcGIS, specifically through edge-tracing techniques. The processing results indicate the percentage of shaded area within a 100 m radius from the measurement points along the mobile route. It is evident that the correlation between shading rate and air temperature along the mobile route is both stronger and negatively correlated; an increase in shading rate contributes to the mitigation of the thermal environment at pedestrian height. Fixed measurement points F1 and F2 consistently lie above the fitted line in the figure, indicating that the cooling effect in the absence of shading is less pronounced.

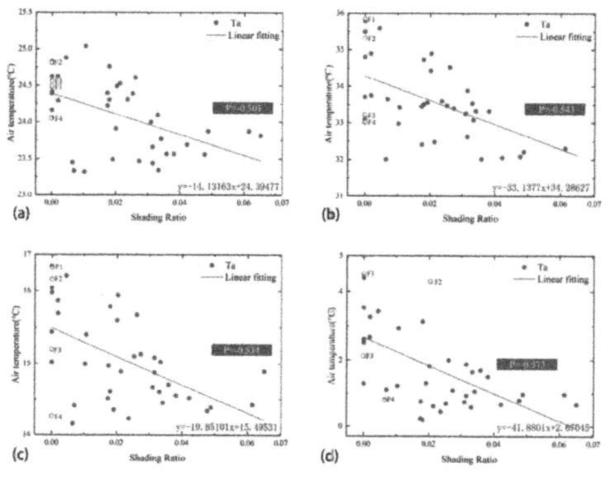

Figure 15. Correlation analysis between shading ratio and Ta ((**a**) for spring, (**b**) for summer, (**c**) for autumn, (**d**) for winter).

3.4.4. Sky View Factor (SVF)

Figure 16 illustrates the correlation analysis between temperature variations throughout the year and the sky view factor. The sky view factor was calculated by processing fisheye photographs with Rayman software. The processing results show the degree of shading of the buildings and environment around different measurement points on the moving route. It is obvious that the correlation between SVF and air temperature at the measurement points on the mobile route is stronger and positive; it indicates that the shading of buildings and trees on the mobile route can significantly reduce the air temperature.

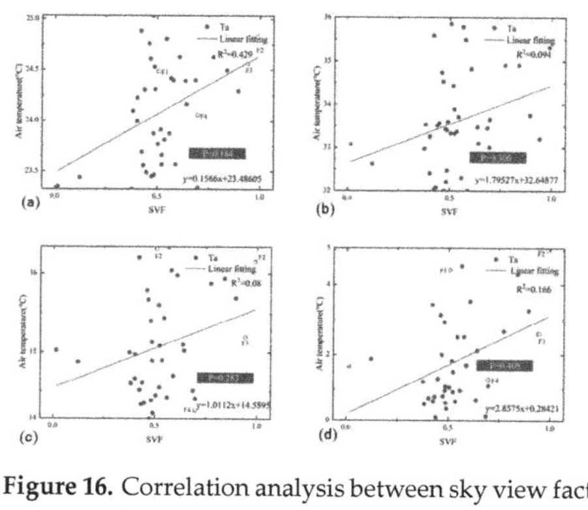

Figure 16. Correlation analysis between sky view factor and Ta ((**a**) for spring, (**b**) for summer, (**c**) for autumn, (**d**) for winter).

3.4.5. Multivariate Fitting Analysis

The annual air temperature data, together with the associated PSF, BD, and SR percentages, underwent a multiple linear regression analysis, resulting in distinct fitting equations for each season, namely Equations (4) to (7) for spring, summer, autumn, and winter, respectively. The r-squared values of 0.372, 0.517, 0.533, and 0.458, respectively, demonstrated a positive association between the ground surface (PSF) and temperature and a negative correlation between the building density (BD) and shading ratio (SR). To further analyze the impact of thermal environment characteristics on thermal weight, the standardized coefficients of each parameter were compared.

$$y = 0.562 * PSF + -1.238BD + -11.285 * SR + 0.697 * SVF + 23.901 \quad (4)$$

$$y = 2.101 * PSF + -5.807 * BD + -31.595 * SR + 0.480 * SVF + 33.864 \quad (5)$$

$$y = 1.105 * PSF + -4.513 * BD + -19.178 * SR + 0.1925 * SVF + 15.424 \quad (6)$$

$$y = 0.929 * PSF + -6.370 * BD + -36.045 * SR + 1.333 * SVF + 2.078 \quad (7)$$

Table 3 illustrates that the standardized coefficients of the SR term much exceed those of the PSF and BD terms, signifying that the influence of shade on air temperature is more pronounced than that of ground and building density. Consequently, the influence of shadowing must be considered in the relationship between mobile points and stationary meteorological stations when implementing spatial and temporal adjustments in mobile measurements.

Table 3. The standardized coefficients for the terms PSF and SR in Equation (4) to Equation (6), respectively.

Terms	Standardized Coefficients for Terms in Equation (4)	Standardized Coefficients for Terms in Equation (5)	Standardized Coefficients for Terms in Equation (6)	Standardized Coefficients for Terms in Equation (7)
PSF	0.193	0.332	0.286	0.123
BD	−0.138	−0.297	−0.379	−0.272
SR	−0.402	−0.517	−0.516	−0.493
SVF	0.259	0.082	0.054	0.190

4. Discussion

4.1. Comprehensive Analysis of Different Underlying Surface Parameters Affecting Slow Lane Thermal Environment

In this study, multiple linear regression analysis was used to reveal the main influencing factors affecting the thermal environment of urban slow lanes, mainly analyzing the mechanisms of permeable underlying surface, building density, shading, and SVF factors on the thermal environment.

Firstly, it was found that in high-density cities, the increase of a permeable underlying surface can mitigate the heat island effect, which is consistent with the findings of Chen, RN [55]. They noted that green spaces, water bodies, and permeable brick pavements could mitigate the urban heat island effect. Our analysis of the effect of permeable understory on the urban thermal environment is inadequate, and the analysis of the results of the study shows that the increase in permeable understory is not able to mitigate the LUHI, so it is necessary to classify the different types of vegetation to analyze their impact on the thermal environment. Secondly, building density also affects changes in the thermal environment. Building density has a warming effect, leading to an increase in the heat island effect within the city [56]; however, in actual high-density cities, building density does not determine the urban thermal environment, but is also affected by other factors [57]. Finally, the shading ratio can significantly reduce air temperature and improve the urban heat island effect in areas with higher SVF values. Higher SVF values provide a more open view of the sky, which affects the thermal environment in localized areas [58]. However, there is no distinction made between the different vegetation types and the constituent shading of buildings to analyze their different shading effects.

4.2. Urban Thermal Environment Optimization Recommendations

In this study, we propose a comprehensive optimization plan for improving the thermal environment of urban pedestrian slow lanes by addressing the results of the effects of different underlying surface parameters on the thermal environment of urban slow lanes.

Firstly, given the finding that shading significantly reduces temperature, it is recommended to increase the density of tree planting in the public space of the studied slow lanes. High-shade-species, such as deciduous trees, should be planted in areas with high heat island intensity, especially in areas with little shading. These trees will provide adequate shading at critical times, reduce air temperatures, and increase humidity through transpiration. In addition, taking into account the difference in the distribution of heat island intensity between summer and winter, and the fact that trees lose their leaves in winter, the trees planted in the slow lanes should alternate between evergreen trees and deciduous trees, so that the ability to ensure the effect of shade in summer can also improve the comfort of pedestrians in winter. A well-designed tree-planting scheme can reduce the surrounding area's temperature by an additional 1.5 °C throughout the year. In downtown Los Angeles, the positive cooling effects of trees during periods of hot weather significantly outweigh the negative impacts during colder periods. By strategically placing trees to maximize shading and considering seasonal variations, cities like Los Angeles can effectively enhance thermal comfort and mitigate the urban heat island effect [59]. The proposed tree-planting strategies offer practical guidance for urban planners and are broadly applicable, providing valuable insights for cities beyond Los Angeles. Secondly, optimizing the layout of buildings is an important regulatory measure. By adjusting the height and location of the building, it will be able to provide more shading at noon and afternoon, effectively reducing direct sunlight. Utilizing the shading effect of buildings can create cooler air for slow lanes and improve the urban heat island effect. Proper use of SVF is also key to optimizing the thermal environment of the slow lanes. By controlling the height and width of tree canopies to maintain adequate sky openings, slow lanes can benefit from good natural lighting and air circulation while avoiding excessive shading that blocks the view of the sky. Finally, improving infrastructure and amenities along slow lanes is also an important optimization measure.

4.3. Limitations of the Study

In line with previous studies, it can be said that this study investigates the air temperature effects of the pedestrian-level on slow lanes at a micro scale using mobile measurements, and our results show that there are differences in the applicability of different temporal correction models, which makes it particularly important to select a temporal correction model that is applicable to one's study area. Secondly, there are differences in the effects of different underlying surface parameters on the thermal environment of the slow lanes, and the shading factor is particularly important to create a more comfortable spatial environment for the slow lane space and to bring comfort to pedestrians travelling in summer. Our study has several limitations. Firstly, the current acquisition of the underlying surface parameters is not detailed enough to distinguish the vegetation types from the permeable underlying surface. In addition, the underlying surface parameters can be further expanded to analyze the mechanism of the influence of the underlying surface parameters on the thermal environment [60]. Secondly, among the effects of shading on the thermal environment, the effects of grass and trees on the thermal environment are different, and there are also differences in the effects of shading composed of buildings and shading composed of vegetation on the thermal environment, so we will further differentiate the effects of shading composed of vegetation and buildings on the thermal environment, and analyze the effects of shading on the thermal environment in depth. We can use the Rhinoceros software and the Grasshopper3 platform to further analyze and study LUHI [61]. We will follow-up on this research with a study on the urban spatial morphology and underlying surface parameters to deeply analyze the specific mechanism of their influence on the thermal environment by using machine learning models, as well as unmanned aerial vehicles.

5. Conclusions

This research examines the urban thermal environment by utilizing the mobile measurement method. Initially, the precision of various temporal correction models is evaluated, followed by the application of temporal correction to the thermal environment parameters derived from mobile measurements. This is integrated with the spatial interpolation technique of ArcGIS to visualize the distribution of heat island intensity along mobile routes, facilitating an analysis of the variations in heat island intensity distribution at different times of the year and the impact of the underlying surface on the thermal environment of these routes. Additionally, we analyzed the impact of shading on the thermal environment. The results show that the "Multiple-distance-underlying surface" temporal correction model exhibits the lowest MAE value for Ta, around 1.144 °C, with an RMSE value of 1.387 °C. The MAER and MSE of Ta for the other three temporal correction models exceed 1.2 °C, surpassing those of the "Multiple-distance-underlying surface" temporal correction model. Consequently, the "Multiple-distance-underlying surface" temporal correction model has superior accuracy compared to alternative models.

Upon comparing the distribution characteristics of LUHI throughout the year, it is evident that the intensity of the heat island along the mobile route varies temporally and spatially. Notably, there is no significant disparity in heat island intensity distribution along the mobile route during spring and autumn, while the most pronounced difference occurs in the summer, followed by winter. The LUHI ranges from 0.1 to 4 in summer and from −0.3 to 3 in winter, with findings indicating that LUHI values in high-rise building areas exhibit both elevated and diminished heat island intensities.

The analysis of measurements at four representative points reveals that shading influences the variations in the thermal environment along the travel route. The air temperatures at F4 exhibit a greater disparity compared to the other three measuring points. A comparison of their average hourly air temperatures indicates that the temperature values at the point located beneath the viaduct are markedly distinct from those at the other three locations, with a maximum average difference in Ta ranging from 3 to 5 °C across various time intervals, and an overall average that is 1.7 to 2.0 °C lower than that of the remaining three measuring points. The sequence of mean Ts across the tests, in ascending order, was F4 < F3 < F1 < F2. The mean surface temperature at F2 was approximately 17.4 °C greater than that at F4, as F4 was obscured by the viaduct, which diminishes shortwave radiation, reaching the ground more effectively than the tree shading at F1.

The analysis of the thermal environmental influences on Ta revealed that the standardized coefficients for shading were considerably greater than those for permeable ground and building density. This indicates that the shading factor exerted a more substantial impact on Ta variations along the mobile measurement route than the permeable underlying surface. The influence of shade on temperature changes was most pronounced in winter, followed by summer. The standardized coefficient for winter shade is −0.572, and for summer shade it is −0.551.

This study investigated the effects of different underlying surface parameters on the thermal environment of urban slow lanes based on a mobile measurement method. The accuracy of different temporal correction models for mobile measurements is evaluated. By comparing the distribution of heat island intensity along the mobile route in different seasons, it can be found that summer and winter exhibit the most significant disparities, with heat island intensity values fluctuating between 0.1 and 4 in summer and between −0.3 and 3 in winter.

In addition, the provision of shading is crucial in the urban planning process. This human-to-human focus paves the way for a more systematic improvement of the urban heat island effect. Therefore, the government should encourage the development of horticulture and green buildings to improve the localized heat island intensity of slow lanes, mitigate the urban heat island effect, and continuously improve the living environment for residents.

Author Contributions: Conceptualization, T.S. and M.L.; methodology, T.S. and M.L.; software, T.S. and M.L.; validation, T.S.; formal analysis, M.L.; investigation, M.L., L.S. and R.C.; resources, T.S.; data curation, T.S. and M.L.; writing—original draft preparation, M.L.; writing—review and editing, T.S. and M.L.; visualization, M.L. and L.S.; supervision, T.S.; funding acquisition, T.S. All authors have read and agreed to the published version of the manuscript.

Funding: The research was supported by the National Nature Science Foundation of China (NO. 52008001), Housing and Urban-Rural Construction Science and Technology Planned Projects of Anhui Province (No. 2022-YF045), and Open Project Program of BIM Engineering Centre of Anhui Province (Grant No. AHBIM2022ZR01).

Institutional Review Board Statement: Not applicable.

Informed Consent Statement: Not applicable.

Data Availability Statement: The data presented in this study are available on request from the corresponding author.

Conflicts of Interest: The authors declare no conflicts of interest.

References

1. Xie, Q.; Sun, Q. Monitoring thermal environment deterioration and its dynamic response to urban expansion in Wuhan, China. *Urban Clim.* **2021**, *39*, 100932. [CrossRef]
2. Wang, M.; Lu, H.; Chen, B.; Sun, W.; Yang, G. Fine-Scale Analysis of the Long-Term Urban Thermal Environment in Shanghai Using Google Earth Engine. *Remote Sens.* **2023**, *15*, 3732. [CrossRef]
3. Kim, S.W.; Brown, R.D. Urban heat island (UHI) intensity and magnitude estimations: A systematic literature review. *Sci. Total Environ.* **2021**, *779*, 146389. [CrossRef]
4. Ding, N.; Zhang, Y.; Wang, Y.; Chen, L.; Qin, K.; Yang, X. Effect of landscape pattern of urban surface evapotranspiration on land surface temperature. *Urban Clim.* **2023**, *49*, 101540. [CrossRef]
5. Guzman, L.A.; Arellana, J.; Castro, W.F. Desirable streets for pedestrians: Using a street-level index to assess walkability. *Transp. Res. Part D* **2022**, *111*, 103462. [CrossRef]
6. Hong, S.; Luo, Y.; Zhu, Y.; Huang, C.; Li, Y.; Sun, C. Childhood respiratory diseases related to indoor and outdoor extreme thermal environment (air temperature) in Shanghai, China. *Indoor Built Environ.* **2024**, *33*, 341–358. [CrossRef]
7. Cao, X.; Tang, B.; Zou, X.; He, L. Analysis on the cooling effect of a heat-reflective coating for asphalt pavement. *Road Mater. Pavement Des.* **2015**, *16*, 716–726. [CrossRef]

8. Cleland, S.E.; Steinhardt, W.; Neas, L.M.; West, J.J.; Rappold, A.G. Urban heat island impacts on heat-related cardiovascular morbidity: A time series analysis of older adults in US metropolitan areas. *Environ. Int.* **2023**, *178*, 108005. [CrossRef]
9. Lee, M.; Kim, S.; Kim, H.; Hwang, S. Pedestrian visual satisfaction and dissatisfaction toward physical components of the walking environment based on types, characteristics, and combinations. *Build. Environ.* **2023**, *244*, 110776. [CrossRef]
10. Zhang, X.; Kasimu, A.; Liang, H.; Wei, B.; Aizizi, Y.; Zhao, Y.; Reheman, R. Construction of Urban Thermal Environment Network Based on Land Surface Temperature Downscaling and Local Climate Zones. *Remote Sens.* **2023**, *15*, 1129. [CrossRef]
11. Atri, M.; Nedae-Tousi, S.; Shahab, S.; Solgi, E. The Effects of Thermal-Spatial Behaviours of Land Covers on Urban Heat Islands in Semi-Arid Climates. *Sustainability* **2021**, *13*, 13824. [CrossRef]
12. Lu, Y.; Yue, W.; Liu, Y.; Huang, Y. Investigating the spatiotemporal non-stationary relationships between urban spatial form and land surface temperature: A case study of Wuhan, China. *Sustain. Cities Soc.* **2021**, *72*, 103070. [CrossRef]
13. Zhou, X.; Chen, H. Experimental Analysis of the Influence of Urban Morphological Indices on the Urban Thermal Environment of Zhengzhou, China. *Atmosphere* **2021**, *12*, 1058. [CrossRef]
14. Ibrahim, A.; Ali, H.; Abuhendi, F.; Jaradat, S. Thermal seasonal variation and occupants' spatial behaviour in domestic spaces. *Build. Res. Inf.* **2019**, *48*, 364–378. [CrossRef]
15. Yang, J.; Nam, I.; Sohn, J.-R. The influence of seasonal characteristics in elderly thermal comfort in Korea. *Energy Build.* **2016**, *128*, 583–591. [CrossRef]
16. Yin, P.; Xie, J.; Ji, Y.; Liu, J.; Hou, Q.; Zhao, S.; Jing, P. Winter indoor thermal environment and heating demand of low-quality centrally heated houses in cold climates. *Appl. Energy* **2023**, *331*, 120480. [CrossRef]
17. Mallick, J. Evaluation of Seasonal Characteristics of Land Surface Temperature with NDVI and Population Density. *Pol. J. Environ. Stud.* **2021**, *30*, 3163–3180. [CrossRef]
18. Fang, Z.; Zhang, F.; Guo, Z.; Zheng, Z.; Feng, X. Investigation into the outdoor thermal comfort on different urban underlying surfaces. *Urban Clim.* **2024**, *55*, 101911. [CrossRef]
19. Cheela, V.R.S.; John, M.; Biswas, W.; Sarker, P. Combating Urban Heat Island Effect—A Review of Reflective Pavements and Tree Shading Strategies. *Buildings* **2021**, *11*, 93. [CrossRef]
20. Zou, Z.; Yan, C.; Yu, L.; Jiang, X.; Ding, J.; Qin, L.; Wang, B.; Qiu, G. Impacts of land use/ land cover types on interactions between urban heat island effects and heat waves. *Build. Environ.* **2021**, *204*, 108138. [CrossRef]
21. Rahman, M.A.; Dervishi, V.; Moser-Reischl, A.; Ludwig, F.; Pretzsch, H.; Rötzer, T.; Pauleit, S. Comparative analysis of shade and underlying surfaces on cooling effect. *Urban For. Urban Green.* **2021**, *63*, 127223. [CrossRef]
22. Shu, B.; Chang, H.T.; Chen, Y.X. Influence of the thermal environment of urban sidewalks under green shading from a human scale. *All Earth* **2024**, *36*, 127223. [CrossRef]
23. Kong, H.; Choi, N.; Park, S. Thermal environment analysis of landscape parameters of an urban park in summer—A case study in Suwon, Republic of Korea. *Urban For. Urban Green.* **2021**, *65*, 127377. [CrossRef]
24. Meili, N.; Acero, J.A.; Peleg, N.; Manoli, G.; Burlando, P.; Fatichi, S. Vegetation cover and plant-trait effects on outdoor thermal comfort in a tropical city. *Build. Environ.* **2021**, *58*, 107733. [CrossRef]
25. Meili, N.; Manoli, G.; Burlando, P.; Carmeliet, J.; Chow, W.T.; Coutts, A.M.; Roth, M.; Velasco, E.; Vivoni, E.R.; Fatichi, S. Tree effects on urban microclimate: Diurnal, seasonal, and climatic temperature differences explained by separating radiation, evapotranspiration, and roughness effects. *Urban For. Urban Green.* **2020**, *58*, 126970. [CrossRef]
26. Manley, G. On the frequency of snowfall in metropolitan England. *Q. J. R. Meteorol. Soc.* **1958**, *84*, 70–72. [CrossRef]
27. Basu, R.; Ostro, B.D. A multicounty analysis identifying the populations vulnerable to mortality associated with high ambient temperature in California. *Am. J. Epidemiol.* **2008**, *168*, 632–637. [CrossRef]
28. Manoli, G.; Fatichi, S.; Schläpfer, M.; Yu, K.; Crowther, T.W.; Meili, N.; Burlando, P.; Katul, G.G.; Bou-Zeid, E. Magnitude of urban heat islands largely explained by climate and population. *Nature* **2019**, *573*, 55. [CrossRef]
29. Mohite, S.; Surawar, M. Impact of urban street geometry on outdoor pedestrian thermal comfort during heatwave in Nagpur city. *Sustain. Cities Soc.* **2024**, *108*, 105450. [CrossRef]
30. Pan, Y.; Li, S.; Tang, X. Investigation of Bus Shelters and Their Thermal Environment in Hot–Humid Areas—A Case Study in Guangzhou. *Buildings* **2024**, *14*, 2377. [CrossRef]
31. Kim, E.J.; Kim, H. Walking-based mobile measurement: Examining its reliability for spatial thermal characteristics in urban environments. *Urban Clim.* **2024**, *58*, 102154. [CrossRef]
32. Jiang, L.; Zhao, H.; Cao, B.; He, W.; Yun, Z.; Cheng, C. A UAV Thermal Imaging Format Conversion System and Its Application in Mosaic Surface Microthermal Environment Analysis. *Sensors* **2024**, *24*, 6267. [CrossRef] [PubMed]
33. Parison, S.; Hendel, M.; Royon, L. A statistical method for quantifying the field effects of urban heat island mitigation techniques. *Urban Clim.* **2020**, *33*, 100651. [CrossRef]

34. Parison, S.; Chaumont, M.; Kounkou-Arnaud, R.; Long, F.; Bernik, A.; Da Silva, M.; Hendel, M. The effects of greening a parking lot as a heat mitigation strategy on outdoor thermal stress using fixed and mobile measurements: Case-study project "tertiary forest". *Sustain. Cities Soc.* **2023**, *98*, 104818. [CrossRef]
35. Ye, X.; Ren, H.; Wang, P.; Duan, Y.; Zhu, J. Urban LST Retrieval From the Ultrahigh Spatial Resolution Remote Sensing Data. *IEEE Geosci. Remote Sens. Lett.* **2024**, *21*, 7000305. [CrossRef]
36. Costanzini, S.; Despini, F.; Beltrami, L.; Fabbi, S.; Muscio, A.; Teggi, S. Identification of SUHI in Urban Areas by Remote Sensing Data and Mitigation Hypothesis through Solar Reflective Materials. *Atmosphere* **2022**, *13*, 70. [CrossRef]
37. Luo, T.; He, Q.; Wang, W.; Fan, X. Response of summer Land surface temperature of small and medium-sized cities to their neighboring urban spatial morphology. *Build. Environ.* **2024**, *250*, 111198. [CrossRef]
38. Schnell, I.; Cohen, P.; Mandelmilch, M.; Potchter, O. Portable-trackable methodologies for measuring personal and place exposure to nuisances in urban environments: Towards a people oriented paradigm. *Comput. Environ. Urban Syst.* **2021**, *86*, 101589. [CrossRef]
39. Pigliautile, I.; D'eramo, S.; Pisello, A.L. Intra-urban microclimate mapping for citizens' wellbeing: Novel wearable sensing techniques and automatized data-processing. *J. Clean. Prod.* **2021**, *279*, 123748. [CrossRef]
40. Sundborg, A. Climatlogical studies in Uppsala with special regard to the temperature conditions in the urban area. *Georaphica* **1951**, 22.
41. Masashi, M.; Minako, N.; Masaki, N.; Masatoshi, N. Study on temperature distribution of an urbanized area using a vehicle with mobile measurement system. *J. Environ. Eng.* **2009**, *74*, 1179–1185.
42. Leconte, F.; Bouyer, J.; Claverie, R.; Pétrissans, M. Using Local Climate Zone scheme for UHI assessment: Evaluation of the method using mobile measurements. *Build. Environ.* **2015**, *83*, 39–49. [CrossRef]
43. Liu, L.; Liu, J.; Lin, Y. Spatial-temporal Analysis of the Urban Heat Island of a Subtropical City by Using Mobile Measurement. *Procedia Eng.* **2016**, *169*, 55–63. [CrossRef]
44. Cao, C.; Yang, Y.; Lu, Y.; Schultze, N.; Gu, P.; Zhou, Q.; Xu, J.; Lee, X. Performance Evaluation of a Smart Mobile Air Temperature and Humidity Sensor for Characterizing Intracity Thermal Environment. *J. Atmos. Ocean. Technol.* **2020**, *37*, 1891–1905. [CrossRef]
45. Emery, J.; Pohl, B.; Crétat, J.; Richard, Y.; Pergaud, J.; Rega, M.; Zito, S.; Dudek, J.; Vairet, T.; Joly, D.; et al. How local climate zones influence urban air temperature: Measurements by bicycle in Dijon, France. *Urban Clim.* **2021**, *40*, 101017. [CrossRef]
46. Chàfer, M.; Tan, C.L.; Cureau, R.J.; Hien, W.N.; Pisello, A.L.; Cabeza, L.F. Mobile measurements of microclimatic variables through the central area of Singapore: An analysis from the pedestrian perspective. *Sustain. Cities Soc.* **2022**, *83*, 103986. [CrossRef]
47. Shimazaki, Y.; Aoki, M.; Nitta, J.; Okajima, H.; Yoshida, A. Experimental Determination of Pedestrian Thermal Comfort on Water-Retaining Pavement for UHI Adaptation Strategy. *Atmosphere* **2021**, *12*, 127. [CrossRef]
48. May, S.; Oliphant, A.J. Characteristics of the Park Cool Island in Golden Gate Park, San Francisco. *Theor. Appl. Climatol.* **2023**, *151*, 1269–1282. [CrossRef]
49. Jin, L.; Pan, X.; Liu, L.; Liu, L.; Liu, J.; Gao, Y. Block-based local climate zone approach to urban climate maps using the UDC model. *Build. Environ.* **2020**, *186*, 107334. [CrossRef]
50. Fung, W.Y.; Lam, K.S.; Nichol, J.; Wong, M.S. Derivation of Nighttime Urban Air Temperatures Using a Satellite Thermal Image. *J. Appl. Meteorol. Climatol.* **2009**, *48*, 863–872. [CrossRef]
51. Zhou, Y.; Luo, Y.; Yi, X.; Lun, F.; Hu, Q.; Huang, N.; Wen, G.; Zhou, H.; Hu, X. Exploring the influence of local urban heat features on park cooling effects: Insights from Chinese cities. *Build. Environ.* **2024**, *262*, 111782. [CrossRef]
52. Liu, L.; Lin, Y.; Wang, D.; Liu, J. Dynamic Spatial-temporal Evaluations of Urban Heat Islands and Thermal Comfort of a Complex Urban District Using an Urban Canopy Model. *J. Asian Archit. Build. Eng.* **2016**, *15*, 627–634. [CrossRef]
53. Yu, Z.; Song, Y.; Song, D.; Liu, Y. Spatial interpolation-based analysis method targeting visualization of the indoor thermal environment. *Build. Environ.* **2020**, *188*, 107484. [CrossRef]
54. Gong, Y.; Li, X.; Liu, H.; Li, Y. The Spatial Pattern and Mechanism of Thermal Environment in Urban Blocks from the Perspective of Green Space Fractal. *Buildings* **2023**, *13*, 574. [CrossRef]
55. Chen, R.; You, X. Reduction of urban heat island and associated greenhouse gas emissions. *Mitig. Adapt. Strateg. Glob. Chang.* **2020**, *25*, 689–711. [CrossRef]
56. Chen, X.; Wang, Z.; Bao, Y.; Luo, Q.; Wei, W. Combined impacts of buildings and urban remnant mountains on thermal environment in multi-mountainous city. *Sustain. Cities Soc.* **2022**, *87*, 104247. [CrossRef]
57. Yang, J.; Shi, B.; Xia, G.; Xue, Q.; Cao, S.-J. Impacts of Urban Form on Thermal Environment Near the Surface Region at Pedestrian Height: A Case Study Based on High-Density Built-Up Areas of Nanjing City in China. *Sustainability* **2020**, *87*, 1737. [CrossRef]

58. Wang, R.; Liu, R.; Chen, Q.; Cheng, Q.; Du, M. Effects of Sky View Factor on Thermal Environment in Different Local Climate Zoning Building Scenarios—A Case Study of Beijing, China. *Buildings* **2023**, *13*, 1882. [CrossRef]
59. Zhu, Y.; Kensek, K.M. Mitigating the Urban Heat Island Effect: The Thermal Performance of Shade-Tree Planting in Downtown Los Angeles. *Sustainability* **2024**, *16*, 8768. [CrossRef]
60. Elmarakby, E.; Khalifa, M.; Elshater, A.; Afifi, S. Tailored methods for mapping urban heat islands in Greater Cairo Region. *Ain Shams Eng. J.* **2022**, *13*, 101545. [CrossRef]
61. Elshater, A.; Abusaada, H.; Alfiky, A.; El-Bardisy, N.; Elmarakby, E.; Grant, S. Workers' Satisfaction vis-à-vis Environmental and Socio-Morphological Aspects for Sustainability and Decent Work. *Sustainability* **2022**, *14*, 1699. [CrossRef]

buildings

MDPI

Article

Estimating Carbon Emissions of Northeast Brazil Railway System

Diogo da Fonseca Soares [1,2,3,*], Sayonara Andrade Eliziário [4], Josicleda Domiciano Galvíncio [2] and Angel Fermin Ramos-Ridao [3]

1 Brazilian Company of Urban Trains, Square Napoleao Laureano, 01 Varadouro, João Pessoa 58010-150, Brazil
2 Department of Geographical Sciences, Federal University of Pernambuco, Avenue Professor Moraes Rego, 1235, Cidade Universitária, Recife 50670-901, Brazil; josicleda.galvincio@ufpe.br
3 Civil Engineering Department, University of Granada, Campus de Fuentenueva, s/n. E.T.S Ingenieros de Caminos, Canales y Puertos, 18071 Granada, Spain; ramosr@ugr.es
4 Department of Renewable Energy Engineering, Federal University of Paraíba, Campus I, Castelo Branco III, João Pessoa 58051-900, Brazil; sayonara@cear.ufpb.br
* Correspondence: diogosoares@cbtu.gov.br; Tel.: +55-83-98701-6734

Abstract: This article addresses the developing of a framework to obtain specific GHG emissions for the railway system and proposes mitigation strategies. To achieve this purpose, a comprehensive life cycle assessment (LCA) method was employed with input data from various sources to analyze the contribution of energy consumption and the emissions of the railway system. This paper included gathering data from an infrastructure operation and maintenance for detailed GHG emissions impact. This study also presents a comparative analysis of the GHG emissions in different urban railway transportation systems in Northeast Brazil, providing valuable contextual insights. As a result of the combination of total GHG emissions analysis from the states of the Northeast Brazil railway system, a total of 11,996.11 metric tons of CO_2 equivalent (tCO_2e) was estimated. The main line traction was a prominent source of the greenhouse gas footprint, especially for the diesel traction systems at Paraiba. The proposed framework shows that significant environmental benefits can be realized with proper decision-making to increase the number of passengers–kilometer transported by rail.

Keywords: railway; GHG emission; electrification; climate change; policies

1. Introduction

In recent decades, the rapid growth of the population and urbanization has led to an increasing demand for alternative transportation services in societies. This emerging trend in mobility requires swift responses from public transportation systems, which must offer reliable, comfortable, and economically and environmentally sustainable services. As economies thrive and cities expand, an efficient transportation infrastructure can play a vital role in mitigating environmental issues, diseases, and fatalities in urban areas [1,2].

The issue of greenhouse gas (GHG) emissions has become one of the most pressing and crucial concerns on a global scale, driven by the growing understanding of its implications for climate change. As an essential element of the worldwide initiatives aimed at addressing climate transformations, multiple sectors have been examined regarding their contribution to GHG emissions. In this context, the railway transportation system has emerged as a fundamental area of investigation, given its significance in urban mobility and freight transport [1,3].

The railway system, although often seen as a more sustainable transportation option compared to road and air modes, is not exempt from significant environmental impacts. The combustion of fossil fuels, such as diesel, to power trains is one of the primary sources of GHG emissions in the railway sector. This generates carbon dioxide (CO_2) and other gasses that contribute to global warming [4].

A comprehensive literature review on greenhouse gas emissions in the railway system unveils a complex interplay of factors contributing to the generation of these gasses. Studies have indicated that the electrification of the railway system, replacing diesel traction with electricity, can substantially decrease GHG emissions. The adoption of cleaner and renewable energy sources for the electricity used in trains can be a crucial step in mitigating emissions [5–15].

Additionally, the effectiveness of operations and the adoption of traffic management technologies can influence GHG emissions. More modern and energy-efficient train models tend to release lower amounts of GHGs per kilometer traveled, in contrast to older and less efficient models. Appropriate maintenance measures and route optimization also prove to be contributing factors in reducing emissions [16–19].

Another relevant aspect is waste management and fugitive emissions, which can occur during the transportation and maintenance processes of railway systems. The proper maintenance of railway tracks and the appropriate management of train components can minimize these emissions [16,17].

The utilization of life cycle assessment (LCA) methods has proven to be a highly useful tool for examining the effect of GHG emissions throughout the entire life cycle of the railway system. These methods enable the analysis of emissions from material production to train operation and maintenance, providing a comprehensive understanding of the emissions involved [18,20–24].

The environmental impact of greenhouse gasses in the railway context highlights the intricate interplay of the factors influencing emissions, underscoring the significance of comprehensive approaches to mitigation. Electrification and enhanced operational efficiency emerge as fundamental elements to attenuate the environmental impact of the GHG emissions from the railway system, thereby contributing to the promotion of sustainability and the fight against climate change [21,25].

While there are studies that delve into environmental performance through life cycle assessment (LCA), there is a shortage of literature specifically examining life cycle assessments in railway systems, particularly concerning greenhouse gas emissions [26]. Consequently, there is an urgent need to research greenhouse gas emissions and environmental impacts in the railway system and identify effective mitigation strategies for these emissions.

The impact of greenhouse gas emissions in the railway system stands as an indispensable tool to underpin effective public mitigation policies. By providing meticulous and precise data on emissions, this work allows for a focused approach to reducing emissions throughout the entire life cycle of the railway system. Such a focus directly contributes to the promotion of environmental sustainability, addressing climate transformations, and building a more resilient and adaptable future [27–30].

Therefore, the objective of this study was providing an analysis of the greenhouse gasses in railway system emissions using a life cycle methodology approach. This analysis employed a life cycle assessment methodology to accurately map the carbon footprints associated with these emissions. As a result, this work calculated the greenhouse gas emissions generated by the metro system in Northeast Brazil, which comprises the states of Paraiba, Pernambuco, Rio Grande do Norte, and Alagoas. This study aimed to compare these emission results with the studies in global railway greenhouse gas emissions. Additionally, it sought to analyze ways to mitigate these greenhouse gas emissions to formulate more effective public and environmental policies within the railway transportation system of these regions.

2. Materials and Methods

2.1. Study Area

This research was conducted on the railway system in Northeast Brazil, specifically focusing on the states of Paraiba, Pernambuco, Rio Grande do Norte, and Alagoas, which are equipped with this transportation infrastructure.

The railway is distributed in different municipalities in the states of Northeast Brazil, as shown in Figure 1: Paraiba (João Pessoa, Cabedelo, Santa Rita, Bayeux); Alagoas (Rio Largo, Setuba, Maceio); Rio Grande do Norte (Ceará-Mirim, Extremoz, Natal, Parnamirim); Pernambuco (Recife, Jaboatão dos Guararapes).

The rail system encompasses an electric metro that operates through an overhead power grid for electrification, along with a diesel train. Spanning over 200 km in length, this extensive transportation network caters to an average daily ridership of 250,000 passengers.

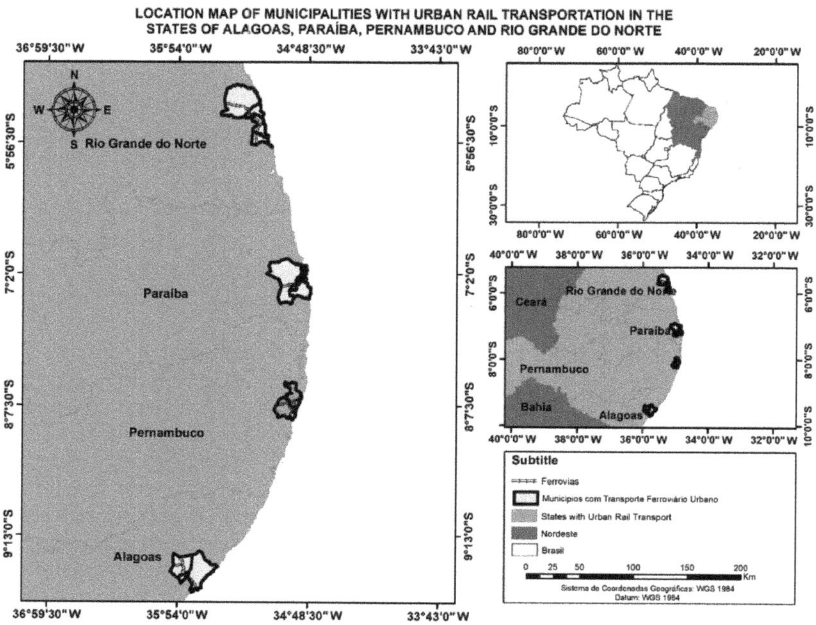

Figure 1. Location map of the northeastern railway system, Brazil.

2.2. Methods

The method employed in this investigation was using the quantitative emissions across the life cycle, with the calculations taking into account the associated resource inputs (materials, fuels, and equipment).

A life cycle assessment (LCA) is an international methodology used to quantify the emissions of specified products or systems throughout their entire lifecycle [31]. However, this methodology was originally defined to be applied to various types of products in a generalized manner. Therefore, conducting an LCA for a railway system can be much more complex than for consumer products in general. Since the urban railway transportation system is not merely a product, commodity, or project, none of the standards can fully apply to it [32]. Modeling may not be sufficient and may require adaptations. Selections concerning system boundary definition, model parameterization, and data selection can significantly impact the calculated results.

For the environmental assessment, a life cycle assessment (LCA) approach was employed using SimaPro 8.4.1 software following the parameters of ISO 14040 and 14044 [32–34]. According to LCA publications choices regarding system boundary definition, model parameterization, and data selection can significantly affect the calculated results [35–38].

In this study, the parameters of the GHG Protocol Product Standard from the World Resources Institute (WRI) and the World Business Council for Sustainable Development (WBCSD) were used. The methodology used was quantitative and based on the specifications of the Brazilian GHG Protocol Program published by the World Resources Institute [39] for corporate greenhouse gas emissions accounting, quantification, and reporting.

The urban railway transportation system comprises tracks, tunnels, viaducts, power facilities, depots, stations, vehicles, and control centers [38,40]. The calculation encompassed the associated resource inputs (materials, fuels, and equipment) and greenhouse gas emissions outputs at each phase, as shown in Figure 2.

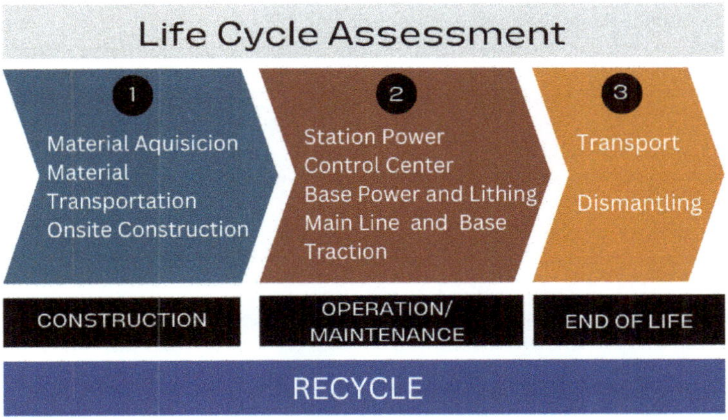

Figure 2. Flowchart of greenhouse gas emissions methodology for LCA railway.

There is an adapted version for the Brazilian product system, but it has not been thoughtfully tailored to the railway system, lacking the diesel railway system currently used in Northeast Brazil in its modeling parameters. This investigation will model this calculation parameter based on the existing indices from the World Resources Institute.

Due to Brazil's northeastern railway system having existed for over 150 years, there are no data available regarding its construction, such as project budget spreadsheets with detailed and complete records of the construction process. As a result, this research will not cover the emissions from the railway's construction phase.

This work displays the total amount of gas emissions based on the year 2022; however, to compare with other international case studies, the results were standardized per passenger–kilometer traveled (PKT) [29].

2.3. Life Cycle Assessment Modeling

Each greenhouse gas has the ability to absorb infrared thermal radiation over 100 years, and this capacity can be compared to the ability of carbon dioxide to perform the same function over the same period of time. For calculating CO_2e, we used the Global Warming Potential (GWP) to compare the global warming impacts based on the emissions estimates of different gasses [41–44], as shown in Table 1.

Table 1. Category of factors for greenhouse gas emissions.

GAS	GWP-100
CO_2	1
CH_4	25
N_2O	298
HFC-125	3500
HFC-134a	1300
HFC-143a	1430
HFC-152a	124
CF_4	7390
C_2F_6	12,200
SF_6	22,880

Source: Adapted [45].

Most of the studies we found were focused into four phases: the production, construction, operation, and maintenance of railway infrastructure [46–49]. They approached the production stage considering all the raw materials and structural components to manufact train car bodyshells and infrastructure projects, where the scarcity of papers about LCA applied on railway vehicles use was shown. This modeling involved calculating the inputs of the associated resources (materials, fuels, and equipment) and the outputs of the greenhouse gas emissions at each phase, categorized by operation, maintenance, transportation, and material procurement from the construction sectors [50,51].

The life cycle greenhouse gas emissions were estimated using Equation (1), as follows:

$$E = (Q1 + Q2 + Q3 + Q4)/L, \quad (1)$$

where

E = greenhouse gas emissions per unit length of the urban railway transportation system (tCO_2e/km);

Q1 = maintenance emission factor;

Q2 = operation emission factor;

Q3 = transportation emission factor;

Q4 = material procurement emission factor;

L = length of the railway track (km).

Calculation of Operation and Maintenance Emission Source

- Fugitive Emission

The calculation of the carbon dioxide equivalent (CO_2e) emissions from refrigeration, air conditioning, and fire extinguishing equipment can be carried out by following the following general steps: identifying all the refrigeration, air conditioning, and fire extinguishing equipment present in the system; identifying the GHGs by determining the specific greenhouse gasses used in each piece of equipment; obtaining the necessary data to perform the calculations, including information about the quantity of GHGs used, the specific emission factors for each GHG, and the equipment efficiency. This model uses the emissions from hydrofluorocarbons (HFCs), perfluorocarbons (PFCs), and sulfur hexafluoride (SF6).

To calculate the GHG emissions from refrigeration and air conditioning equipment and fire extinguishers using a life cycle stage approach, Equation (2) is as follows:

$$E = (EUN + EUE + EUD) * GWP, \quad (2)$$

where

E = emissions in CO_2e (kg);

EUN = emissions from the installation of new units; gas is used to charge the new equipment minus the equipment's capacity (the difference corresponds to losses to the atmosphere);

EUE = gas added to the existing units as maintenance by the organization or supplier (does not include the pre-charging performed by the manufacturer);

EUD = emissions from the disposal of old units; the capacity of the discarded unit minus the amount of recovered gas (the difference corresponds to losses to the atmosphere);

GWP = the Global Warming Potential of the greenhouse gasses (Table 1).

- Rail transportation

The rail fleet in Northeast Brazil is characterized by three types of trains: locomotives, light rail vehicles (VLT), and electric trains. The first two utilize commercial diesel oil as fuel. In Brazil, due to regulation, some fossil fuels are legally required to have a percentage of biofuel incorporated before being sold to the final consumer (ANP Resolution No. 909/2022). Therefore, the calculation is performed separately for pure fossil diesel oil and biofuel B10, with 90.00% fossil fuel diesel oil.

Initially, the values of fossil fuel and biofuel are separated. Then, the emission factors for the use of fossil fuels and biofuel in mobile sources are used for greenhouse gas conversion, as shown in Table 2.

Table 2. Emission factors for fossil fuel diesel oil and biodiesel.

Fuel	Kg CO_2/L	Kg CH_4/L	Kg N_2O/L
Fossil Diesel Oil	2.681	1.385×10^{-4}	1.385×10^{-4}
Biodiesel	2.499	3.316×10^{-4}	1.99×10^{-5}

Source: Adapted [39].

- Emissions in metric tons of CO_2

$$E = ECOF \times GWP\_CO_2 + ECHF\ (t) \times GWP\_CH_4 + ENOF(t) \times GWP, \quad (3)$$

where

E(t) = total emissions (tCO_2e);

ECOF = CO_2 emissions (t) fossil (the sum of the total fossil fuel quantity (L or m^3) × the fossil fuel emission factor (kg CO_2/L)/1000;

ECHF(t) = CH_4 emissions (t) (the sum of the total commercial fuel quantity (L or m^3) × the commercial fuel emission factor (kg CH_4/L)/1000;

ENOF(t) = N_2O emissions (t) (the sum of the total commercial fuel quantity (L or m^3) × the commercial fuel emission factor (kg N_2O/L)/1000.

- Emission of biogenic CO_2

$$EB(t) = FEB \times SQB, \quad (4)$$

where

EB(t) = biogenic CO_2 emissions (tCO_2);

FEB = biofuel emission factor;

SQB = sum of total biofuel quantity present in commercial fossil fuel, mandated by Brazilian law, in L.

In addition to CO_2 emissions, other greenhouse gasses can be emitted during fuel combustion. Therefore, if it is necessary to calculate emissions of other GHGs, the use of GHG emission factors is important.

- Electric Energy

Due to the source of electric energy in Brazil mainly coming from large hydroelectric, thermal power plants, nuclear energy, and renewable energy, there is a monthly variation in the energy sources and they complement each other in case of shortages. The small variation in the monthly CO_2 average emission factors of the national grid presents a relationship with seasonal fluctuations of the variable power generation of renewable sources. Furthermore, these intensities also change every year, and the Ministry of Science and Technology provides the conversion table annually.

For greater accuracy, a calculation of the greenhouse gas emissions from the monthly consumed electric energy was performed. The emission factors for the electricity generation, varying each month supplied from Brazilian Interconnected System (SIN), are shown in Table 3.

- Greenhouse Gas Emission Calculation Equation for Electricity Purchase

$$E(t) = \sum_{n=1}^{\infty}\left(\frac{CEM\_jan \times FEM\_jan}{1000} + \frac{CEM\_feb \times FEM\_feb}{1000} + \cdots + \frac{CEM\_dec \times FEM\_dec}{1000}\right) \quad (5)$$

where

$E(t)$ = mass of CO_2 emissions from electricity (tCO_2);

CEM_m = quantity of electricity that is consumed (MWh) in the month;

FEM_m = emission factor of SIN in the month.

- Calculation of Emission Source for Transportation (Logistics)

The calculation of the carbon dioxide equivalent (CO_2e) emissions from road transportation that utilize greenhouse gasses (GHGs) for the movement of equipment, personnel, and materials for railway operation and maintenance is carried out by identifying the types of transportation and the types of fuels (diesel, gasoline, ethanol, natural gas, etc.) used by road vehicles. Then, the values of the emission factors of each fossil fuel and biofuel are separated and applied for GHG conversion in mobile sources, as per Table 4.

Table 3. Emission factors for electricity generation.

Parameters for the 2019 Inventory	
Month	**FE of SIN (tCO_2/MWh)**
Jan	0.0355
Feb	0.0667
Mar	0.0530
Apr	0.0514
May	0.0482
Jun	0.0426
Jul	0.0906
Aug	0.1070
Sep	0.1024
Oct	0.1040
Nov	0.1078
Dec	0.0919

Source: Adapted [52].

Table 4. Global Warming Potential (GWP) of greenhouse gasses.

Fuel	**Unit**	**Emission Factors (kg CO_2/L)**		
		CO_2	**CH_4**	**N_2O**
Automotive Gasoline (pure)	L	2.212	0.0008	0.00026
Diesel Oil (pure)	L	2.603	0.0001	0.00014
Compressed Natural Gas (CNG)	m^3	1.999	0.0034	0.00011
Liquefied Petroleum Gas (LPG)	kg	2.9325	0.0029	0.00001
Aviation Kerosene	L	2.52	0.00002	0.00007
Aviation Gasoline	L	2.25	0.00002	0.00006
Lubricants	L	2.7175	0.0001	0.00014
Fuel Oil	L	3.1	0.0004	0.00002
Hydrous Ethanol	L	1.457	0.0004	0.00001
Biodiesel (B100)	L	2.431	0.0003	0.00002
Anhydrous Ethanol	L	1.526	0.0002	0.00001

Source: Adapted [44].

The GHG emissions from CO_2, CH_4, and N_2O of the commercial fuel and biofuel were calculated using the conversion factors for CH_4 (kg/L) and for N_2O (kg/L) by type, year, and fleet fuel. In the same way, to estimate the GHG emission, the consumption data per type of fuel we used Equations (3) and (4).

3. Results

3.1. *Raw Material Production, Transportation, and On-Site Construction*

The structure of urban railway transportation in Northeast Brazil dates back more than 150 years, as the Brazilian government strategically chose to repurpose the inactive freight railway to build the urban railway. Therefore, there is no real information available about the construction of the urban railway in Northeast Brazil. Due to this reason, the inventory of greenhouse gas (GHG) emissions from the construction of the railway is not discussed in this case study.

Based on other studies like the article "Calculation of life-cycle greenhouse gas emissions of urban rail transit systems: A case study of Shanghai Metro" [29] the emissions from the construction of a railway track represent less than 5% of the entire life cycle of a railway system. Therefore, the lack of construction data does not have a significant impact on the model.

3.2. Operation and Maintenance

The energy consumption of the operation and maintenance phases can be divided into eight sources: fugitive emissions, station energy and lighting, electric line traction, diesel line traction, base energy and lighting (workshop and administration), control center energy and lighting, and transportation and materials [21,31,53,54]. To serve as a reference for future life cycle assessment (LCA) studies on urban railway transportation systems, the scope of each source is described below, in Table 5.

Table 5. Energy consumption data.

Source	Energy Consumption	Emissions (tCO_2e)	CO_2-Biogenic
Electric (building)	9455.21	700.60	-
Logistic transportation (diesel)	30,923	11.23	11.23
Logistic transportation (gasoline)	86,472	184.31	45.05
Fugitive emissions	1397.98	1035.45	-
Electric train traction	51,663.668	3774.70	-
LRH (diesel)–train traction	19,78746	5648.34	555.20

Considering data availability and accessibility, the year 2022 was chosen as the base year to estimate the overall emission level of the operation and maintenance phases. According to the data obtained from the Brazilian urban train company, the total energy and resources consumed in logistic transportation, including car and truck entries and exits, account for the lowest among all the related sources as are presented in Table 5. The total energy used in building energy as station power and light are also presented in Table 5, with their associated emissions. We summarized all the air conditioning gasses and fire extinguishers used in the railway system in 2022, as fugitive emissions. The table provides the traction energy consumption for electric and diesel trains, respectively. This traction energy represents around 83% of the greenhouse gas emission (GHG) of the railway system.

Since the railway system in Northeast Brazil relies significantly on diesel-powered trains, the emissions from electric energy do not constitute the majority of emissions, in contrast to most other railway systems globally, which are predominantly powered by electricity, as demonstrated in Figure 3.

Although the emissions do directly originate from the electricity consumption during operation, the life cycle assessment (LCA) calculations incorporated the greenhouse gas (GHG) emissions from the processes of electricity generation and transmission. In the year 2022, the average annual electricity consumed for powering trains in the Northeast Brazilian railway system during the operation phase amounted to 51.7 million kWh. The GHG emissions from the electricity used to power the trains were calculated to be 3774.70 metric tons of CO_2 equivalent (tCO_2e).

In the same year, the diesel-powered train system consumed 1,978,746 L of diesel fuel for its operation. The GHG emissions resulting from this consumption were calculated to be 5703.54 metric tons of CO_2e.

By adding emissions from other energy resources, the total annual GHG emissions from the operation and maintenance activities amounted to 11,996.11 metric tons of CO_2e.

The provided data on greenhouse gas (GHG) emissions in the rail system of Pernambuco, Paraiba, Alagoas, and Rio Grande do Norte, as shown in Figure 4, presents an interesting view of the different emission sources associated with railway operations and related infrastructure.

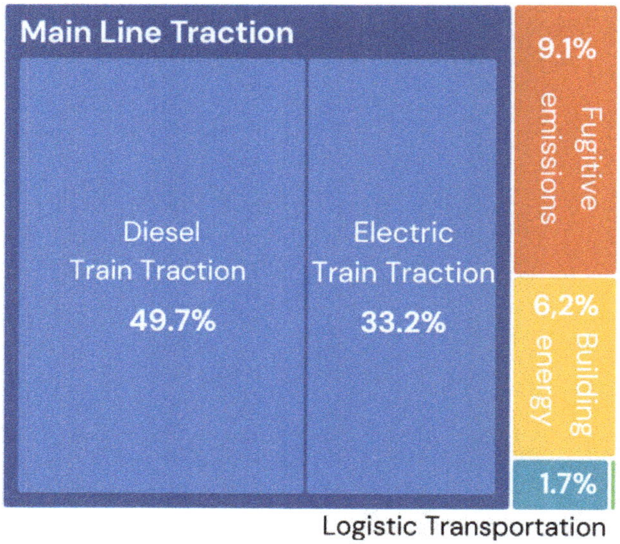

Figure 3. Composition of consumption and emissions in 2021.

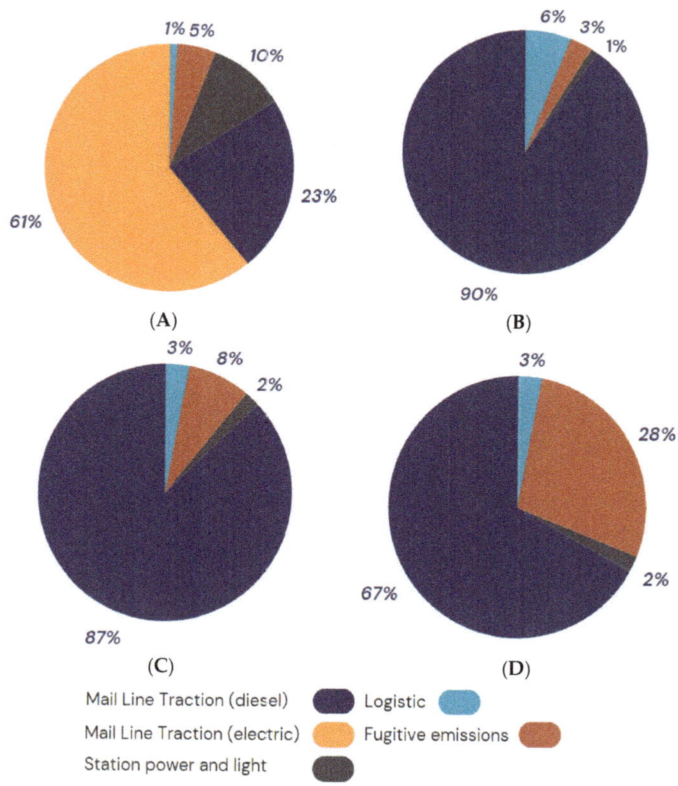

Figure 4. Composition of consumption and GHG emissions of rail system. (**A**) Rail system of Pernambuco. (**B**) Rail system of Paraiba. (**C**) Rail system of Alagoas. (**D**) Rail system of Rio Grande do Norte.

It is evident that GHG emissions vary considerably among the states and between different emission sources. The "Main line traction" category is a prominent source of GHG emissions, both for electric and diesel traction systems. In Pernambuco, for example, electric traction significantly contributes to emissions as the trains are powered by both electric and diesel energy, while in Paraiba, Alagoas, and Rio Grande do Norte, diesel traction is the main source due to the use of diesel-powered light rail vehicles. The emissions from diesel traction can be impacted by factors such as engine efficiency and fleet composition [20,40,55–58].

In summary, these data provide a valuable starting point for assessing the GHG emissions associated with railway operations in different cities. They highlight the need for a comprehensive and context-specific approach to addressing emissions, considering local sources and challenges. By focusing on clean technologies, operational efficiency, and strategic planning, it is possible to significantly reduce the GHG emissions in these railway operations.

The comparison between the states demonstrates the diversity of the energy choices for train traction. While Pernambuco opted for electric traction and thus recorded high emissions associated with electricity consumption, Paraiba, Alagoas, and Rio Grande do Norte relied more on diesel traction, reflected in the emissions from this fossil fuel. Besides CO_2, diesel combustion also emitted fine particles like soot. These particles not only impact air quality but can also absorb solar radiation and affect the climate [59–64]. The burning of diesel also releases nitrogen oxides (NOx), which contribute to the formation of acid rain and tropospheric ozone (a component of smog). Tropospheric ozone is a pollutant and a short-lived greenhouse gas (GHG) [11,65–71].

For the railway systems in the states of Paraiba, Alagoas, and Rio Grande do Norte, which predominantly use diesel traction, the implementation of emission reduction technologies like diesel oxidation catalysts (DOC) and selective catalytic reduction (SCR) systems help minimize NOx and particulate emissions [34,65,72–75]. More efficient engines and fuels with lower sulfur content also contribute to emissions reduction [65,71,76–80], as does transitioning to cleaner energy sources such as electrification (electric vehicles), renewable fuels like biodiesel [13,79,80], and hydrogen technologies [81–84]. In other words, electrification is essential to reduce the GHG emissions from the railway systems in these states, depending on the primary energetic sources used for producing electricity.

The emissions from station power and lighting are factors to consider in all cities. While these emissions may seem smaller compared to traction emissions, they contribute to the overall emission profile of the region.

The "Logistics" category underscores the importance of the emissions related to the transport of equipment, personnel, and materials needed for railway operation and maintenance. This highlights the relevance of logistic efficiency in reducing total emissions. The energy efficiency of vehicles and the logistics operations used for transporting railway system materials in this region should have a more efficient system to consume less fuel and, thus, emit fewer GHGs per unit of transported load [85,86].

The provided data on the greenhouse gas (GHG) emissions in tons of CO_2e from the diesel traction in different railway states in the northeast—Paraiba, Alagoas, Rio Grande do Norte, and Pernambuco, as shown in Figure 5—are essential for understanding the environmental impact of these transportation systems and can serve as a basis for formulating more effective environmental policies in the railway transportation systems of these regions [87–89].

The quantities of the GHG emissions associated with diesel traction vary significantly among the mentioned states. Pernambuco has the highest emissions (578.971 metric tons of CO_2e) followed by Alagoas (554.797 metric tons of CO_2e), Rio Grande do Norte (498.887 metric tons of CO_2e), and Paraíba (346.091 metric tons of CO_2e). This variation is influenced by various factors, such as the length of the routes, where Paraiba has a longer route than the other regions, and the volume of railway traffic, which is higher in Pernambuco due to its status as a major metropolitan area in Northeast Brazil. The energy efficiency of the trains is not a factor since all the light rail vehicles used in this region have the same technology and were manufactured by the same factory. An important factor that could also lead to higher emissions between states is the topography of the region; factors like curves and inclines, among others, could increase GHG emissions [89]. However, this work will not conduct an in-depth study of these factors, as this is one of the limitations of this research.

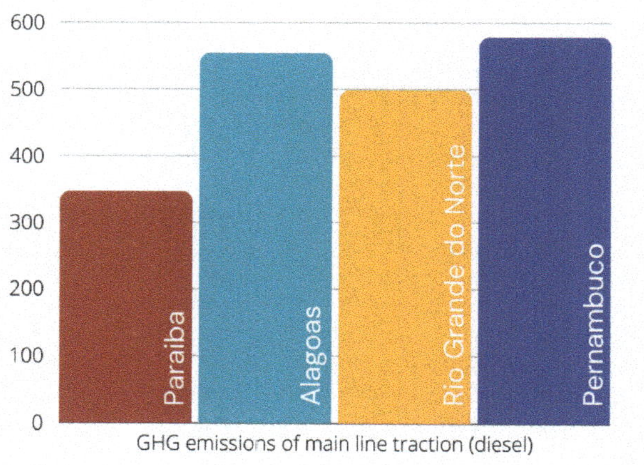

Figure 5. GHG emissions of main line traction (diesel).

3.3. Dismantling and Recycling

Urban railway transportation was developed relatively late in Northeast Brazil, with the opening of the first system in 1985 through repurposing some trains from the existing railway network and the acquisition of new ones. Currently, there is no real data available regarding the dismantling and recycling of urban railway transportation infrastructure in Northeast Brazil. For example, only in 2019 was an industry agreement introduced in Brazil to implement an e-waste reverse logistics system [90]. In addition, there is a compelling need to effectively manage construction and demolition waste due to the limitations of the existing legal framework in Brazil [91]. As a result, dismantling is not discussed in this case study.

The current economy does not incentivize a move towards a more circular economy. However, it is essential to create cleaner and more sustainable rail routes, fully integrated with other environmentally friendly modes of transport. There is clearly a gap in the deployment of supporting infrastructure (such as bridges, tunnels, tracks, yards, electrification, and railway circuits), assets (such as trains, rolling stock, traction systems, locomotives/carriages, and signaling and communication systems) and associated built environments (such as railway stations, mixed-use commercial buildings, and depots), taking into account the concept of the circular economy. According to [4], for this environment to exist, governments/procurement bodies should state that investment decisions will now be made on how circular economic value propositions are thus supporting the concept of circular economic competition. Government intervention should encourage behaviors that support the transition to a circular economy and create a level playing field for transport systems through a rolling strategy that covers short, medium, and long-term goals, grounded in a holistic systems approach [1].

3.4. Limitations

The absence of detailed construction data can hinder the accurate estimation of GHG emissions, which, in turn, can lead to an incomplete understanding of the overall impact of railways on climate change. This can be an obstacle to obtaining a holistic view of the life cycle and conducting accurate comparisons between different modes of transportation or sustainability interventions [92]. Geographic factors, such as ground softness, can play a crucial role in determining the GHG emissions associated with the earthworks in railway infrastructure construction. The inherent characteristics of products and systems, such as stations, tunnels, and materials like cement (e.g., ready-mix concrete), PHC posts, and steel, often result in their development through unique, one-off projects. Additionally, over 30% of the total GHG emissions from earthworks were linked to the use of equipment such as dump trucks, caterpillar excavators, and tractors, with the emissions heavily influenced by the ground conditions. This approach can lead to outputs with significant variability and uncertainty [25]. The study [93] indicated the increase in emissions and heavy metal in the air, soil, and water in the construction phase of the railway line, but the impacts decreased due to the increase in the utilization ratio of the railway during operation. Landgraf and Horvath [94] studied an analysis based on country-specific supply chains and production processes. Within railway infrastructure, the track (including rails, fasteners, sleepers and ballast) is the main contributor to GHG emissions, comprising 55% of the total in Australia. Furthermore, the carbon footprint of the electrification process depends on the energy mix used to generate electricity. Therefore, it is important to acknowledge that LCA is not limited solely to the construction or operational and maintenance phases; the choice of processes also plays a crucial role in evaluating the environmental impact of railway systems. To address this limitation, researchers can adopt strategies such as tailoring specific cases, conducting within their own defined scopes and system boundaries, using indirect data, modeling, and simulating to estimate the GHG emissions in the construction phase. In this research conducted on the railway system in Pernambuco, the assessment of greenhouse gas emissions (GHG) exclusively focused on the maintenance and operational phases because available studies and databases about construction are missing. This suggests that while the lack of construction data may create a limitation in the comprehensive life cycle analysis of GHG emissions, the focus on operational and maintenance phases, which generally have a larger contribution to the total emissions, still offers valuable insights into the sustainability of the railway system.

4. Discussion

4.1. Comparative System

This study aims to emphasize the importance of the comparative analysis of traction emissions, station operations, and infrastructure in different urban railway transport systems around the world. It is important to note that the availability of related studies is limited, and the comparative analysis was conducted within the same scope whenever possible, referencing the data and information found in the specialized literature and various case studies with varying local parameters.

In this study, the greenhouse gas (GHG) emissions of the urban railway transport systems in Northeast Brazil, Shanghai Metro, Delhi Metro, and the California corridor (San Diego to Los Angeles, San Francisco, and Sacramento) were compared. Furthermore, the reasons for the differences in data between these regions were explored. To facilitate the comparison, the emission intensities of each city were standardized per passenger–kilometer traveled (PKT).

This comparative analysis allowed for the identification of variations in the GHG emissions among different urban railway transport systems. These differences can be attributed to factors such as the energy source used, the train energy efficiency, the population density, the transport demand profile, and the specific infrastructure of each city. This information is valuable for understanding sustainable transportation practices and policies and can provide insights to enhance efficiency and reduce the emissions in railway systems [18].

By standardizing emission intensities per PKT, fairer comparisons can be made and areas for improvement in each railway system can be identified. This approach considers both the number of passengers transported and the distance traveled, enabling a more equitable assessment of environmental performance [42].

Therefore, this comparison of emissions standardized per PKT in different urban railway systems provides an important view of the disparities and offers valuable information for the development of more sustainable and emission-efficient transportation strategies.

According to [29], in 2012, Shanghai's metro system generated 637,011.3 tCO_2e from electricity consumption for traction. With a total passenger volume of around 2.3 billion and an average travel distance of 12.0 km, the traction emissions amounted to 23.4 g CO_2e per passenger–kilometer traveled (PKT). In the same context, Delhi has a total passenger volume of 651 million passengers and emitted 232,161.7 tCO_2e from traction emissions in 2011, which means 24.3 g CO_2e/PKT. On the other hand, in 2009, the traction emissions of California's high-speed rail system were 31.8 g CO_2e/PKT. It is important to emphasize that this depends on ridership levels. Chester and Horvath, in 2010, reported a range from well under 100 g CO_2e/PKT to over 700 g CO_2e/PKT for the total emissions of the California HSR system, which varied according to the level of travelers. By comparison, Brazil's Northeast system in 2022 had rail transport traction emissions of 9.5 tCO_2e with a total passenger volume of approximately 2,316,000 passengers over an average distance of 194 km with 21.14 gCO_2e/PKT. In 2016, Andrade analyzed the Rio de Janeiro Metro system and calculated emissions of 13.90 g of carbon dioxide per passenger–kilometer traveled [95].

Figure 6 depicts the variation in emissions among the railway transport systems in different world regions. This suggests that factors such as energy source, system efficiency, population density, and other local aspects have a significant impact on the emissions associated with railway transportation.

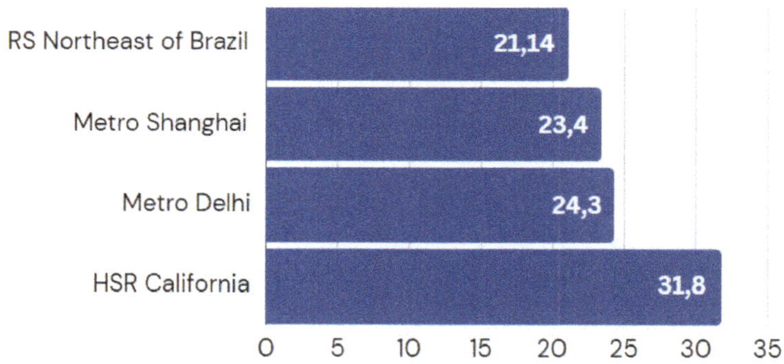

Figure 6. GHG emission (gCO_2/PKT).

The comparison of greenhouse gas (GHG) emissions per passenger–kilometer (gCO_2/PKT) among different railway transport systems provides valuable insights into the environmental efficiency of these systems in different regions of the world [29].

The comparison of the gCO_2/PKT values indicates that metro systems tend to be more emission efficient than light rail systems. This difference can be attributed to the variations in the technology used, the vehicle capacity, and the electrification systems, such as the use of fossil energy for traction in certain trains, as in the case of Northeast Brazil.

It is interesting to note that despite regional differences, the Delhi and Shanghai metro systems exhibit similar gCO_2/PKT values. This might suggest that both systems are adopting similar strategies to reduce emissions, such as the electrification of the main line traction and the adoption of more efficient technologies.

Additionally, metro systems like Delhi's and Shanghai's, which likely have more extensive electrification systems, present lower emissions. It was possible to consider that while construction generates a greater impact than maintenance to the permanent foundation in some countries, the opposite occurs in Northeast Brazil, because most vehicles are brought in only to be assembled. This makes it clear where efforts must be directed to ensure the reduced source of emissions and impacts. In any case, this study sheds light on the importance of electrification in emissions reduction and promoting sustainability in railway systems.

There are several factors that can influence GHG emissions, such as the number of train cars, the type of traction energy (which can vary based on the country's energy matrix), the infrastructure, urban development, and the geographical characteristics of the region, among others. These factors were not extensively studied in this article, serving as limitations to this research.

The gCO_2/PKT values can also be influenced by the local context, including the availability of renewable energy sources, sustainable transportation policies, and investments in infrastructure.

In addition, to compare the greenhouse gas (GHG) emissions per passenger–kilometer (gCO_2/PKT), the extent to which the speed of the system is comparable to other transport systems should still be questioned. This is due to the fact that the functional unit does not take the temporal dimension into account, which limits the significance of comparing train systems. At higher velocities, higher occupation rates over the same distance can be achieved, varying the levels of infrastructure utilization. For example, Beckert et al. [96] suggest changing the functional unit from PKT to PKT/h to make the other systems more comparable.

4.2. Mitigation of GHG Emissions

The mitigation of greenhouse gas (GHG) emissions in the railway system is a topic of great relevance in the context of environmental concerns and emission reduction goals. Based on recently published articles, several promising approaches and strategies to mitigate GHG emissions in the railway sector can be observed.

The electrification of railway lines has been highlighted as one of the main strategies to reduce GHG emissions [20,97]. In the case of the railway system in Northeast Brazil, which consists of both electric and diesel parts, transitioning diesel-powered trains to electric trains powered by renewable energy sources such as solar energy, installed on train rooftops to reduce traction electric energy consumption [10], or the use of wind energy have demonstrated effectiveness in significantly reducing carbon emissions. Additionally, the use of batteries and energy storage systems can enable the use of clean energy even on non-electrified sections [5–9].

The research and development of technologies that improve train efficiency can directly result in GHG emission reduction. Lighter trains [33,98], aerodynamic designs, and energy recovery systems [8,9,99] can minimize fuel consumption and, consequently, carbon emissions.

The implementation of intelligent traffic and operation management systems [100] can optimize train flow, reducing the need for sudden stops and accelerations. This not only saves energy but also contributes to more efficient and lower-emission operations.

Investments in the maintenance and modernization of railway infrastructure can result in smoother tracks, reducing train friction and energy consumption, or even utilizing friction for energy recycling [101–104]. Additionally, modernizing stations and signaling systems can improve operational efficiency [86,105]. The railway sector has a favorable impact on connectivity, saving time and influencing economic and social aspects. The railway system offers highly efficient solutions for densely populated and congested urban areas such as São Paulo. Investing in rail transport is essential for quality of life, as it reduces the commuting time and makes it easier to transport people from regions with lower property costs to areas characterized by limited land area and high population density. Programs promoting the use of these systems are fundamental, since increasing the number of passengers reduces the emissions calculated by life cycle analysis (LCA). However, robust and coordinated investment in transport infrastructure is necessary, as accessibility increases when the system is comprehensive and affordable.

Integration between different modes of transportation, such as rail, buses, and bicycles, can reduce the dependency on private vehicles and, consequently, GHG emissions.

Urban planning focused on accessibility, such as research evaluating the accessibility in high-speed metro systems and light rail vehicles with wheelchair accessibility in public transport, can encourage more people to adopt low-emission transportation modes [106–112].

If Brazil is to expand the rail system, optimize existing infrastructure and target future public investments efficiently, it will need a coordinated strategy that connects states with relevant economic ties. The iron mining sector in Brazil already makes extensive use of this type of transport, and other sectors could be incorporated. The predominance of the road system for heavy transport is a reality, but with the economy growth, the current system will be saturated and new investments will make the rail sector an important alternative, especially if electrified.

The adoption of emission reduction technologies, such as exhaust gas purification systems and particle filters, can contribute to the reduction in emissions in diesel-powered trains. These technologies can minimize the emission of harmful air pollutants.

Empowering operators and maintenance personnel with sustainable and efficient practices can also contribute to emission reduction [113]. Economical driving techniques and proper train maintenance can optimize energy consumption [23].

Finally, it is crucial to ensure that rail network construction and maintenance projects comply with all the applicable environmental requirements, minimizing impacts on the environment. Policies that consider the use of sustainable materials and the recycling of waste should be implemented. The government and future projects need to focus on the importance of recycling routes, the promotion of a circular economy, and the use of sustainable sources of materials, especially those considered critical and strategic.

5. Conclusions

This research aimed to calculate greenhouse gas (GHG) emissions through a life cycle analysis of the railway system in Northeast Brazil. Additionally, a comparative analysis was conducted to the global context to understand the emission levels of urban railway transport systems worldwide. This analysis also aimed to provide guidance for future emission reduction efforts.

The results obtained revealed that the total GHG emissions over the life cycle, considering the construction length of the entire railway system in Northeast Brazil, were 11,996.11 metric tons of CO_2 equivalent (tCO_2e).

The Brazilian urban railway system differs from the rest of the world, as a significant portion of the system is still diesel-powered. The quantities of GHG emissions associated with diesel traction vary significantly among the mentioned states, where Paraiba has the greatest contribution. Nevertheless, when comparing the operation and maintenance phase emissions, the traction emissions of the Brazilian Northeast system were 21.14 gCO_2e/PKT, which were competitive with those of the Delhi Metro and the California high-speed rail corridor, based on a much larger passenger volume in Shanghai. However, there was still significant potential for energy savings in the operation phase, especially in the stations designed to be excessively grandiose and ornamented with energy-intensive facilities. A more reasonable ventilation structure for metro stations and energy-saving lighting, such as LED lights, can effectively reduce their emissions.

These results provide an important understanding of the GHG emissions of the railway system in Northeast Brazil and serve as a basis for identifying the areas for improvement and guiding future actions towards reducing GHG emissions in this context. The analysis of such data can serve as a basis for formulating more effective environmental policies in the railway transport systems of these regions. An awareness of the predominant emission sources in each city can guide the efforts to mitigate environmental impact.

Data like these can highlight the importance of transitioning to cleaner energy sources, such as electric railway traction, low-carbon biofuels, photovoltaic energy, improvements in railway infrastructure, and sustainable stations to reduce the environmental impact of railway transport.

The railway sector, as part of the transportation sector, plays a significant role in meeting the environmental and emissions reduction targets set by various governments and international agreements. However, the railway supply chain is not incentivized to work towards a circular economy. An awareness of the emissions associated with diesel traction is crucial to guide sustainability policies and efforts.

Furthermore, the development of a more comprehensive estimation tool and an applicable reference standard for various areas, equipment types, structures, and techniques would be highly beneficial. This would facilitate the collection and analysis of consistent and comparable data, enabling a more accurate assessment of the GHG emissions in different railway systems.

In summary, the analysis of GHG emissions data in railways provides valuable insights for sustainability-oriented decision-making. These insights can steer investments towards cleaner technologies, efficient transportation strategies, and collaborations between the public and private sectors to achieve environmental goals and reduce the carbon footprint of railway transport.

Author Contributions: Conceptualization, S.A.E.; Methodology, D.d.F.S.; Validation, D.d.F.S.; Investigation, D.d.F.S.; Resources, D.d.F.S.; Writing—original draft, D.d.F.S. and S.A.E.; Writing—review & editing, D.d.F.S., S.A.E.; Supervision, A.F.R.-R., J.D.G. and S.A.E.; Project administration, D.d.F.S.; Funding acquisition, D.d.F.S. and A.F.R.-R. All authors have read and agreed to the published version of the manuscript.

Funding: This research received no external funding.

Data Availability Statement: The data presented in this study are openly available on Sistema de Registro Nacional de Emissões (SIRENE) website, in a publicly accessible repository: https://www.gov.br/mcti/pt-br/acompanhe-o-mcti/sirene (accessed on 27 August 2024).

Conflicts of Interest: Author Diogo da Fonseca Soares was employed by the company Brazilian Company of Urban Trains, Square Napoleao Laureano. The remaining authors declare that the research was conducted in the absence of any commercial or financial relationships that could be construed as a potential conflict of interest.

References

1. Kaewunruen, S.; Sresakoolchai, J.; Peng, J. Life cycle cost, energy and carbon assessments of Beijing-Shanghai high-speed railway. *Sustainability* **2020**, *12*, 206. [CrossRef]
2. Lin, J.; Li, H.; Huang, W.; Xu, W.; Cheng, S. A Carbon Footprint of High-Speed Railways in China: A Case Study of the Beijing-Shanghai Line. *J. Ind. Ecol.* **2019**, *23*, 869–878. [CrossRef]
3. Shi, X.Y.; Qiao, X.Q.; Ni, J.M.; Zheng, Y.Y. Experimental and CFD study on combustion noise and pollutant emissions of diesel engine with split injections of common rail system. *Shanghai Jiaotong Daxue Xuebao/J. Shanghai Jiaotong Univ.* **2010**, *44*, 1566–1572.
4. Trevisan, L.; Bordignon, M. Screening Life Cycle Assessment to compare CO_2 and Greenhouse Gases emissions of air, road, and rail transport: An exploratory study. *Procedia CIRP* **2020**, *90*, 303–309. [CrossRef]
5. Gee, A.M.; Dunn, R.W. Analysis of Trackside Flywheel Energy Storage in Light Rail Systems. *IEEE Trans. Veh. Technol.* **2015**, *64*, 3858–3869. [CrossRef]
6. Zhu, F.; Yang, Z.; Lin, F.; Xin, Y. Decentralized Cooperative Control of Multiple Energy Storage Systems in Urban Railway Based on Multiagent Deep Reinforcement Learning. *IEEE Trans. Power Electron.* **2020**, *35*, 9368–9379. [CrossRef]
7. Khodaparastan, M.; Dutta, O.; Saleh, M.; Mohamed, A.A. Modeling and Simulation of DC Electric Rail Transit Systems With Wayside Energy Storage. *IEEE Trans. Veh. Technol.* **2019**, *68*, 2218–2228. [CrossRef]
8. Zhu, F.; Yang, Z.; Zhao, Z.; Lin, F. Two-Stage Synthetic Optimization of Supercapacitor-Based Energy Storage Systems, Traction Power Parameters and Train Operation in Urban Rail Transit. *IEEE Trans. Veh. Technol.* **2021**, *70*, 8590–8605. [CrossRef]
9. Krueger, H.; Fletcher, D.; Cruden, A. Vehicle-to-Grid (V2G) as line-side energy storage for support of DC-powered electric railway systems. *J. Rail Transp. Plan. Manag.* **2021**, *19*, 100263. [CrossRef]
10. Carneiro, M.A.; Soares, D.D.F. Solar photovoltaic assistance system study for a Brazilian light rail vehicle. *U.Porto J. Eng.* **2020**, *6*, 35–45. [CrossRef]
11. Zare, A.; Stevanovic, S.; Jafari, M.; Verma, P.; Babaie, M.; Yang, L.; Rahman, M.M.; Ristovski, Z.D.; Brown, R.J.; Bodisco, T.A Analysis of cold-start NO_2 and NOx emissions, and the NO_2/NOx ratio in a diesel engine powered with different diesel-biodiesel blends. *Environ. Pollut.* **2021**, *290*, 118052. [CrossRef] [PubMed]
12. Kurczyński, D. Effect of the rme biodiesel on the diesel engine fuel consumption and emission. *Commun. Sci. Lett. Univ. Zilina* **2021**, *23*, B308–B316. [CrossRef]

13. Thongchai, S.; Lim, O. Macroscopic spray behavior of a single-hole common rail diesel injector using gasoline-blended 5% biodiesel. *Energies* **2020**, *13*, 2276. [CrossRef]
14. Li, F.; Lin, Q. Radial development of pentanol-biodiesel fuel spray in a high-pressure common-rail system. *Phys. Fluids* **2021**, *33*, 095115. [CrossRef]
15. Zhu, H.Y.; Fang, J.H.; Zhang, W.G.; Huang, Z. Spray characteristics of biodiesel within a common rail system. *K. Cheng Je Wu Li Hsueh Pao/J. Eng. Thermophys.* **2011**, *32*, 1785–1788.
16. Aminzadegan, S.; Shahriari, M.; Mehranfar, F.; Abramović, B. Factors affecting the emission of pollutants in different types of transportation: A literature review. *Energy Rep.* **2022**, *8*, 2508–2529. [CrossRef]
17. Chen, Z. Socioeconomic Impacts of high-speed rail: A bibliometric analysis. *Socio-Econ. Plan. Sci.* **2022**, *85*, 101265. [CrossRef]
18. Da Fonseca-Soares, D.; Eliziário, S.A.; Galvinicio, J.D.; Ramos-Ridao, A.F. Life-Cycle Greenhouse Gas (GHG) Emissions Calculation for Urban Rail Transit Systems: The Case of Pernambuco Metro. *Appl. Sci.* **2023**, *13*, 8965. [CrossRef]
19. Gulcimen, S.; Aydogan, E.K.; Uzal, N. Life cycle sustainability assessment of a light rail transit system: Integration of environmental, economic, and social impacts. *Integr. Environ. Assess. Manag.* **2021**, *17*, 1070–1082. [CrossRef]
20. Ramos da Silva, T.; Moura, B.; Monteiro, H. Life Cycle Assessment of Current Portuguese Railway and Future Decarbonization Scenarios. *Sustainability* **2023**, *15*, 11355. [CrossRef]
21. Hausberger, L.; Cordes, T.; Gschösser, F. Life Cycle Assessment of High-Performance Railway Infrastructure, Analysis of Superstructures in Tunnels and on Open Tracks. *Sustainability* **2023**, *15*, 7064. [CrossRef]
22. Keiser, D.; Schnoor, L.H.; Pupkes, B.; Freitag, M. Life cycle assessment in aviation: A systematic literature review of applications, methodological approaches and challenges. *J. Air Transp. Manag.* **2023**, *110*, 102418. [CrossRef]
23. Merchan, A.L.; Belboom, S.; Léonard, A. Life cycle assessment of rail freight transport in Belgium. *Clean Technol. Environ. Policy* **2020**, *22*, 1109–1131. [CrossRef]
24. Gebler, M.; Cerdas, J.F.; Thiede, S.; Herrmann, C. Life cycle assessment of an automotive factory: Identifying challenges for the decarbonization of automotive production—A case study. *J. Clean. Prod.* **2020**, *270*, 122330. [CrossRef]
25. Lee, J.Y.; Lee, C.K.; Chun, Y.Y. Greenhouse gas emissions from high-speed rail infrastructure construction in Korea. *Transp. Res. Part D Transp. Environ.* **2020**, *87*, 102514. [CrossRef]
26. Da Fonseca-Soares, D.; Galvinicio, J.D.; Eliziário, S.A.; Ramos-Ridao, A.F. A Bibliometric Analysis of the Trends and Characteristics of Railway Research. *Sustainability* **2022**, *14*, 13956. [CrossRef]
27. Thondoo, M.; Marquet, O.; Márquez, S.; Nieuwenhuijsen, M.J. Small cities, big needs: Urban transport planning in cities of developing countries. *J. Transp. Health* **2020**, *19*, 100944. [CrossRef]
28. Hamurcu, M.; Eren, T. Strategic planning based on sustainability for urban transportation: An application to decision-making. *Sustainability* **2020**, *12*, 3589. [CrossRef]
29. Li, Y.; He, Q.; Luo, X.; Zhang, Y.R.; Dong, L. Calculation of life-cycle greenhouse gas emissions of urban rail transit systems: A case study of Shanghai Metro. *Resour. Conserv. Recycl.* **2018**, *128*, 451–457. [CrossRef]
30. Chang, T.; Yang, D.; Yang, Y.; Huo, J.; Wang, G.; Xiong, C. Impact of urban rail transit on business districts based on time distance: Urumqi light rail. *J. Urban Plan. Dev.* **2018**, *144*, 04018024. [CrossRef]
31. Lazzerini, G.; Lucchetti, S.; Nicese, F.P. Green House Gases(GHG) emissions from the ornamental plant nursery industry: A Life Cycle Assessment(LCA) approach in a nursery district in central Italy. *J. Clean. Prod.* **2016**, *112*, 4022–4030. [CrossRef]
32. *ISO 14040:2006*; Environmental Management—Life Cycle Assessment—Principles and Framework. ISO: Geneva, Switzerland, 2006; pp. 1–20.
33. *ISO 14044:2006*; Environmental Management—Life Cycle Assessment—Requirements and Guidelines. ISO: Geneva, Switzerland, 2006; pp. 1–46.
34. *SimaPro*, Version 8.4.1; PRe Sustainability: Amersfoort, The Netherlands, 2018. Available online: https://support.simapro.com/s/ (accessed on 12 March 2022).
35. Liljenstrom, C.; Toller, S.; Akerman, J.; Bjorklund, A. Annual climate impact and primary energy use of Swedish transport infrastructure. *Eur. J. Transp. Infrastruct. Res.* **2019**, *19*, 77–116. [CrossRef]
36. Kimball, M.; Chester, M.; Gino, C.; Reyna, J. Assessing the Potential for Reducing Life-Cycle Environmental Impacts through Transit-Oriented Development Infill along Existing Light Rail in Phoenix. *J. Plan. Educ. Res.* **2013**, *33*, 395–410. [CrossRef]
37. Puig-Samper Naranjo, G.; Bolonio, D.; Ortega, M.F.; García-Martínez, M.-J. Comparative life cycle assessment of conventional, electric and hybrid passenger vehicles in Spain. *J. Clean. Prod.* **2021**, *291*, 125883. [CrossRef]
38. Olugbenga, O.; Kalyviotis, N.; Saxe, S. Embodied emissions in rail infrastructure: A critical literature review. *Environ. Res. Lett.* **2019**, *14*, 21. [CrossRef]
39. WRI Brasil. Greenhouse gas protocol calculation tool for forestry in Brazil. Technical note. Janeiro de 2020. 32p. Available online: https://wribrasil.org.br/sites/default/files/ghg_protocolo-florestastechnicalnote.pdf (accessed on 23 November 2024).
40. Giunta, M.; Leonardi, G. Framework for Life Cycle Railway Sustainability Assessment: A Methodological Approach Based on Advanced Methods and Tools. In Proceedings of the Computational Science and Its Applications—ICCSA 2024 Workshops, Hanoi, Vietnam, 1–4 July 2024; pp. 233–244.
41. Rempelos, G.; Preston, J.; Blainey, S. A carbon footprint analysis of railway sleepers in the United Kingdom. *Transp. Res. Part D Transp. Environ.* **2020**, *81*, 24. [CrossRef]

42. Chang, Y.; Lei, S.H.; Teng, J.J.; Zhang, J.X.; Zhang, L.X.; Xu, X. The energy use and environmental emissions of high-speed rail transportation in China: A bottom-up modeling. *Energy* **2019**, *182*, 1193–1201. [CrossRef]
43. Wang, L.; Jiao, G.H.; Lu, H.S.; Wang, Q.Q.; IOP. Environmental Load of Iron Ore Transportation, Bayan Obo Mine, China. In Proceedings of the International Conference on Air Pollution and Environmental Engineering (APEE), Hong Kong, China, 26–28 October 2018.
44. Zhang, H.; Li, S.; Wang, Z.; Li, M.; Yu, C.; Ren, S.; Sun, K.; Li, G.; Wang, J. Emission performance of gasoline homogeneous charge induced ignition in multi-cylinder diesel engine. *Nongye Gongcheng Xuebao/Trans. Chin. Soc. Agric. Eng.* **2017**, *33*, 108–114.
45. EPA. Greenhouse Gas Inventory Guidance: Direct Emissions from Mobile Combustion Source. 2020. Available online: https://www.epa.gov/sites/default/files/2020-12/documents/mobileemissions.pdf (accessed on 13 August 2023).
46. Banar, M.; Özdemir, A. An evaluation of railway passenger transport in Turkey using life cycle assessment and life cycle cost methods. *Transp. Res. Part D Transp. Environ.* **2015**, *41*, 88–105. [CrossRef]
47. Pons, J.J.; Villalba Sanchis, I.; Insa Franco, R.; Yepes, V. Life cycle assessment of a railway tracks substructures: Comparison of ballast and ballastless rail tracks. *Environ. Impact Assess. Rev.* **2020**, *85*, 106444. [CrossRef]
48. Iyer, K.; Wennhagea, P.; Åkermo, M. Life-cycle Assessment of a Composite Railway bogie frame. *Procedia CIRP* **2024**, *122*, 988–993. [CrossRef]
49. Soleimani-Chamkhorami, K.; Garmabaki, A.H.S.; Kasraei, A.; Famurewa, S.M.; Odelius, J.; Strandberg, G. Life cycle cost assessment of railways infrastructure asset under climate change impacts. *Transp. Res. Part D Transp. Environ.* **2024**, *127*, 104072. [CrossRef]
50. Chang, B.; Kendall, A. Life cycle greenhouse gas assessment of infrastructure construction for California's high-speed rail system. *Transp. Res. Part D Transp. Environ.* **2011**, *16*, 429–434. [CrossRef]
51. Montoya-Torres, J.; Akizu-Gardoki, O.; Iturrondobeitia, M. Measuring life-cycle carbon emissions of private transportation in urban and rural settings. *Sustain. Cities Soc.* **2023**, *96*, 104658. [CrossRef]
52. MCTI. Ministério da Ciência, Tecnologia e Inovação: Fatores de emissão de CO2 pela geração de energia elétrica no Sistema Interligado Nacional do Brasil—Ano Base 2019. Available online: https://www.gov.br/mcti/pt-br/acompanhe-o-mcti/sirene/dados-e-ferramentas/fatores-de-emissao (accessed on 12 March 2022).
53. Disney, J.; Rossiter, W.; Smith, D.J. Nottingham Express Transit: The role of green innovation in the drive for sustainable mobility through improved public transport. *Int. J. Entrep. Innov.* **2018**, *19*, 56–68. [CrossRef]
54. Bates, T.B.; Jablonski, P. San Diego, California, Trolley's new green line: Early success for distinctive service. *Transp. Res. Rec.* **2007**, *2006*, 40–49. [CrossRef]
55. Qi, Y.Y.; Dai, H.Y.; Gao, H.; Gan, F. Dynamic analysis of high-speed train with considering drive system. *Zhendong Gongcheng Xuebao/J. Vib. Eng.* **2019**, *32*, 176–183.
56. Chu, W.J.; Zhang, X.C.; Chen, J.H.; Xu, B. An ELM-Based Approach for Estimating Train Dwell Time in Urban Rail Traffic. *Math. Probl. Eng.* **2015**, *2015*, 473432. [CrossRef]
57. Jackson, D.J.; Brander, M. The risk of burden shifting from embodied carbon calculation tools for the infrastructure sector. *J. Clean. Prod.* **2019**, *223*, 739–746. [CrossRef]
58. Mao, J.; Xiang, J.; Yu, C.; Liu, X.; Gong, K. Wear and stability analysis of the train-curve rail system with an elastic anti-friction guard rail. *Dyna* **2015**, *90*, 532–539. [CrossRef]
59. Fan, L.; Bai, Y.; Ma, X.; Song, E. Analysis upon fuel injection quantity variation of common rail system for diesel engines. *J. Mech. Sci. Technol.* **2016**, *30*, 3365–3377. [CrossRef]
60. Li, Q.; Backes, F.; Wachtmeister, G. Application of canola oil operation in a diesel engine with common rail system. *Fuel* **2015**, *159*, 141–149. [CrossRef]
61. Shao, L.M.; An, S.J.; Chang, H.B. Calculation and analysis of combustion and emission of a diesel engine with high pressure common rail system. *Neiranji Gongcheng/Chin. Intern. Combust. Engine Eng.* **2005**, *26*, 1–4.
62. Zhang, Z.; Zhao, F.; Deng, J.; Li, L.; Shen, Y. Characteristics of the multiple injection diesel spray employed common rail system *J. Therm. Sci. Technol.* **2013**, *8*, 106–119. [CrossRef]
63. Kuti, O.A.; Zhu, J.; Nishida, K.; Wang, X.; Huang, Z. Characterization of spray and combustion processes of biodiesel fuel injected by diesel engine common rail system. *Fuel* **2013**, *104*, 838–846. [CrossRef]
64. Zheng, J.B.; Miao, X.L.; Hong, J.H.; Wang, X.Y.; Chen, X.Y.; Ju, Y.S.; Bi, F.Y. Combustion delay of premixed combustion in diesel engine. *Neiranji Xuebao/Trans. CSICE (Chin. Soc. Intern. Combust. Engines)* **2011**, *29*, 241–247.
65. Mohamed Ibrahim, M.; Varuna Narasimhan, J.; Ramesh, A. Comparison of the predominantly premixed charge compression ignition and the dual fuel modes of operation with biogas and diesel as fuels. *Energy* **2015**, *89*, 990–1000. [CrossRef]
66. Krasinski, A.; Gradon, L. Comprehensive system for diesel fuel cleaning. *Filtr. Sep.* **2016**, *53*, 33–36. [CrossRef]
67. Liu, L.; Yang, F.Y.; Ouyang, M.G.; Fang, C.; Jiang, Z.J.; Hao, S.G. Control and optimization of diesel/LNG system of a 4G200 diesel engine. *Jilin Daxue Xuebao (Gongxueban)/J. Jilin Univ. (Eng. Technol. Ed.)* **2015**, *45*, 414–419.
68. Alcántara-Carmona, A.; Sáez-Bastante, J.; Dorado, M.P. Corrosion testing of a diesel engine common rail system using various types of biodiesel. *Renew. Energy Power Qual. J.* **2017**, *1*, 427–432. [CrossRef]
69. Song, G.; Li, J.; Hu, L.; Jiang, Z. Development of magnetoelectric speed sensor for high pressure common rail system. *Yi Qi Yi Biao Xue Bao/Chin. J. Sci. Instrum.* **2006**, *27*, 290–294.

70. Ferrari, A.; Salvo, E. Determination of the transfer function between the injected flow-rate and high-pressure time histories for improved control of common rail diesel engines. *Int. J. Engine Res.* **2017**, *18*, 212–225. [CrossRef]
71. Gong, W.G.; Lou, J.P.; Ping, Y.S.; Zhu, G.H.; Wang, Q. Development and application of Model D19 Diesel Engine for passenger car. *Neiranji Gongcheng/Chin. Intern. Combust. Engine Eng.* **2013**, *34*, 5–12.
72. Donnici, G.; Rocchi, I.; Pezzuti, E. Aircraft diesel engine turbocompound optimized. *Far East J. Electron. Commun.* **2014**, *13*, 137–154.
73. Shin, J.; Hong, S.; Park, I.; Sunwoo, M. Common rail pressure control algorithm for passenger car diesel engines using quantitative feedback theory. *Trans. Korean Soc. Mech. Eng. B* **2014**, *38*, 107–114. [CrossRef]
74. Li, F.; Fu, W.; Yi, B.; Song, L.; Liu, T.; Wang, X.; Wang, C.; Lei, Y.; Lin, Q. Comparison of macroscopic spray characteristics between biodiesel-pentanol blends and diesel. *Exp. Therm. Fluid Sci.* **2018**, *98*, 523–533. [CrossRef]
75. Zhang, X.; Wei, J.; Lü, G.; Song, C. A Comparison of Number and Size Distribution and Mass of Particulates for Biodiesel and Diesel in Post-Injection Combustion Process. *Ranshao Kexue Yu Jishu/J. Combust. Sci. Technol.* **2017**, *23*, 261–267.
76. Ferrari, A.; Novara, C.; Paolucci, E.; Vento, O.; Violante, M.; Zhang, T. Design and rapid prototyping of a closed-loop control strategy of the injected mass for the reduction of CO_2, combustion noise and pollutant emissions in diesel engines. *Appl. Energy* **2018**, *232*, 358–367. [CrossRef]
77. Liu, X.; Zhang, H.M.; Ji, L.W.; Li, S.A. Development of common rail system and performance researching of electronic-controlled diesel engine. *Neiranji Gongcheng/Chin. Intern. Combust. Engine Eng.* **2005**, *26*, 1–5.
78. Matsumoto, K. Development of diesel common rail system and production Technology. *Seimitsu Kogaku Kaishi/J. Jpn. Soc. Precis. Eng.* **2009**, *75*, 140–141. [CrossRef]
79. Thongchai, S.; Lim, O. Investigation of the combustion characteristics of gasoline compression ignition engine fueled with gasoline-biodiesel blends. *J. Mech. Sci. Technol.* **2018**, *32*, 959–967. [CrossRef]
80. Vargas, F.; Pérez, A.; Delgado, R.; Hernández, E.; Suástegui, J.A. Performance analysis of a compression ignition engine using mixture biodiesel palm and diesel. *Sustainability* **2019**, *11*, 4918. [CrossRef]
81. Fang, W.; Huang, B.; Kittelson, D.B.; Northrop, W.F. Dual-Fuel diesel engine combustion with hydrogen, gasoline, and ethanol as fumigants: Effect of diesel injection timing. *J. Eng. Gas Turbines Power* **2014**, *136*, 081502. [CrossRef]
82. Fernández-Ríos, A.; Santos, G.; Pinedo, J.; Santos, E.; Ruiz-Salmón, I.; Laso, J.; Lyne, A.; Ortiz, A.; Ortiz, I.; Irabien, Á.; et al. Environmental sustainability of alternative marine propulsion technologies powered by hydrogen—A life cycle assessment approach. *Sci. Total Environ.* **2022**, *820*, 153189. [CrossRef] [PubMed]
83. Ibrahim, M.M.; Ramesh, A. Experimental investigations on a hydrogen diesel homogeneous charge compression ignition engine with exhaust gas recirculation. *Int. J. Hydrogen Energy* **2013**, *38*, 10116–10125. [CrossRef]
84. Ciancetta, F.; Ometto, A.; D'ovidio, G.; Masciovecchio, C. Modeling, Analysis and Implementation of an Urban Electric Light-Rail Train Hydrogen Powered. *Int. Rev. Electr. Eng.* **2019**, *14*, 237–245. [CrossRef]
85. Okyere, S.; Yang, J.; Aminatou, M.; Tuo, G.; Zhan, B. Multimodal transport system effect on logistics responsive performance: Application of ordinal logistic regression. *Eur. Transp. Trasp. Eur.* **2018**, *68*, 1–17.
86. Perez-Martinez, P.J.; Miranda, R.M.; Andrade, M.F. Freight road transport analysis in the metro Sao Paulo: Logistical activities and CO_2 emissions. *Transp. Res. Part A Policy Pract.* **2020**, *137*, 16–33. [CrossRef]
87. Holler Branco, J.E.; Bartholomeu, D.B.; Alves Junior, P.N.; Caixeta Filho, J.V. Evaluation of the economic and environmental impacts from the addition of new railways to the brazilian's transportation network: An application of a network equilibrium model. *Transp. Policy* **2022**, *124*, 61–69. [CrossRef]
88. Holian, M.J.; Kahn, M.E. Household demand for low carbon policies: Evidence from california. *J. Assoc. Environ. Resour. Econ.* **2015**, *2*, 205–234. [CrossRef]
89. Wen, L.; Wang, A. System dynamics model of Beijing urban public transport carbon emissions based on carbon neutrality target. *Environ. Dev. Sustain.* **2022**, *25*, 12681–12706. [CrossRef]
90. Dias, P.; Palomero, J.; Cenci, M.P.; Scarazzato, T.; Bernardes, A.M. Electronic waste in Brazil: Generation, collection, recycling and the covid pandemic. *Cleaner Waste Systems* **2022**, *3*, 100022. [CrossRef]
91. Brandao, R.; Edwards, D.J.; Melo, A.C.S.; Macedo, A.N. Reverse supply chain practices for construction and demolition waste in the Brazilian Amazon: A multi-stakeholder view. *J. Mater. Cycles Waste Manag.* **2023**, *25*, 2245–2261. [CrossRef]
92. Ingrao, C.; Scrucca, F.; Matarazzo, A.; Arcidiacono, C.; Zabaniotou, A. Freight transport in the context of industrial ecology and sustainability: Evaluation of uni-and multi-modality scenarios via life cycle assessment. *Int J Life Cycle Assess* **2021**, *26*, 127–142. [CrossRef]
93. Bilgili, L.; Kuzu, S.L.; Çetinkaya, A.Y.; Kumar, P. Evaluation of railway versus highway emissions using LCA approach between the two cities of Middle Anatolia. *Sustain. Cities Soc.* **2019**, *49*, 101635. [CrossRef]
94. Landgraf, M.; Horvath, A. Embodied greenhouse gas assessment of railway infrastructure: The case of Austria. *Environ. Res. Infrastruct. Sustain.* **2021**, *1*, 025008. [CrossRef]
95. De Andrade, C.E.S.; D'Agosto, M.A. Energy use and carbon dioxide emissions assessment in the lifecycle of passenger rail systems: The case of the Rio de Janeiro Metro. *J. Clean. Prod.* **2016**, *126*, 526–536. [CrossRef]
96. Beckert, P.; Pareschi, G.; Ehwald, J.; Sacchi, R.; Bauer, C. Fast as a plane, clean as a train? Prospective life cycle assessment of a hyperloop system. *Resour. Environ. Sustain.* **2024**, *17*, 100162. [CrossRef]

97. Chen, L.; Jung, J.; Sohn, G. Multi-Scale Hierarchical CRF for Railway Electrification Asset Classification from Mobile Laser Scanning Data. *IEEE J. Sel. Top. Appl. Earth Obs. Remote Sens.* **2019**, *12*, 3131–3148. [CrossRef]
98. Dalkic, G.; Balaban, O.; Tuydes-Yaman, H.; Celikkol-Kocak, T. An assessment of the CO_2 emissions reduction in high speed rail lines: Two case studies from Turkey. *J. Clean. Prod.* **2017**, *165*, 746–761. [CrossRef]
99. Soffker, C.; Gralla, C. "Smart" power supply for urban rail systems thanks to Hesop. *Elektr. Bahnen* **2016**, *114*, 10–16.
100. Masirin, M.I.M.; Salin, A.M.; Zainorabidin, A.; Martin, D.; Samsuddin, N. Review on Malaysian Rail Transit Operation and Management System: Issues and Solution in Integration. *IOP Conf. Ser. Mater. Sci. Eng.* **2017**, *226*, 012029. [CrossRef]
101. Mayba, I.A.; Glazunov, D.V. Optimization of Tribotechnical Characteristics of Wheel–Rail Friction Modifiers. *J. Frict. Wear* **2020**, *41*, 517–520. [CrossRef]
102. Gerlici, J.; Gorbunov, M.; Kravchenko, K.; Domin, R.; Kovtanets, M.; Lack, T. Slipping and Skidding Occurrence Probability Decreasing by Means of the Friction Controlling in the Wheel-Braking Pad and Wheel-Rail Contacts. *Manuf. Technol.* **2017**, *17*, 17–18. [CrossRef]
103. Cui, X.; He, Z.; Huang, B.; Chen, Y.; Du, Z.; Qi, W. Study on the effects of wheel-rail friction self-excited vibration and feedback vibration of corrugated irregularity on rail corrugation. *Wear* **2021**, *477*, 203854. [CrossRef]
104. Sánchez, J.C.; Jaramillo, J.A.; Santa, J.F.; Toro, A. Twin-disc assessment of the effect of top-of-rail friction modifiers on the tribological response of er8-r370ht pairs for use in wheel-rail systems. *Rev. Latinoam. Metal. Mater.* **2019**, *39*, 2–15.
105. Gassner, A.; Lederer, J.; Kanitschar, G.; Ossberger, M.; Fellner, J. Extended ecological footprint for different modes of urban public transport: The case of Vienna, Austria. *Land Use Policy* **2018**, *72*, 85–99. [CrossRef]
106. Yao, E.; Cheng, X.; Liu, S.; Zhang, R. Accessibility-based forecast on passenger flow entering and departing existing urban railway stations. *Tiedao Xuebao/J. China Railw. Soc.* **2016**, *38*, 1–7.
107. Bonotti, R.; Rossetti, S.; Tiboni, M.; Tira, M. Analysing Space-Time Accessibility Towards the Implementation of the Light Rail System: The Case Study of Brescia. *Plan. Pract. Res.* **2015**, *30*, 424–442. [CrossRef]
108. Moyano, A.; Martínez, H.S.; Coronado, J.M. From network to services: A comparative accessibility analysis of the Spanish high-speed rail system. *Transp. Policy* **2018**, *63*, 51–60. [CrossRef]
109. Liu, Z.; Zhu, A.X.; Zhang, W.; Ren, M. An improved potential-based approach to measuring the daily accessibility of HSR. *Transp. Res. Part A Policy Pract.* **2021**, *150*, 271–284. [CrossRef]
110. Janic, M. Light rail rapid transit systems for more sustainable ground accessibility of airports. *Transp. Plan. Technol.* **2011**, *34*, 569–592. [CrossRef]
111. Hong, L.; Ouyang, M.; Xu, M.; Hu, P. Time-varied accessibility and vulnerability analysis of integrated metro and high-speed rail systems. *Reliab. Eng. Syst. Saf.* **2020**, *193*, 106622. [CrossRef]
112. Ferreira, A.F.; Leite, A.D.; Pereira, L.D.F.; Neves, J.M.D.J.; Oliveira Pinheiro, M.G.D.; Chang, S.K.J. Wheelchair accessibility of urban rail systems: Some preliminary findings of a global overview. *IATSS Res.* **2021**, *45*, 326–335. [CrossRef]
113. Carneiro, M.A.; Da Fonseca-Soares, D.; Pereira, L.H.M.; Ramos-Ridao, A.F. An Approach for Water and Energy Savings in Public Buildings: A Case Study of Brazilian Rail Company. *Sustainability* **2022**, *14*, 15921. [CrossRef]

buildings

MDPI

Article

Layout Optimization of Residential Buildings to Improve the Outdoor Microclimate of Neighborhoods Along an Urban Bay: A Case Study of Shantou's Inner Bay, China

Wenqing Liu [1,*], Chang Miao [2], Lei Xiao [1], Junhang Mai [3] and Yingzi Wang [1,*]

[1] Department of Civil Engineering and Intelligent Construction, College of Engineering, Laboratory of Vernacular Architecture Surveying and Mapping, Shantou University, Shantou 515063, China; rayshaw674@gmail.com

[2] Key Laboratory of Cold Region Urban and Rural Human Settlement Environment Science and Technology, School of Architecture and Design, Harbin Institute of Technology, Harbin 150006, China; changmiao0211@gmail.com

[3] School of Architecture, South China University of Technology, Guangzhou 510641, China; stmichaelhang@163.com

* Correspondence: wqliu@stu.edu.cn (W.L.); yzwang@stu.edu.cn (Y.W.)

Abstract: In summer, the urban heat island effect causes unbearable warmth in Shantou City, especially in the urban areas along the Inner Bay with densely populated neighborhoods. An investigation of the layout patterns of 100 residential neighborhoods along Shantou's Inner Bay was conducted, leading to the establishment of four types and nine sub-types of idealized residential neighborhood models. Their wind speed, relative humidity, and air temperature were simulated in the ENVI-met software (version No.5.0.1). The simulation results show that high-rise buildings in the front areas play a decisive role in the overall microclimate environment. Accordingly, three principal drawbacks regarding neighborhood layout for thermal climate adaptation were extracted. Furthermore, by comparing the simulation results before and after modifying the layout of high-rise buildings, three spatial strategies to strengthen the humidification and cooling effect of sea–land breezes to optimize the outdoor microclimatic environment of neighborhoods were proposed, and these strategies were subsequently verified in the Golden-Harbor neighborhood.

Keywords: residential buildings; outdoor microclimate; layout optimization; urban bay

1. Introduction

1.1. Background

Over the past 50 years, more than 11,000 weather, climate, and water-related disasters have been reported globally [1]. Data from UN-Habitat's *World Cities Report 2020* suggest that urbanization will continue to accelerate over the next decade. By 2030, it is projected that the urban population will rise from 56.2% to 60.4% [2]. The ongoing rise in urban populations is expanding residential areas, which, in turn, amplifies the urban heat island (UHI) effect. This effect, a consequence of urbanization, is increasingly affecting urban climate change [3–5]. The UHI effect will have a long-term impact on people. When combined with extremely high temperatures, the thermal comfort of urban areas will be worsened, which may even increase mortality from heat-related diseases [6,7]. The above environmental and health problems could be alleviated by exploring the correlation between space form and local microclimate from the perspective of urban design and by proposing optimization strategies [8–12].

Amid global climate change and the UHI effect, China has released the *Green Building Evaluation Standard* (GB/T 50378-2019) [13] and the Carbon Peaking and Carbon

Neutrality Goals. The microclimate adaptation of urban neighborhoods has become a focus, inspiring scholars to identify the affecting and controlling factors of the microclimate of urban neighborhoods.

Recent studies have shown that building layout has a significant impact on the outdoor microclimate of residential areas, especially in terms of thermal comfort [14]. The outdoor microclimate can be improved by altering the morphological parameters of residential buildings, such as the building height, plot ratio, building coverage, and volume ratio of buildings and green plants [15,16]. In addition, the layout optimization of residential buildings also has the potential to reduce energy consumption and carbon emissions [17,18]. These studies focus on inland areas; relevant research on residential areas close to the inner coast is insufficient. Thus, this study aims to optimize building layouts to improve the microclimate of neighborhoods along Shantou's Inner Bay and similar areas.

1.2. Research Object and Theoretical Framework

Shantou is a coastal city in southeast China with an inner bay urban area that has become one of the core areas for urban development and construction since China's reform and opening up, and a comparison of satellite maps shows that neighborhoods in the area have surged in recent decades (Figure 1). The numerous high-rise residential buildings along the coastline of Shantou's Inner Bay have altered the outdoor microclimate of traditional coastal neighborhoods, and their negative influence has even extended to the broader urban area. The impact of the layout of new residential buildings on the microclimate of coastal urban areas and the corresponding optimization have emerged as critical contemporary issues.

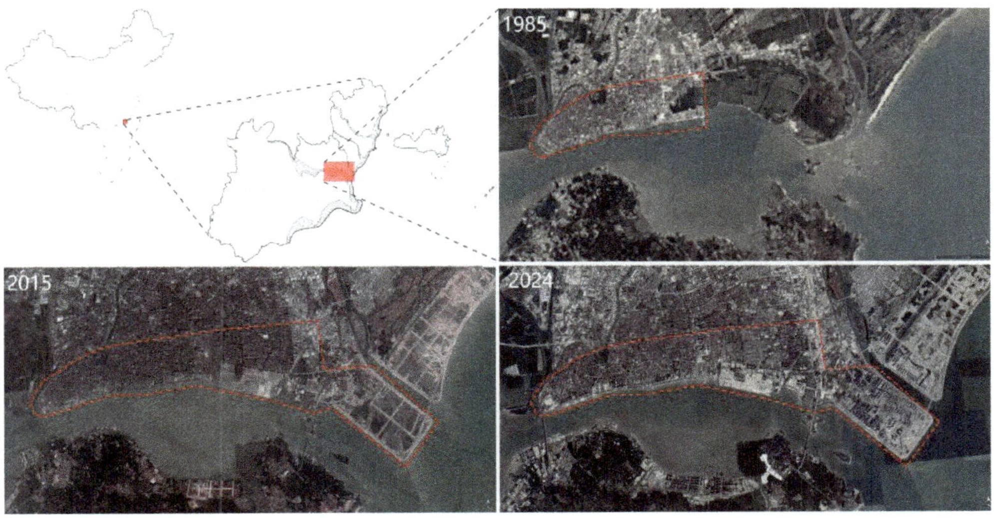

Figure 1. Comparison of satellite maps of built areas along Shantou's Inner Bay.

As shown in the Figure 2, this study includes four stages. In the first stage, according to climate statistics showing summer high-temperature days in the past 20 years for Shantou, it can be found that the availability of sea–land breeze in this area has been weakened, and the large built areas generated in the process of rapid urbanization have worsened the urban climate. In the second stage, by investigating sufficient samples, the neighborhood morphology along the Inner Bay coast was classified, and the meteorological data of a typical high-temperature day in summer were adopted as a parameter source to construct microclimate simulation models in the ENVI-met software. In the third stage, the outdoor microclimate difference for each type was compared and analyzed, and layout patterns suitable for the humidification and cooling effect of coastal sea–land breeze were found. In the fourth stage, the optimization strategy for the layout of neighborhoods along the Inner Bay of Shantou was proposed, and the microclimate of out areas after layout optimization was simulated again to verify the effectiveness.

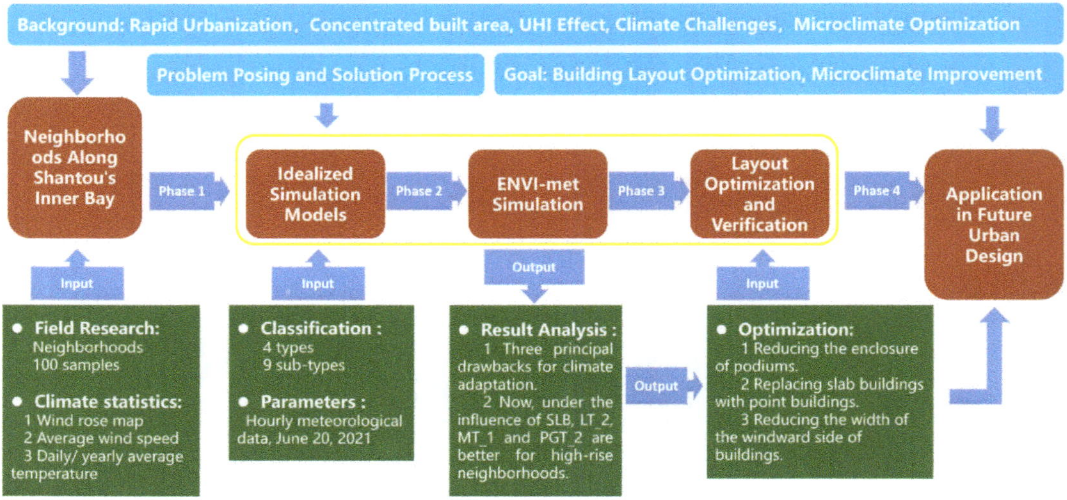

Figure 2. The methodological framework of this paper.

2. Urban Climate and Layout of Residential Neighborhoods Along Shantou's Inner Bay

2.1. Urban Climate of Shantou's Inner Bay

2.1.1. Sea–Land Breeze (SLB)

Changes in the lower urban interface and the urban spatial form caused by urbanization will affect the circulation characteristics of the sea–land breeze (SLB) [19–22]. Exploring the SLB may become an important way to mitigate the UHI effect. There is obviously an SLB in the coastal area of Shantou's Inner Bay. In 2016, there were 64 recorded days of SLB activities. The sea wind typically began at around 13:00, lasting for 4.8~9.5 h, with an average wind speed of 2.8 m/s, peaking most frequently at 16:00. Conversely, the land wind predominantly started at 01:00, with a duration of 5.5~8.6 h and an average wind speed of 2.4 m/s, peaking most often at 02:00. The SLB circulation was most pronounced in summer, with the longest duration and the highest wind speeds [23]. A case study of Jinjiang City adjacent to Shantou revealed that the SLB speed was negatively correlated with the UHI intensity in summer. In addition, SLB can cool down a city and weaken UHI intensity [24].

As the UHI effect mostly occurs in summer, its mitigation should be more targeted to the summer climate [25]. Using the ladybug and grasshopper plug-in of the Rhino software (version No.8.2), in combination with *the Chinese standard meteorological dataset* [26], wind speeds and directions were selected for the days when the air temperature in Shantou City surpassed 28 °C during the months of June to September. A wind rose diagram was then generated with the data (Figure 3). It shows that when the air temperature is higher than 28 °C, the east–southeast wind has the highest frequency, followed by the southeast wind. This indicates that the prevalent wind in hot summer is perpendicular to the residential blocks along Shantou's Inner Bay, which could improve microclimate obviously.

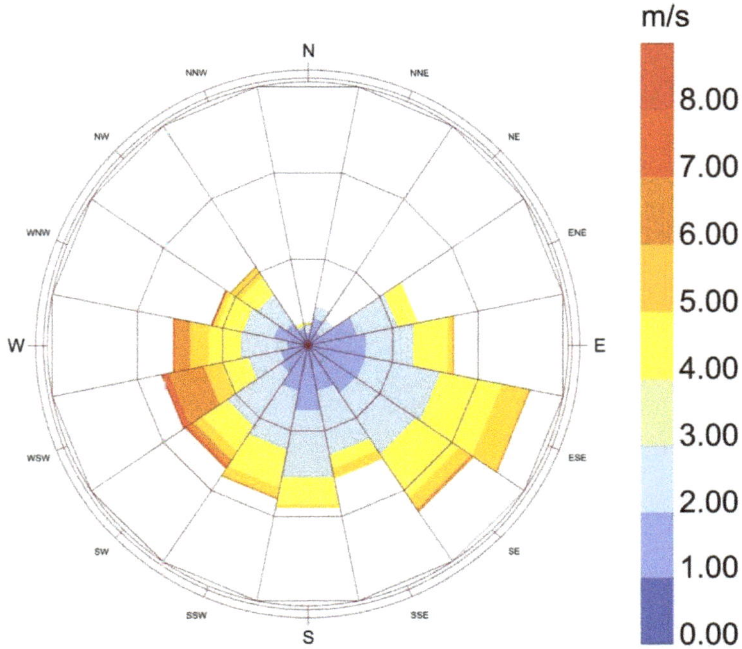

Figure 3. Wind rose map of high temperatures in Shantou City from June to September.

Based on the observation data of Shantou Meteorological Station (No. 59316, 116°41′ E, 23°24′ N) from 2000 to 2020 in *the daily value dataset of the surface climate data in China* "https://data.cma.cn/ (accessed on 10 May 2021)", the mean wind speed and temperature of Shantou City over the past 20 years were analyzed (Figures 4–6). Figure 3 presents a significant downward trend, with the annual average decreasing from 2.53 m/s in 2000 to 1.67 m/s in 2020. The urbanization of Shantou City has accelerated real estate development in coastal areas, and the emergence of numerous high-rise neighborhoods on the north and south sides of Shantou's Inner Bay has obstructed the SLB flow and adversely affected the mean wind speed. Compared to the case study of Jinjiang City [24], the decreasing wind speed in the coastal area of Shantou City will also weaken the microclimate-improving ability of SLB.

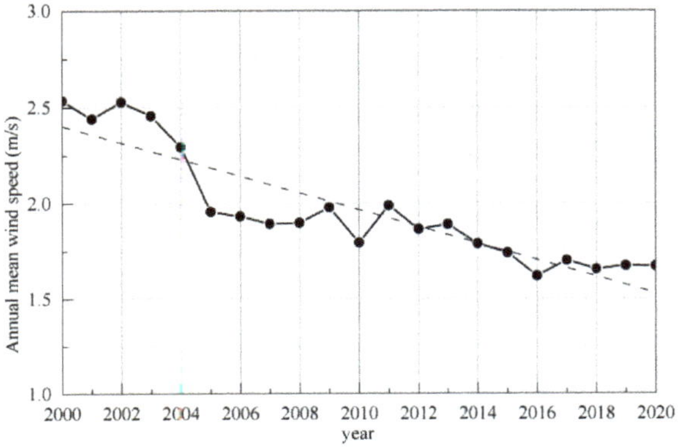

Figure 4. Statistics for average wind speed changes from 2000 to 2020 in Shantou City.

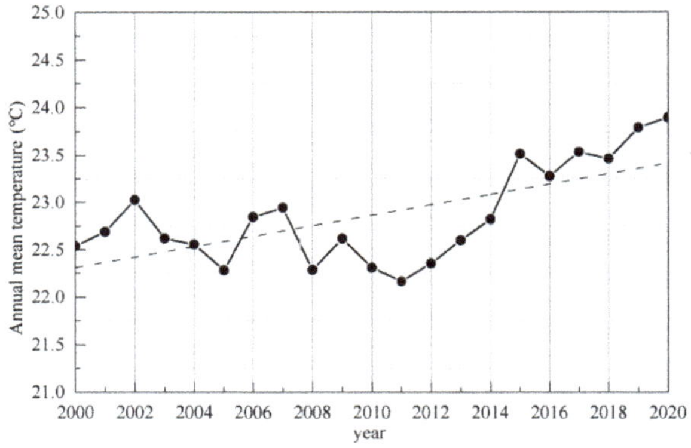

Figure 5. Annual average temperature statistics for Shantou from 2000 to 2020.

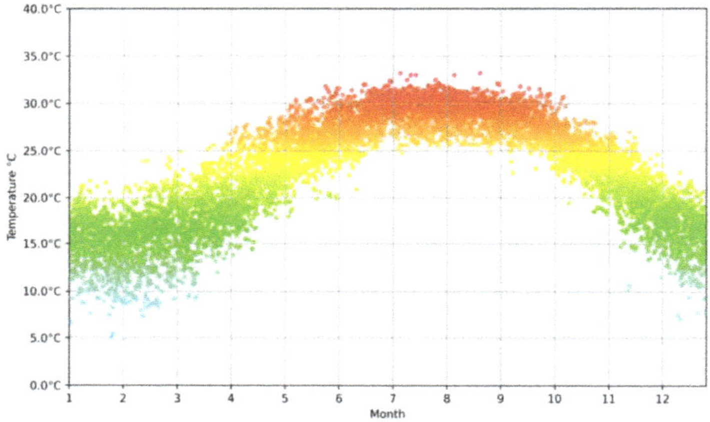

Figure 6. Daily average temperature statistics for Shantou from 2000 to 2020.

2.1.2. Land Surface Temperature

The annual mean temperature of Shantou City has seen a general upward trend in the past 20 years, increasing by 1.35 °C from 22.54 °C in 2000 to 23.89 °C in 2020 (Figure 5). Starting in 2011, the annual mean temperature has shown a continuous increase [27], indicating that the UHI effect in Shantou City has been aggravated over the last decade, necessitating urgent mitigation. The daily mean temperature in Shantou City from June to September is always higher than 25 °C and sometimes even higher than 35 °C (Figure 6). According to China's Measures for the Administration of Measures to Prevent Heat Stroke and Cooling, "scorching weather" is when the daily maximum temperature reaches or exceeds 35 °C, which makes people feel uncomfortable. "Scorching weather" in Shantou City is concentrated from June to September, making the outdoor microclimate of neighborhoods in this period the focus of this study.

2.2. Building Layout of Residential Neighborhoods Along Shantou's Inner Bay

Given the close connection between the layout pattern of urban neighborhoods and the outdoor microclimate, a layout tailored to the local climate is an effective way to improve microclimate, involving factors such as local wind speed, surface temperature, and thermal comfort [28–31]. Since the 1980s, Chinese cities have experienced rapid urban expansion, suburbanization, and urban sprawl [32–34]. Due to commercial real estate development, the spatial forms of Chinese urban neighborhoods and building layout patterns have gradually become homogeneous [35–37]. This has also happened in Shantou City. In this study, neighborhoods within 3 km of Shantou's Inner Bay were observed through satellite images and field research, and 100 neighborhoods were investigated (Figure 7).

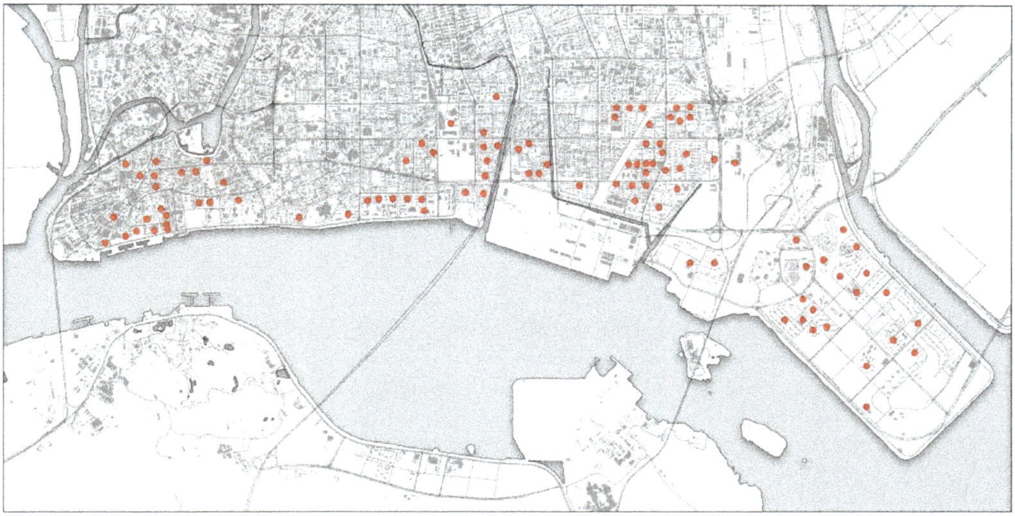

Figure 7. Distribution of neighborhoods within 3 km of Shantou's Inner Bay.

According to the above investigation of 100 neighborhoods along Shantou's Inner Bay, four layout patterns for residential buildings, i.e., linear type (63%), enclosed type (10%), point-group type (11%), and mixed type (16%), were extracted (Figure 8). Their relationships and distance from the coastline of Shantou's Inner Bay are shown in Figure 9. Within 0.5 km of the coastline, the proportion of linear types and enclosed types is highest; the mixed-type layouts are mainly found in a range of more than 2 km from the coastline; and the point-group type is distributed more evenly, with the highest proportion in a range of 1.5–2 km from the coastline. Moreover, the floor numbers of the surveyed neighborhoods include three types, i.e., multi-story (4–6 stories), mid-high-rise (7–11 stories), and high-rise (more than 12 stories), and their distribution in the Inner Bay is shown in Figure 9. Multi-story and mid-rise neighborhoods are mainly distributed within 0.5~1 km and 1.5 km of the coastline, respectively, while high-rise neighborhoods have the largest proportion in the area closest to the coastline.

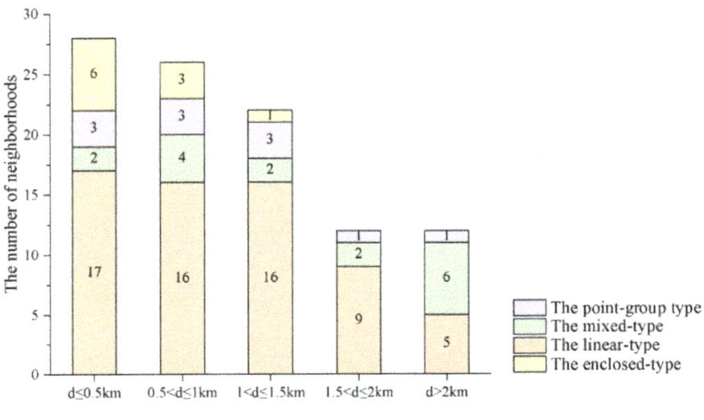

Figure 8. Distribution of different building layouts.

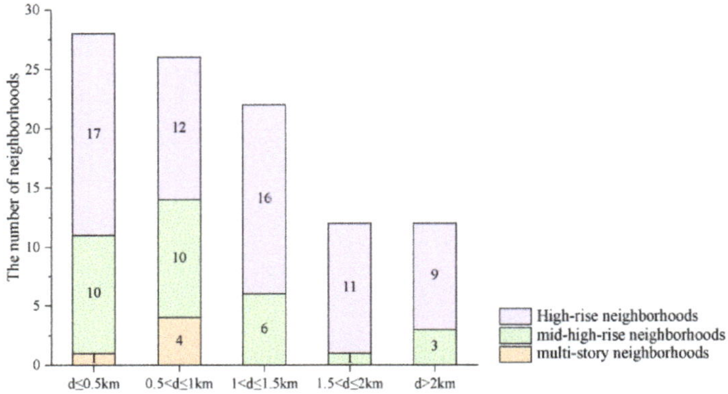

Figure 9. Distribution of different building heights.

3. Model Establishment and Simulation Parameter Settings

3.1. Spatial Models of Neighborhoods

The survey reveals that an urban spatial pattern of "Inner Bay–high-rise neighborhoods–multi-story neighborhoods" has emerged along Shantou's Inner Bay. High-rise neighborhoods exhibit a variety of spatial forms, including linear, enclosed, point-group, and mixed types. In contrast, multi-story neighborhoods predominantly feature a linear layout. To analyze these patterns, nine types of idealized models were developed and implemented within the ENVI-met software (Figure 10). ENVI-met is an urban climate modeling tool for simulating the microclimatic effects of buildings, vegetation, and other objects in urban design. It can be used to simulate and analyze urban microclimate simulations, the heat island effect, wind–heat environment and comfort, and air pollutant dispersion and has shown application value in sustainable building construction, healthy living environments, and urban space optimization. It has yielded reliable results in many urban research cases [38–41]. In idealized models of high-rise neighborhoods, the parameters are land area (2.8 hm^2), floor area ratio (3.5), the size of linear buildings (45*12 m), the size of point buildings (30*30), the height of a single floor (3 m), and the number of floors (20); for multi-story neighborhoods, they are land area (1.05 hm^2), the size of the building (30*12 m), building height (3 m), and the number of floors (8). The model area contains $80 \times 135 \times 30$ grids, with a resolution of 3 m.

The linear type is the main form of high-rise neighborhoods along Shantou's Inner Bay, involving three forms, LT_1, LT_2, and LT_3 (Figure 10a–c). The enclosed-type layout includes two types (ET_1 and ET_2) (Figure 10d,e). ET_1 is surrounded by high-rise buildings, while ET_2 is surrounded by a continuous commercial building of 2–3 floors and high-rise residential buildings. The point-group type is formed by point-type buildings, and its layout includes two types (PGT_1 and PGT_2) (Figure 10f,g). The buildings in the PGT_1 array are aligned but staggered in PGT_2. The spatial layout of the mixed-type layout, which consists of point-type and linear buildings, includes two forms (MT_1 and MT_2) (Figure 10h,i), with MT_1 mixed horizontally and MT_2 mixed vertically.

3.2. Simulation Parameter Settings

The simulation was conducted for the date of 20 June 2021, using hourly meteorological data from the National Meteorological Data Center, and this choice was mainly based on the following reasons. Firstly, this day is a typical meteorological day in simulation research, and it meets the research background of "high-temperature days in summer" for this study. More specifically, considering the longitude and latitude of Shantou, this day was the hottest day near the summer solstice of 2021, when our research started, with a maximum temperature of 35.9 °C and clear skies. Secondly, the simulation period is from

10:00 to 16:00, during which, the most abundant sunshine and the most serious urban heat island effect can be captured. In summary, taking the high-temperature period of a typical summer high-temperature day as the simulation period can highlight the real challenges faced by the outdoor microclimate of neighborhoods along Shantou's Inner Bay.

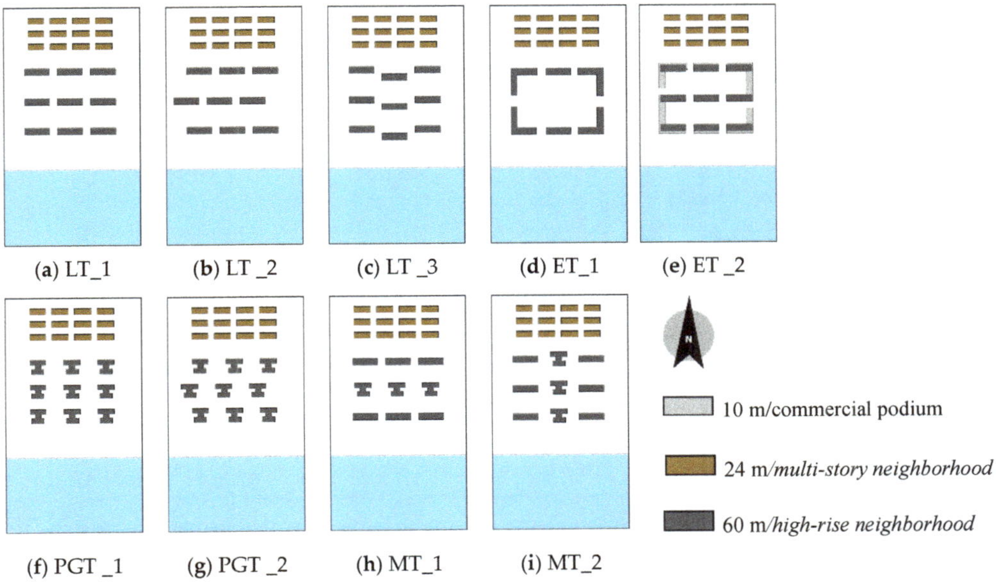

Figure 10. Plans for four types of idealized models.

Table 1 provides the hourly temperature and relative humidity. The boundary conditions for the simulation were set to Simple Forcing mode. The incoming wind direction was set to be the south wind perpendicular to the main facade according to wind rose analysis (Figure 2). Table 2 shows the parameter settings in ENVI-met. According to China's *Evaluation Standard for Green Buildings* (GB/T 50378-2019) [13] and *Thermal Environment Design Standard for Urban Residential Areas* (JGJ 286-2013) [42], air temperature, relative humidity, and wind speed were taken as the main outdoor thermal environment evaluation indexes, and simulation calculations were carried out for the outdoor microclimate environment of the nine residential layout forms in summer high temperatures at the pedestrian level (1.5 m).

Table 1. Hourly meteorological data of Shantou City on 20 June 2021.

Time	Air Temperature (°C)	Relative Humidity (%)	Time	Air Temperature (°C)	Relative Humidity (%)
00:00	30.4 °C	75%	12:00	34.7 °C	58%
01:00	30.3 °C	75%	13:00	35.7 °C	53%
02:00	30.1 °C	75%	14:00	35.9 °C	52%
03:00	29.8 °C	76%	15:00	34.5 °C	60%
04:00	29.7 °C	75%	16:00	33.5 °C	62%
05:00	29.7 °C	75%	17:00	33.3 °C	65%
06:00	29.6 °C	75%	18:00	32.6 °C	69%
07:00	30.0 °C	72%	19:00	31.8 °C	72%
08:00	30.4 °C	72%	20:00	31.2 °C	75%
09:00	31.4 °C	68%	21:00	31.0 °C	73%
10:00	32.7 °C	64%	22:00	30.8 °C	74%
11:00	33.7 °C	62%	23:00	30.6 °C	72%

Table 2. Simulation parameter settings for ENVI-met.

The ENVI-Met Parameter	Value
Simulation date	20 June 2021
Starting time	10:00
Time duration	6 h
Wind speed at the inlet boundary	2.09 m/s
Wind direction	180° (south wind)
Surface roughness	0.10

This singular analysis during the same time frame is based on the following considerations. Firstly, the chosen date and time for this singular analysis are representative of extreme climate conditions with a maximum temperature of 35.9 °C, which could be defined as "scorching weather" according to China's Measures for the Administration of Measures to Prevent Heat Stroke and Cooling. It represents the most challenging condition for neighborhoods along Shantou's Inner Bay in summer, and the optimization strategy based on this extreme situation can also be applied to the most common summer weather conditions in Shantou. Secondly, this study is designed to optimize building layouts to improve the microclimate of neighborhoods along Shantou's Inner Bay and similar areas. Our focus is on how to optimize the layout of different types of buildings (four types and nine sub-types) to adapt to the same high-temperature climate in summer, so we adopted a singular analysis with the same meteorological parameters based on a different layout pattern. Thirdly, due to computational and time constraints, a singular analysis with the same meteorological parameters based on a different layout pattern is the most practical approach to addressing the research question without compromising the quality of the study.

4. Outdoor Microclimate Simulation for Different Neighborhood Patterns

4.1. Air Temperature Simulation

Figure 11 illustrates the simulation results for air temperature at 1.5 m aboveground within the high-rise neighborhood for each building layout when the daily maximum temperature is reached at 14:00. The air temperatures under these layouts vary significantly.

Among the three linear-type layouts (Figure 11a–c), the air temperature of the high-rise neighborhood in LT_3 is the highest (Figure 11c), and it is lowest in LT_2 (Figure 11b). In addition, the air temperature between the first and second rows of buildings along the coastline in LT_1 and LT_3 is lower than that between the second and third rows (Figure 10a,c), while it is the opposite in LT_2 (Figure 11b). This is probably because when the sea breeze is perpendicular to the main facade, the staggered layout of the buildings can change the direction and length of ventilation corridors, and the wind shadow zone is smaller, leading to a wider sea breeze cooling area. Therefore, the LT_2 layout can reduce the overall air temperature inside the neighborhood.

In both enclosed-type layouts, the air temperature of the high-rise neighborhood in ET_1 is lower than in ET_2 (Figure 11d,e). The high-rise buildings on the east and west sides of ET_1 feature relatively poor living quality, so ET_1 is rarely adopted in actual neighborhoods. ET_2 has continuous commercial podiums on the east and west sides, which, compared to ET_1, reduces the area of the centralized open space. Despite maintaining similar plot ratios, this design leads to weaker air circulation and a relatively high air temperature within the neighborhood.

In both point-group-type layouts (Figure 11f,g), the air temperature of the high-rise neighborhood in PGT_2 is lower than PGT_1. The reason might be akin to factors influencing the linear-type layouts. In addition, simulations comparing PGT_1 with LT_1 and PGT_2 with LT_2 show that point-type buildings, when plot ratios are consistent, effectively reduce the overall air temperature within the neighborhood.

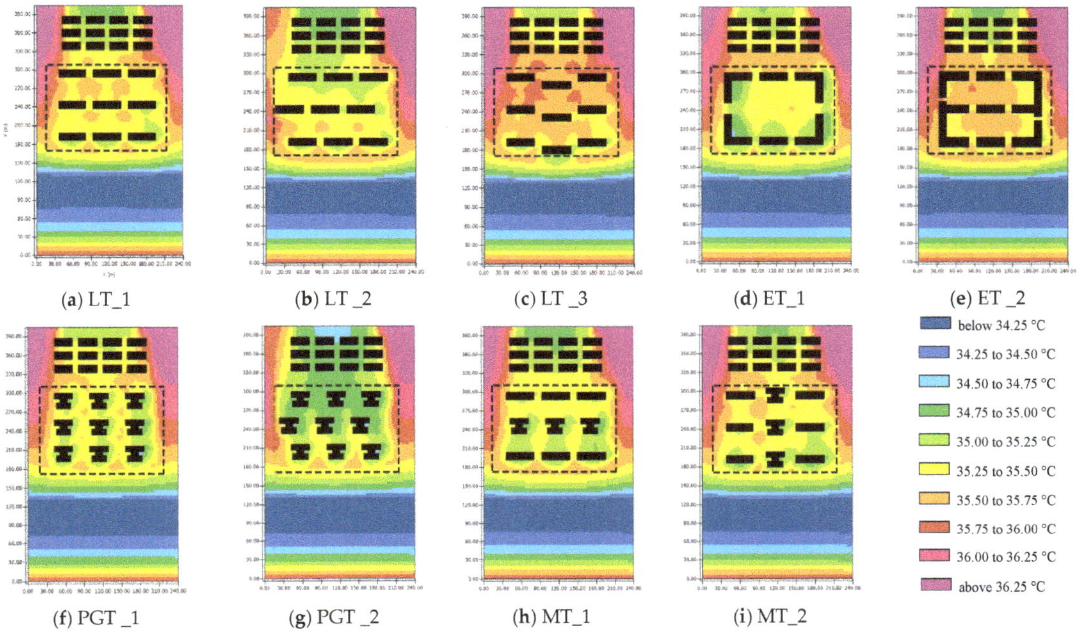

Figure 11. Simulation results for air temperature at 14:00.

In both mixed-type layouts (Figure 11h,i), the air temperature of the high-rise neighborhood in MT_1 is lower than MT_2. The reason may be that when the sea breeze is perpendicular to the main facade, the different widths of the front-row and back-row buildings could change the width of ventilation corridors and reduce the wind shadow area, thereby contributing to air circulation inside the neighborhood and expanding the sea breeze cooling area. Therefore, the horizontal mixed layout (LT_1) reduces the overall air temperature inside the neighborhood.

Figures 12 and 13 plot the air temperature diagrams inside the high-rise neighborhoods of each layout at a height of 1.5 m.

In the linear-type layout, there are obvious differences in the peaks of the air temperature distribution curves. The air temperature at the peak of LT_2 is lower than the other two and concentrated in a range of 35.13~36.22 °C, with an average temperature of 35.39 °C. This indicates that the overall air temperature of high-rise neighborhoods under the staggered layout is lower and balanced. In the enclosed-type layout, ET_1 exhibits lower peaks in the air temperature distribution curves, primarily ranging from 34.74 to 36.12 °C, with an average of 35.34 °C. This suggests that the presence of a large, centralized open space within the neighborhood aids in lowering the air temperature. In the two point-group-type layouts, the peaks of their air temperature distribution curves are obviously different. The air temperature at the peak of PGT_2 is lower and mainly distributed in a range of 34.66~36.18 °C, with an average temperature of 35.22 °C. This reveals that when the buildings are all point types, the staggered layout also helps to reduce and balance air temperature. In the two mixed-type layouts, their air temperature distribution curves are similar, mainly distributed from 34.75 to 36.12 °C, with an average temperature of 35.37 °C, and the distribution curve of MT_1 is lower and concentrated relatively.

In general, PGT_2 has the lowest average temperature, followed by ET_1 and MT_1. However, as previously noted, the enclosed-type layout is less commonly applied in practice. When the sea breeze is perpendicular to the main facade, PGT_2 and MT_1 are recommended.

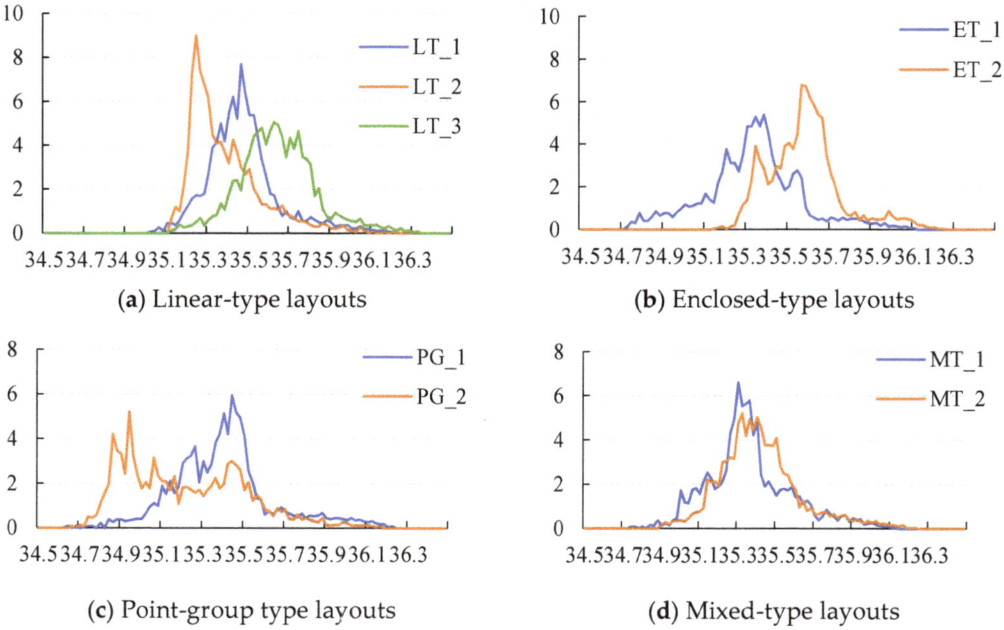

Figure 12. Distribution statistics for air temperature inside the high-rise neighborhoods.

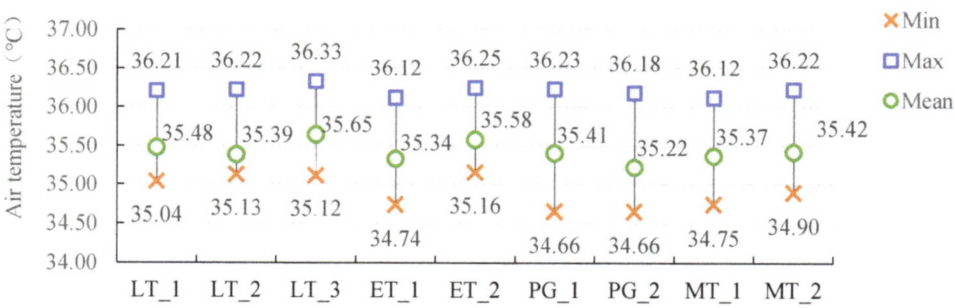

Figure 13. Interval point statistics for air temperature inside the high-rise neighborhoods.

4.2. Relative Humidity Simulation

Figure 14 displays the relative humidity simulation at 1.5 m height for each layout during the day's peak temperature at 14:00. Relative humidity within the high-rise neighborhoods decreases with distance from the coast, with variations observed between layouts.

Among the three linear-type layouts, LT_2 has the highest relative humidity, while LT_3 has the lowest. In addition, the relative humidity between the first and second rows of buildings is higher than that between the second and third rows in both LT_1 and LT_3 along the coastline, whereas the relative humidity in LT_2 is more balanced. This indicates that the staggered arrangement of buildings contributes to a higher penetration range for the humidification effect when sea breeze is perpendicular to the main facade. The reason may be that the staggered arrangement of buildings changes the direction and length of ventilation corridors, thereby increasing the flow range of sea breeze after it enters the neighborhood.

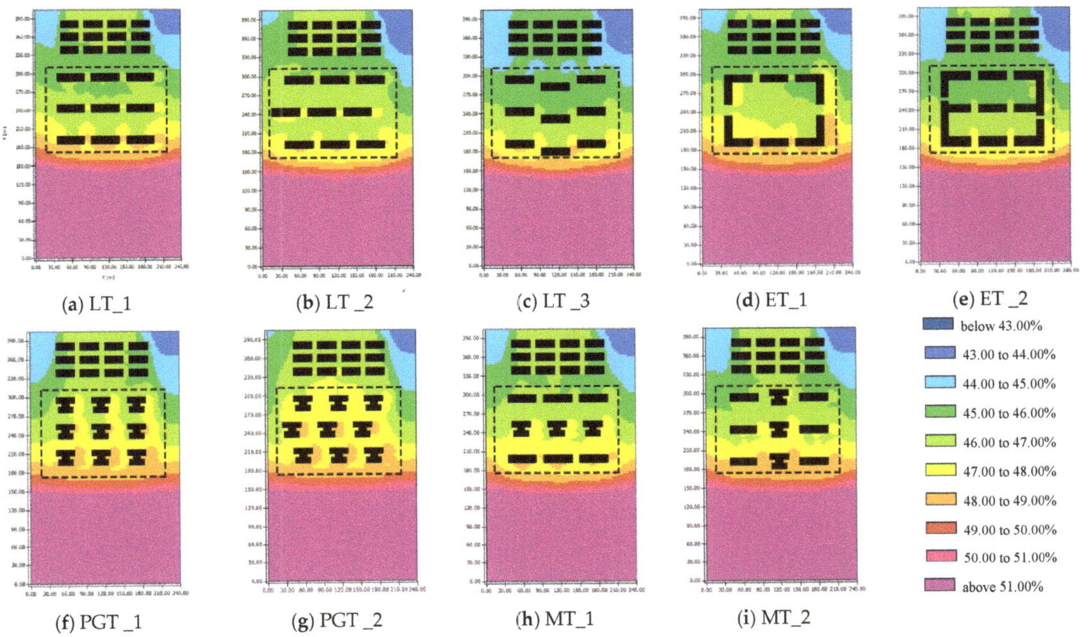

Figure 14. Simulation result for relative humidity at 14:00.

In the two enclosed-type layouts, the relative humidity of ET_1 is higher. The large open areas inside the ET_1 improve the humidification effect of sea breeze. However, ET_1 is rarely adopted in actual neighborhoods. The comparison results for ET_2 and ET_1 show that the relative humidity decreases after the addition of continuous commercial podiums, especially in the area between the second and third rows of buildings.

In point-group layouts, the relative humidity on the southeast side of buildings is elevated compared to other regions. This can be attributed to the creation of building shadows in these areas at 14:00, which significantly reduces direct solar radiation. Consequently, the evaporation of moisture is diminished in these shadowed areas. The relative humidity is higher in PGT_2 compared to PGT_1. The reason may be that the layout of buildings in PGT_2 increases the flow area of sea breeze and improves the penetration range of humidification. In addition, a comparison between the point-group layout and the linear layout reveals that with a similar plot ratio, the point-type buildings facilitate an increase in relative humidity.

In mixed-type layouts, the simulation results for MT_1 and MT_2 are similar. By comparing the simulation results for MT_1, MT_2, and LT_1, it is found that the humidification effect of sea breeze could be improved by the adoption of point-type buildings when the sea breeze is perpendicular to the main facade.

Figures 15 and 16 plot relative humidity diagrams inside the high-rise neighborhoods of each layout at a height of 1.5 m.

In linear-type layouts, the relative humidity curves of the three layouts all appear to have multiple peaks, indicating a gradient in the distribution of relative humidity, where the peaks of LT_2 are more concentrated and mainly distributed from 45.85% to 48.22%, with a highest average relative humidity of 46.70%. In the enclosed-type layouts, multiple peaks appear in the relative humidity curves of both layouts, in which the peak of PT_1 is higher and mainly distributed from 45.31% to 48.77%, with an average relative humidity of 46.87%. In the point-group layouts, there is an obvious difference between the peaks of the two layouts; the peak of PGT_2 is higher and mainly distributed from 45.05% to 49.01%, with an average of 47.60%. This indicates that the humidifying effect of sea breeze could be enhanced by point-type buildings with a staggered arrangement. In mixed-type layouts, the relative humidity curves of the two layouts both show multiple peaks with the same average relative humidity, 46.97%, which is higher compared to LT_1. This suggests that replacing slab-type buildings with point-type buildings could partially improve the relative humidity inside the neighborhood.

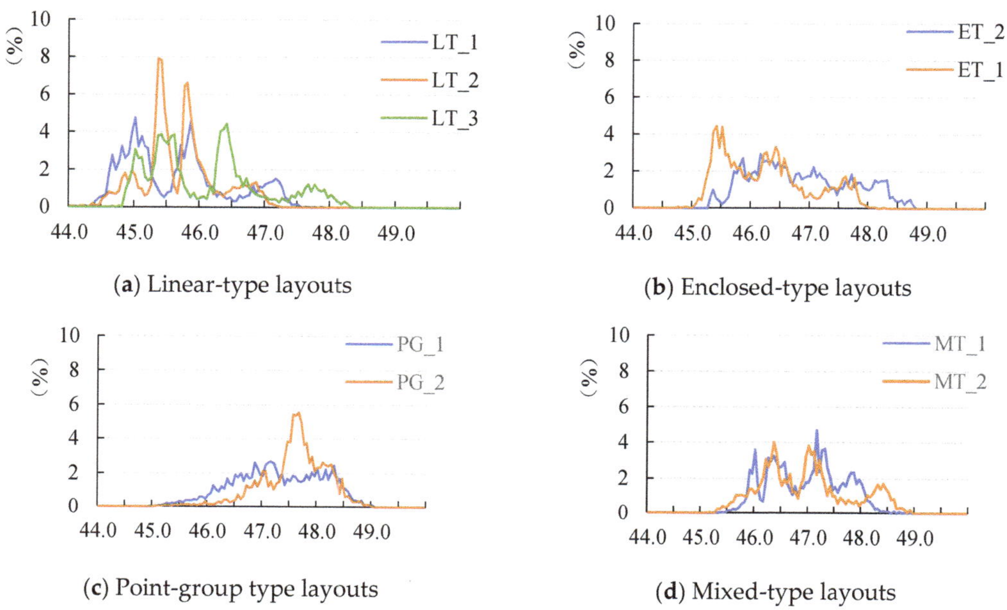

Figure 15. Distribution statistics for relative humidity inside the high-rise neighborhoods.

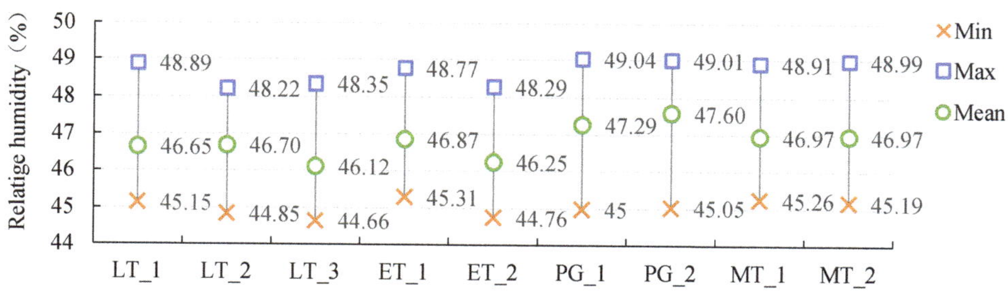

Figure 16. Interval point statistics for relative humidity inside the high-rise neighborhoods.

Overall, PGT_2 exhibits the highest relative humidity, followed by PGT_1 and the two mixed-type layouts. When the sea breeze is perpendicular to the main facade, it is recommended that a higher and more balanced level of relative humidity within the high-rise neighborhood be achieved with more point-type buildings.

4.3. Wind Speed Simulation

The simulation results for wind speed at a height of 1.5 m for each layout during the daily maximum temperature at 14:00 are shown in Figure 17. When the sea breeze is perpendicular to the main facade, corner wind forms on the east and west sides of buildings, and a ventilation corridor with a high wind speed forms between buildings. Conversely, wind shadow areas with reduced wind speeds can be observed on opposite sides, particularly when the front-row and back-row buildings are in direct alignment with each other.

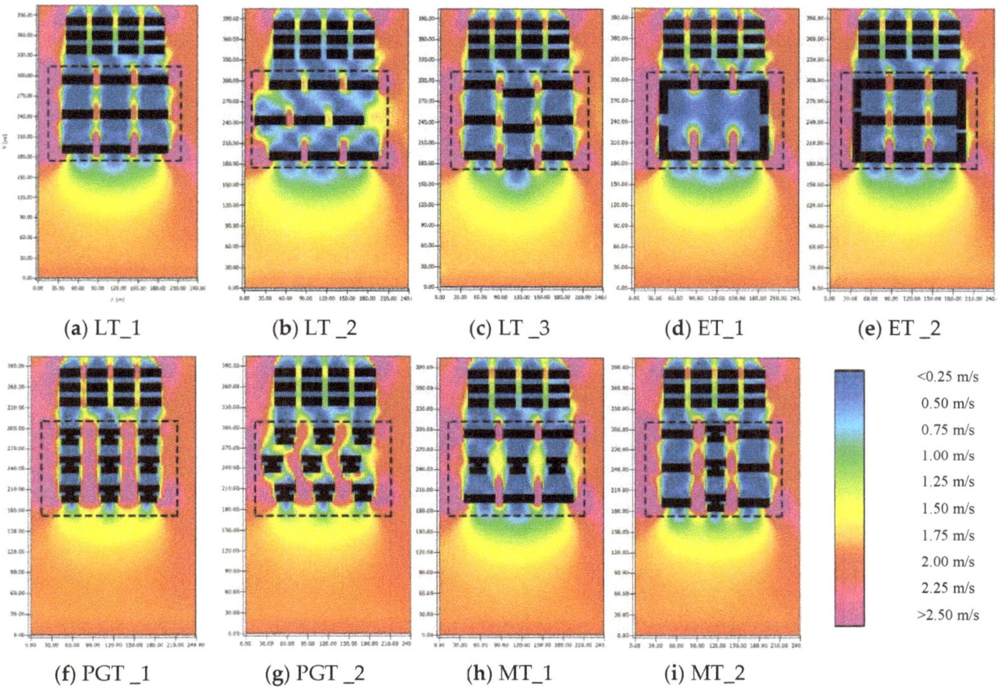

Figure 17. Simulation results for wind speed at 14:00.

Among the three linear-type layouts, the wind speeds of high-rise neighborhoods in LT_2 are higher, and those of LT_1 and LT_3 are both lower. The wind speed between the pediments of the first row of buildings is more than 2.5 m/s. The second row of buildings in LT_2 changes the direction of the ventilation corridor, increasing the flow range of the sea breeze and wind speed between the front and rear buildings and decreasing the wind speed in the ventilation corridor between the pediments of back-row buildings. The wind environment is better inside the neighborhood in LT_2.

In the two enclosed-type layouts, the wind speed of the high-rise neighborhood in ET_1 is higher. The wind speeds flowing into the neighborhoods in ET_1 and ET_2 both exceed 2.5 m/s, but the wind speed flowing out of the neighborhood in ET_1 is higher and still more than 2.5 m/s. The simulation results for ET_1 show that the sea breeze passes through the first row of buildings at a high speed and then decreases rapidly, thus preventing the formation of a continuous ventilation corridor. By comparing the simulation results for ET_2 and LT_1, it can be found that the continuous commercial podiums will prevent the inflow and outflow of wind and reduce the wind speed inside the neighborhood. Hence, enclosed-type layouts are not recommended.

In the point-group layouts, PGT_2 exhibits higher wind speeds within the high-rise neighborhood. PGT_1 features continuous ventilation corridors between building pediments, with wind speeds above 2.5 m/s, yet the distribution is unbalanced. In PGT_2, the staggered arrayed buildings change the direction of the ventilation corridor and increase the airflow range. Consequently, the excessive wind speed is reduced, making the wind environment better.

In the two mixed-type layouts, MT_1 has higher wind speeds within high-rise neighborhoods compared to MT_2. In MT_2, while the average wind speed in the ventilation corridor is higher, the area of low wind speed is more extensive. Conversely, in MT_1, there is a significant reduction in the average wind speed of the ventilation corridor, yet the overall wind speed remains higher, indicating a superior wind environment in MT_1.

Figures 18 and 19 present the wind speed distribution in the simulation results for layout modes at 1.5 m inside the high-rise neighborhood.

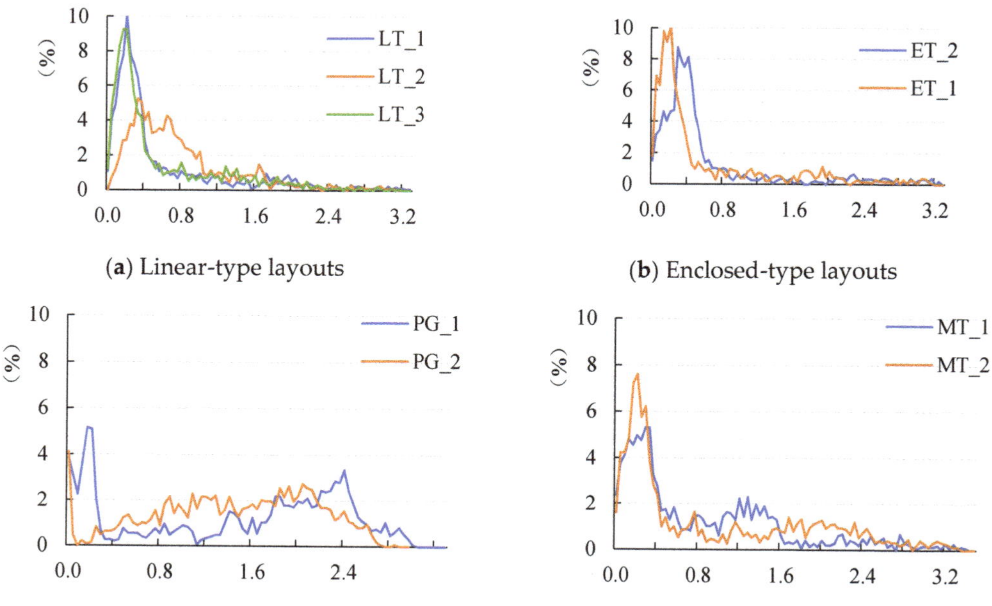

(**a**) Linear-type layouts (**b**) Enclosed-type layouts (**c**) Point-group type layouts (**d**) Mixed-type layouts

Figure 18. Distribution statistics for wind speed inside the high-rise neighborhoods.

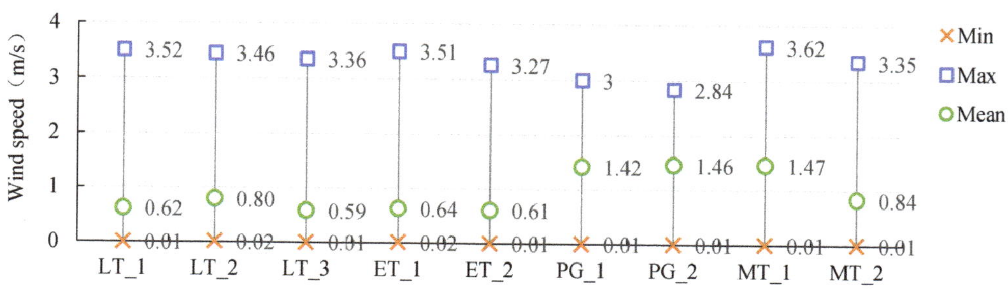

Figure 19. Interval point statistics for wind speed inside the high-rise neighborhoods.

In the linear-type layouts, LT_1 and LT_3 exhibit similar wind speed distribution curves, both featuring extensive zones of low wind speeds. LT_2, on the other hand, has a wind speed primarily ranging from 0.02 to 3.46 m/s, averaging 0.80 m/s, which is the highest among the linear-type layouts. LT_2's distribution curve exhibits a higher peak as well. In addition, the proportion of low-wind-speed zones in LT_2 is also smaller, so the wind environment of the high-rise neighborhood in LT_2 is better compared to the other two. In the enclosed-type layout, the wind speed curves of the two layout modes both have more low-wind-speed zones. The wind speed of ET_1 is mainly distributed in a range of 0.02–3.51 m/s, with an average value of 0.64 and a higher curve peak. This indicates that a large open-space area inside the neighborhood could increase the wind speed. In the point-group type layout, PGT_2 has a more consistent wind speed curve, leading to a balanced distribution, averaging 1.46 m/s within a range of 0.01–2.84 m/s, which surpasses that of PGT_1. When adopting point-type buildings in neighborhoods, the left–right staggered layout improves the wind environment. In the mixed-type layout, the wind speed of MT_1 is mainly distributed in a range of 0.01–3.62 m/s, with a higher average value of 1.47 m/s and a smaller proportion of low speeds. This suggests that, in a mixed layout of slab and point buildings, increasing the width of the ventilation corridor inside could improve the wind environment.

Collectively, MT_1 has the highest average wind speed, followed by PGT_1 and PGT_2. However, the wind speed inside the neighborhood of PGT_1 is unbalanced, and a large area with extreme wind speed makes pedestrians feel uncomfortable. In summary, when the sea breeze is perpendicular to the main facades, MT_1 and PGT_2 are recommended for high-rise neighborhoods.

5. Spatial Strategies for Optimizing the Microclimate of Neighborhoods

5.1. Reducing the Enclosure of Podiums

According to this investigation on the morphology of residential neighborhoods along Shantou's Inner Bay, most of the high-rise buildings are surrounded by two or three layers of podiums along streets. As shown in Figure 20, these neighborhoods are almost completely closed. The continuous commercial podiums increase commercial interests while affecting the wind and heat environment at pedestrian height inside the neighborhood. Meanwhile, enclosed podiums prevent the inflow and outflow of sea breeze inside high-rise neighborhoods and even change the outdoor microclimate of back-row multi-story neighborhoods.

Figure 20. High-rise neighborhoods with continuous commercial podiums.

Based on the above simulation, a comparison of the simulation results for LT_1 and ET_2 (Table 3) reveals an average temperature decrease of 0.1 °C and an average relative humidity increase of 0.40% within the high-rise neighborhood after the removal of podiums around it. Similarly, within the multi-story neighborhood, the average temperature drops by 0.09 °C, accompanied by a 0.37% rise in average relative humidity.

Table 3. Comparison of air temperature and relative humidity of LT_1 and ET_2.

Layout	High-Rise Neighborhood		Multi-Story Neighborhood	
	Average Air Temperature (°C)	Average Relative Humidity (%)	Average Air Temperature (°C)	Average Relative Humidity (%)
LT_1	35.48	46.65	35.25	45.97
ET_2	35.58	46.25	35.34	45.60

The difference in air temperature and wind speed between LT_1 and ET_2 is shown in Figure 21. The air temperature difference graph indicates that the air temperature in most areas of the high-rise neighborhood decreases after reducing continuous podiums, especially in the middle area. The air temperature near the podiums on the east and west sides increases, mainly because the building shadow of the podiums can reduce direct solar radiation around the shadow area. Overall, the air temperatures in both high-rise and multi-story neighborhoods decrease after the removal of continuous podiums. The wind speed difference graph shows that the removal of continuous podiums increases wind speed in most regions, with particularly notable gains on the east and west sides within the high-rise neighborhood. The difference graph indicates that the continuous podium on the east and west sides could prevent the inflow and outflow of air, while the small podiums on the north and south sides increase the lengths of ventilation corridors and wind speeds.

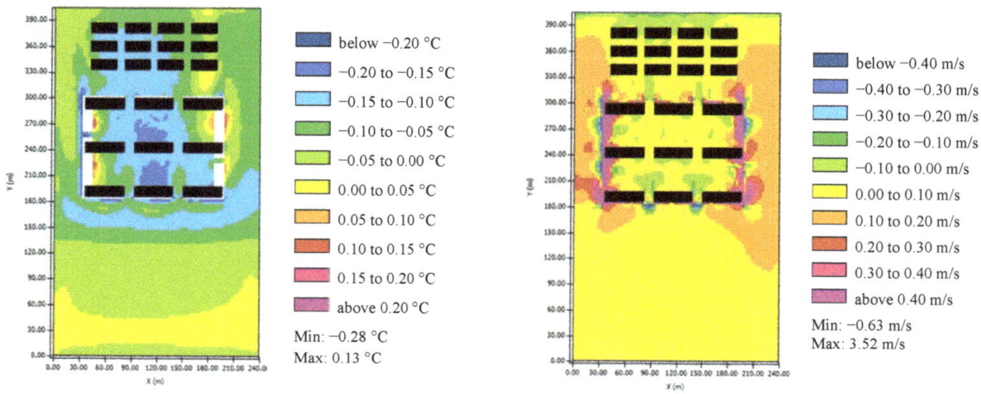

(**a**) Differences in air temperature (**b**) Differences in wind speed

Figure 21. Differences in air temperature and wind speed between LT_1 and ET_2.

In summary, to improve the wind and heat environment in the high-rise neighborhoods along Shantou's Inner Bay, we recommend removing the continuous podiums on the east and west sides and retaining the podiums on the north and south sides appropriately.

5.2. *Replacing Slab Buildings with Point Buildings*

The windward and wind shadow areas of point buildings are smaller than those of slab buildings, and the width of ventilation corridors also increases for point buildings. In this case, replacing slab buildings with point buildings could effectively increase wind speed inside high-rise neighborhoods. A comparison of wind speeds in LT_1 (Figure 17a) and MT_1 (Figure 17h) shows that replacing some slab buildings with point buildings in a high-rise neighborhood decreases the area of low wind speeds and increases instances of wind speed acceleration. Similar results can be found in the comparison of LT_2 (Figure 17b) and PGT_2 (Figure 17g).

When the sea breeze enters the high-rise neighborhood, the better wind environment helps to increase its humidification range and cooling effect. Table 4 compares the air temperature and relative humidity of high-rise and multi-story neighborhoods under four layouts. After replacing some slab buildings with point buildings, the average air temperature of the high-rise and multi-story neighborhoods decreases by 0.11 °C vs. 0.06 °C, and the average relative humidity increases by 0.32% vs. 0.14%. Upon replacing all slab buildings with point buildings, the average air temperature of the high-rise and multi-story neighborhoods decreases by 0.17 °C vs. 0.19 °C, and the average relative humidity increases by 0.9% vs. 0.72%. In summary, the replacement of slab buildings with point buildings could improve the outdoor microclimate environment of high-rise and multi-story neighborhoods.

Table 4. Comparison of air temperature and relative humidity of four layouts.

Layout	High-Rise Neighborhood		Multi-Story Neighborhood	
	Average Air Temperature (°C)	Average Relative Humidity (%)	Average Air Temperature (°C)	Average Relative Humidity (%)
LT_1	35.48	46.65	35.25	45.97
MT_1	35.37	46.97	35.19	46.11
LT_2	35.39	46.70	35.13	46.01
PGT_2	35.22	47.60	34.94	46.73

5.3. Reducing the Width of the Windward Side of Buildings

Most high-rise neighborhoods along Shantou's Inner Bay are arrayed continuously in a linear-type layout. The first row of buildings is notably tall and lengthy, creating a "wind barrier" that results in a substantial wind shadow zone behind it. To study the influence of reducing the length of the "wind barrier" on the outdoor microclimate, simulations were carried out after splitting slab buildings with similar building plot ratios. A comparison of wind speeds (Figure 22a) indicates that when the length of the "wind barrier" decreases, the area of wind shadow behind the high-rise buildings reduces, and the number of ventilation corridors inside the neighborhood increases. Consequently, the overall wind environment in the neighborhood is optimized. A comparison of air temperatures (Figure 22b) shows that when the length of the windward side of buildings is reduced, the air temperature of both the high-rise neighborhood and the back-row multi-story neighborhood decreases, indicating an improvement in conditions compared to the previous state.

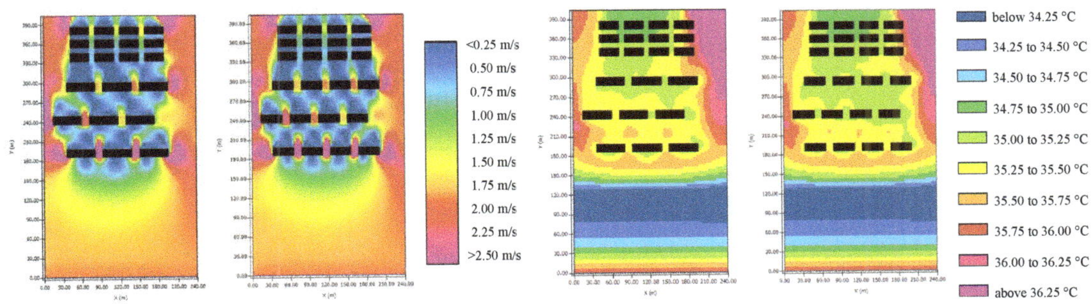

(**a**) wind speed distribution (**b**) air temperature distribution

Figure 22. Comparison of wind speed and air temperature after reducing the windward length.

A comparison of the simulation results in Table 5 reveals that after reducing the length of the "wind barrier", the average temperature inside high-rise and multi-story neighborhoods decreases by 0.06 °C vs. 0.03 °C, and the average relative humidity increases by 0.21% vs. 0.1%. In summary, for neighborhoods along the Inner Bay, reducing the length of the "wind barrier" results in a reduction in areas of low wind speed within the high-rise neighborhood. This change also increases the number of ventilation corridors, which, in turn, aids in enhancing the outdoor microclimate environment for high-rise and multi-story neighborhoods.

Table 5. Comparison of air temperature and relative humidity after reducing the windward length.

Layout	High-Rise Neighborhood		Multi-Story Neighborhood	
	Average Air Temperature (°C)	Average Relative Humidity (%)	Average Air Temperature (°C)	Average Relative Humidity (%)
LT_2	35.39	46.70	35.13	46.01
LT_2_1	35.33	46.91	35.10	46.11

6. Discussion

Based on an investigation of 100 neighborhoods along Shantou's Inner Bay, idealized layout models of neighborhoods were created and simulated with the ENVI-met software. Through a comparison analysis of wind speed, air temperature, and relative humidity of each model, measures to optimize the layout of neighborhoods were proposed. The following conclusions are drawn:

(1) In the context of rapid urbanization, affected by an intensified UHI, Shantou City has seen a steady rise in air temperatures over the last decade. Shantou's Inner Bay has obvious humidification and cooling effects in its coastal areas. These effects are influenced by the layout patterns of buildings within the coastal blocks.

(2) Areas along the Inner Bay have a higher commercial value, leading to a high concentration of residential and commercial space. "Inner Bay–high-rise neighborhoods–multi-story neighborhoods" constitutes the main distribution pattern of neighborhoods along Shantou's Inner Bay. This urban pattern exhibits three principal drawbacks for climate adaptation: firstly, the continuous and expansive facade of the front-row high-rise buildings impedes the natural SLB circulation; secondly, the cooling effect of SLB at the pedestrian level is substantially curtailed by the pervasive and enclosed commercial podiums encircling the high-rise areas; thirdly, the multi-story neighborhoods suffer from inadequate ventilation, predominantly due to the confined spaces between buildings and the blockage caused by the preceding high-rise structures.

(3) By changing the building layout of high-rise neighborhoods, the main distribution patterns can be refined into four types and nine sub-types of idealized models, each distinguished by its impact on the outdoor microclimate. The simulation results for wind speed, air temperature, and relative humidity show that when the linear-type layout is adopted, the outdoor microclimate of LT_2 is better than that of LT_1 and LT_3; when the enclosed-type layout is adopted, the outdoor microclimates of ET_1 and ET_2 are both worse; when the mixed-type layout is utilized, the outdoor microclimate of MT_1 is better compared to MT_2; a point-group type layout in the high-rise neighborhood results in a superior outdoor microclimate compared to the other three types, with PGT_2 demonstrating even better performance than PGT_1. The simulation results also showed that the outdoor area of PGT_2 exhibited the highest relative humidity and the lowest average temperature. In addition, MT_1 and PGT_2 show higher values and balanced distributions of average wind speeds. From this, it can be concluded that when the plate-structure buildings were partially or completely replaced by point-structure buildings, the outdoor microclimate of neighborhoods improved obviously. However, from the perspective of an indoor living environment and house-type designs, plate-structure buildings show more convenience and comfort than point-structure buildings. How to weigh decision-making factors other than outdoor microclimate should be further discussed in future studies.

The above analysis focuses on a single simulation; it does not capture the full range of temporal variability in weather conditions, which could also affect the outdoor microclimate of neighborhoods, so the outcomes may not account for seasonal fluctuations in thermal demand and comfort. In addition, without multiple simulations under different parameters, this study is not enough to explain the sensitivity of neighborhood layouts under different climatic conditions.

(4) Through further comparative simulations of wind speed, air temperature, and relative humidity, the following three spatial optimization measures for high-rise neighborhoods can be proposed to improve the outdoor microclimate: firstly, efforts should be made to reduce the number of consecutive podiums on the east and west sides and cut down the enclosure degree of facades at pedestrian height; secondly, slab-type buildings should be partially replaced with point-type buildings while avoiding the layout pattern of aligning front-row and back-row buildings; thirdly, the width and continuity of the windward surfaces of the front buildings should be reduced, and the ground space inside should be opened by cutting off the first or second floor of buildings.

In contrast to earlier research, the strategies derived from these idealized models are specifically aimed at enhancing the summer climate in coastal regions. The residential buildings along the shoreline of Shantou's Inner Bay are mostly oriented in the south–north direction, and the dominant wind direction during high summer temperatures tends to be southerly, which have been both considered in the idealized models. Future research should place greater emphasis on the impact of various building orientations. Furthermore, this paper primarily concentrates on examining the humidification and cooling effects of SLB in a variety of building layouts, somewhat weakening the consideration of greening and paving strategies. While this approach is beneficial for guiding building design, it may fall short in providing comprehensive insights into the design of neighborhood landscapes.

In the course of the investigation, it was observed that high-rise neighborhoods are characterized by more extensive green spaces and a higher greening rate. The air temperatures recorded by instruments in these areas were found to be lower compared to those in multi-story neighborhoods. However, the findings from the simulation of idealized models revealed a contrary trend, indicating that the air temperatures in high-rise neighborhoods tend to be generally higher than those in multi-story neighborhoods. This discrepancy may be attributable to the larger open spaces in high-rise areas, which are more susceptible to absorbing solar radiation, consequently influencing the air temperatures at pedestrian levels. Future research should take into account a broader range of factors that impact the outdoor microclimate. Additionally, this study suggests that optimizing the layout of high-rise neighborhoods has a limited effect on the microclimate of adjacent back-row multi-story neighborhoods. Therefore, the exploration of more effective strategies also remains an open area for future investigation.

Author Contributions: Conceptualization, W.L. and Y.W.; methodology, W.L.; software, L.X., W.L.; validation, W.L., L.X. and C.M.; formal analysis, C.M.; investigation, L.X.; resources, L.X.; data curation, C.M.; writing—original draft preparation, W.L.; writing—review and editing, W.L. and J.M.; visualization, J.M.; supervision, Y.W.; project administration, W.L.; funding acquisition, Y.W. All authors have read and agreed to the published version of the manuscript.

Funding: This study was supported by the Social Science Fund for Youth of the Chinese Ministry of Education [23YJCZH142] and the Shantou University Scientific Research Fund [NTF21028].

Data Availability Statement: The data presented in this study are available on request from the corresponding author.

Conflicts of Interest: The authors declare no conflict of interest.

References

1. WMO. *The Atlas of Mortality and Economic Losses from Weather, Climate and Water Extremes (1970–2019)*; World Meteorological Organization: Geneva, Switzerland, 2021.
2. UN-Habitat. *World Cities Report 2020: The Value of Sustainable Urbanization*; UN-Habitat: Nairobi, Kenya, 2020.
3. Wai, K.; Wang, X.; Lin, T.; Wong, M.; Zeng, S.; He, N.; Ng, E.; Lau, K.; Wang, D. Observational Evidence of a Long-term Increase in Precipitation Due to Urbanization Effects and Its Implications for Sustainable Urban Living. *Sci. Total Environ.* **2017**, *599–600*, 647–654. [CrossRef] [PubMed]
4. Wang, M.Y.; Xu, H.Q. Analyzing the influence of urban forms on surface urban heat islands intensity in Chinese mega cities. *J. Geo-Inf. Sci.* **2018**, *20*, 1787–1798. [CrossRef]

5. Tam, B.Y.; Gough, W.A.; Mohsin, T. The impact of urbanization and the urban heat island effect on day to day temperature variation. *Urban Clim.* **2014**, *12*, 1–10. [CrossRef]
6. Kim, S.W.; Brown, R.D. Urban heat island (UHI) intensity and magnitude estimations: A systematic literature review. *Sci. Total Environ.* **2021**, *779*, 146389. [CrossRef] [PubMed]
7. Mills, D.; Schwartz, J.; Lee, M.; Sarofim, M.; Jones, R.; Lawson, M.; Duckworth, M.; Deck, L. Climate change impacts on extreme temperature mortality in select metropolitan areas in the United States. *Clim. Chang.* **2015**, *131*, 83–95. [CrossRef]
8. Banerjee, S.; Y, G.C.N.; Yik, S.K.; Dzyuban, Y.; Crank, P.J.; Yi, R.P.X.; Chow, W.T. Analysing impacts of urban morphological variables and density on outdoor microclimate for tropical cities: A review and a framework proposal for future research directions. *Build. Environ.* **2022**, *225*, 109646. [CrossRef]
9. Paramita, B.; Suparta, W. Alteration of Urban Microclimate in Bandung, Indonesia Based on Urban Morphology. *Geogr. Tech.* **2019**, *14*, 213–220. [CrossRef]
10. Middel, A.; Häb, K.; Brazel, A.J.; Martin, C.A.; Guhathakurta, S. Impact of urban form and design on mid-afternoon microclimate in Phoenix Local Climate Zones. *Landsc. Urban Plan.* **2014**, *122*, 16–28. [CrossRef]
11. Kamal, A.; Abidi, S.M.H.; Mahfouz, A.; Kadam, S.; Rahman, A.; Hassan, I.G.; Wang, L.L. Impact of Urban Morphology on Urban Microclimate and Building Energy Loads. *Energy Build.* **2021**, *253*, 111499. [CrossRef]
12. Allegrini, J.; Dorer, V.; Carmeliet, J. Influence of morphologies on the microclimate in urban neighbourhoods. *J. Wind. Eng. Ind. Aerodyn.* **2015**, *144*, 108–117. [CrossRef]
13. *GB/T 50378-2019*; Assessment Standard for Green Building. China Architecture & Building Press: Beijing, China, 2019.
14. Li, H.; Wu, J.; Zhao, Y.; Huang, J.L.; Li, Z.; Ruan, Y. Influence Analysis of Building Layouts on Micro- environment of Residence Community. *Build. Energy Effic.* **2012**, *44*, 57–63. [CrossRef]
15. Wang, Y.; Zhou, D.; Wang, Y.; Fang, Y.; Yuan, Y.; Lv, L. Comparative study of urban residential design and microclimate characteristics based on ENVI-met simulation. *Indoor Built Environ.* **2019**, *28*, 1200–1216. [CrossRef]
16. Zhang, M.; You, W.; Qin, Q.; Peng, D.; Hu, Y.; Gao, Z.; Buccolieri, R. Investigation of typical residential block typologies and their impact on pedestrian- level microclimate in summers in Nanjing, China. *Front. Archit. Res.* **2022**, *11*, 278–296. [CrossRef]
17. Wei, X. Research on Reducing Carbon Consumption in Residential Community Spaces as Influenced by Microclimate Environments. *J. Urban Plan. Dev.* **2021**, *147*, 04021037. [CrossRef]
18. Yu, H.; Zhang, H.; Qiu, L.; Lai, D. Evaluation of Renovation Strategies for Existing Residential Districts in Shanghai based on Coupling Calculation of Microclimate and Energy Consumption. *Energy Build.* **2023**, *298*, 113569. [CrossRef]
19. Wen, W.-J.; Shen, T.-L.; Ding, Z.-Y.; Song, J.-H. Numerical experiment on the effect of urbanization upon summer land-sea breezes in the coastland of guangxi. *J. Trop. Meteorol.* **2010**, *16*, 263–270.
20. Wang, Y.; Di Sabatino, S.; Martilli, A.; Li, Y.; Wong, M.S.; Gutiérrez, E.; Chan, P.W. Impact of land surface heterogeneity on urban heat island circulation and sea-land breeze circulation in Hong Kong. *J. Geophys. Res. Atmos.* **2017**, *122*, 4332–4352. [CrossRef]
21. Shen, L.; Zhao, C.; Ma, Z.; Li, Z.; Li, J.; Wang, K. Observed decrease of summer sea-land breeze in Shanghai from 1994 to 2014 and its association with urbanization. *Atmos. Res.* **2019**, *227*, 198–209. [CrossRef]
22. Yamamoto, Y.; Ishikawa, H. Influence of urban spatial configuration and sea breeze on land surface temperature on summer clear-sky days. *Urban Clim.* **2020**, *31*, 100578. [CrossRef]
23. Huang, H.; Lv, Y.; Tan, H. Analysis of Sea-Land Breezes Characteristics in the Coastal Region of Shantou, Guangdong. *Guangdong Meteorol* **2018**, *40*, 14–17. [CrossRef]
24. Chen, C.; Ye, D.; Qiu, X.; Ye, X.; Zhang, Y.; Zhu, X. The impact of sea-land breeze on urban heat island in Jinjiang. *J. Meteorol. Sci.* **2020**, *40*, 257–263. [CrossRef]
25. Yamamoto, Y. Measures to Mitigate Urban Heat Islands. Environmental and Energy Research Unit. *Quaterly Rev.* **2006**, *18*, 65–83.
26. China Meteorological Administration, Tsinghua University. *Special Meteorological Dataset for Thermal Environment Analysis of Chinese Buildings*; China Architecture & Building Press: Beijing, China, 2005.
27. Guo, L.B.; Jiang, W.S.; Li, F.Q.; Wang, X.L. Analyses of urban island effect in Shantou city of Guangdong Province in recent ten years. *Period. Ocean. Univ. China* **2022**, *52*, 19–30. [CrossRef]
28. Clarice, B.D.S.; Fayzullaev, F.F.; Dunichkin, I.V. Influence of planning decisions on the local climate of residential building of the Krasnodar region and the Primorsk region. *Vestn. MGSU* **2019**, *14*, 954–966. [CrossRef]
29. Lin, H.; Yin, S.; Xie, C.; Lin, Y. Research-Integrated Pedagogy with Climate-Responsive Strategies: Vernacular Building Renovation Design. *Buildings* **2022**, *12*, 1294. [CrossRef]
30. Su, Y.; Li, Z.; Meng, Q.; Zhao, Q.; Wu, X.; Liu, Q.; Wu, Z.; Wang, C. Simulation and optimization of thermal comfort in residential areas based on outdoor morphological parameters. *Build. Environ.* **2023**, *245*, 110837. [CrossRef]
31. Liu, S.; Kwok, Y.T.; Ren, C. Investigating the impact of urban microclimate on building thermal performance: A case study of dense urban areas in Hong Kong. *Sustain. Cities Soc.* **2023**, *94*, 104509. [CrossRef]
32. Deng, F.F.; Huang, Y. Uneven Land Reform and Urban Sprawl in China: The Case of Beijing. *Prog. Plan.* **2004**, *61*, 211–236. [CrossRef]
33. Yue, W.; Liu, Y.; Fan, P. Measuring Urban Sprawl and Its Drivers in Large Chinese Cities: The Case of Hangzhou. *Land Use Policy* **2013**, *31*, 358–370. [CrossRef]
34. Lin, L.; Chen, X.; Moudon, A.V. Measuring the Urban Forms of Shanghai's City Center and Its New Districts: A Neighborhood-Level Comparative Analysis. *Sustainability* **2021**, *13*, 8481. [CrossRef]

35. He, S.; Wu, F. Socio-spatial impacts of property-led redevelopment on China's urban neighbourhoods. *Cities* **2007**, *24*, 194–208. [CrossRef]
36. Lu, D. Travelling urban form: The neighbourhood unit in China. *Plan. Perspect.* **2006**, *21*, 369–392. [CrossRef]
37. Ye, L. Urban Regeneration in China: Policy, Development, and Issues. *Local Econ.* **2011**, *26*, 337–347. [CrossRef]
38. Perini, K.; Chokhachian, A.; Dong, S.; Auer, T. Modeling and simulating urban outdoor comfort: Coupling ENVI-Met and TRNSYS by grasshopper. *Energy Build.* **2017**, *152*, 373–384. [CrossRef]
39. Tsoka, S.; Tsikaloudaki, A.; Theodosiou, T. Analyzing the ENVI-met microclimate model's performance and assessing cool materials and urban vegetation applications–A review. *Sustain. Cities Soc.* **2018**, *43*, 55–76. [CrossRef]
40. Forouzandeh, A. Prediction of surface temperature of building surrounding envelopes using holistic microclimate ENVI-met model. *Sustain. Cities Soc.* **2021**, *70*, 102878. [CrossRef]
41. Elraouf, R.A.; Elmokadem, A.; Megahed, N.; Eleinen, O.A.; Eltarabily, S. Evaluating urban outdoor thermal comfort: A validation of ENVI-met simulation through field measurement. *J. Build. Perform. Simul.* **2022**, *15*, 268–286. [CrossRef]
42. *JGJ 286-2013*; Design Standard for Thermal Environment of Urban Residential Areas. China Architecture & Building Press: Beijing, China, 2013.

buildings

MDPI

Article

Adaptive Optimization of Wind Environment in Coastal Village Spatial Forms of Western Guangdong

Yue Pang [1], Zhanxun Liang [1], Peisheng Xie [2] and Li Li [1,*]

[1] College of Architecture and Urban Planning, Guangzhou University, Guangzhou 510006, China; pangyue@gzhu.edu.cn (Y.P.); 2112209059@e.gzhu.edu.cn (Z.L.)
[2] Guangdong Provincial Architectural Design Research Institute Group Co., Ltd., Guangzhou 510000, China; xiepeisheng0419@outlook.com
* Correspondence: lili_ar@gzhu.edu.cn

Abstract: Naozhou Island is located in a subtropical marine monsoon climate, with frequent windy days throughout the year, which has a significant impact on the residents' lives. The spatial form of local traditional villages has adapted to the local wind environment through long-term practical exploration. This study aims to quantitatively analyze this layout to explore the patterns of its climate adaptability, thereby providing guidance for modern village construction. The research method primarily involves using CFD software (2019) to analyze the spatial form parameters of the village, namely village scale, planar form, building density, and orientation, along with their effects on average wind speed, wind speed amplification factor, and wind field coefficient under normal and extreme wind conditions. The results show that an appropriate planar form can enhance the wind adaptability of the village, while village scale and building density significantly affect the wind environment. However, the orientation of the village does not have a significant impact on wind field changes due to the discontinuity of the street system. These patterns of wind adaptability can assist in the planning and design of future coastal villages to enhance the wind environment regulation and disaster resilience of island villages.

Keywords: wind adaptation; coastal village planning; spatial morphology analysis; computational fluid dynamics (CFD); village microclimate; wind environment simulation

1. Introduction

Wind, as a natural phenomenon, plays an important role in multiple fields. From improving air quality [1] to alleviating the urban heat island effect [2], and to the fine design of vibration control in buildings or machinery [3], the influence of wind is ubiquitous. Particularly in island environments, wind not only shapes unique natural landscapes but also profoundly affects the living conditions and construction methods of local residents. Island inhabitants demonstrate a deep understanding and adaptability to the wind environment when faced with unique climatic conditions. This adaptability is particularly evident in architecture, manifested in low and less dense coastal village buildings, wide street dimensions, and other spatial forms that facilitate air circulation and reduce the hazards posed by extreme weather such as typhoons [4,5]. This wind adaptability not only reflects the harmonious coexistence of human wisdom and the natural environment but also provides valuable experience for modern architectural design.

Currently, research on the impact of spatial morphology on wind environments has made significant progress in urban areas [6–12], while studies in rural areas, especially coastal villages, remain relatively weak. Previous research on rural areas has mostly focused on landscape layout [13,14], village site selection, and feng shui theory [15,16], lacking in-depth exploration of the relationship between spatial morphology and wind adaptability. With the deepening implementation of the rural revitalization strategy, the living standards of villagers have significantly improved [17], and the sustainability of rural living environments has gained attention, making the demand for related research more urgent.

In existing research, many scholars have begun to focus on the relationship between the spatial morphology of inland villages and wind adaptability, achieving certain research results. For example, Xu et al. extracted the morphologies of 17 typical villages in Tianjin and used CFD to simulate the impact of different planar forms on village ventilation conditions and residents' health risks. Through statistical data analysis, they derived the optimal rural development morphology [18]. Zang et al. conducted quantitative analysis by selecting typical node spaces, considering the impact of village planar morphology and building density on wind speed, and proposed optimization and renovation suggestions to assist new village development [19]. Yao et al. explored how the location and spatial morphology of villages adapt to the local environment to promote the formation of a comfortable local microclimate [20]. Sun et al. used CFD to simulate and analyze the winter wind environment of four typical courtyard layouts, providing a theoretical basis for micro-level studies of village courtyard wind environments [21]. Gou et al. aimed to improve the energy efficiency of residential buildings in this region of China, exploring the relationship between spatial morphology and energy consumption in villages located in regions with hot summers and cold winters [22].

Scholars often use CFD numerical simulations and field measurements to explore the relationship between spatial morphological indicators such as building density, height, street orientation, aspect ratio, and village plan form with the wind environment [23–26]. However, most research subjects focus on inland villages under normal wind conditions, while studies related to island villages that are easily affected by typhoons remain a weak link. Considering the impact of wind on buildings under severe weather conditions is equally important.

The Pacific Northwest is the source of 36% of the world's typhoons [24]. According to publicly available data from the China Meteorological Administration, since 1949, a total of 264 typhoons have made landfall in Guangdong, occurring 277 times, ranking 1st in both the number of typhoons and landfall occurrences nationwide. Among the typhoons that made landfall in Guangdong, the Zhanjiang area is the most favored. During the peak typhoon season each year, residents of nearby islands, such as Naozhou Island, are troubled by these extreme weather events, making it a typical island affected by such climatic conditions.

Therefore, this paper takes the coastal village of Naozhou Island in western Guangdong as the research object, using CFD methods to analyze the impacts on the coastal village under two different wind conditions: normal wind and extreme wind. Through this study, we aim to reveal the patterns of how traditional layouts affect the wind environment, thereby providing a scientific basis for planners and architects to develop more effective strategies to enhance the wind adaptability of traditional villages. In addition, this research can also provide references for policy decision making and practical implementation in urban development, promoting sustainable and resilient design practices.

2. Materials and Methods

2.1. Research Subjects

Naozhou Island is an island located in the southeastern part of Zhanjiang City, Guangdong Province, formed approximately 200,000 to 500,000 years ago by underwater volcanic eruptions. It is also the largest volcanic island in China, bordered by Donghai Island to the north, Leizhou Bay to the west, and the South China Sea to the southeast, with a total area of about 56 square kilometers (Figure 1). Naozhou Island is situated in a subtropical monsoon zone, with prevailing easterly and southeasterly winds from April to September. From October to March of the following year, northerly and northeasterly winds dominate (Figure 2). Additionally, the island is often affected by typhoons to varying degrees during the summer. These typically occur from May to November, with the majority happening from July to September [27]. This results in two distinctly different states of wind conditions along the coastal areas throughout the year. One is a relatively stable and common state, known as the normal state; the other is an extreme and abnormal state, characterized by strong winds and heavy rainfall brought by typhoons and other extreme weather conditions.

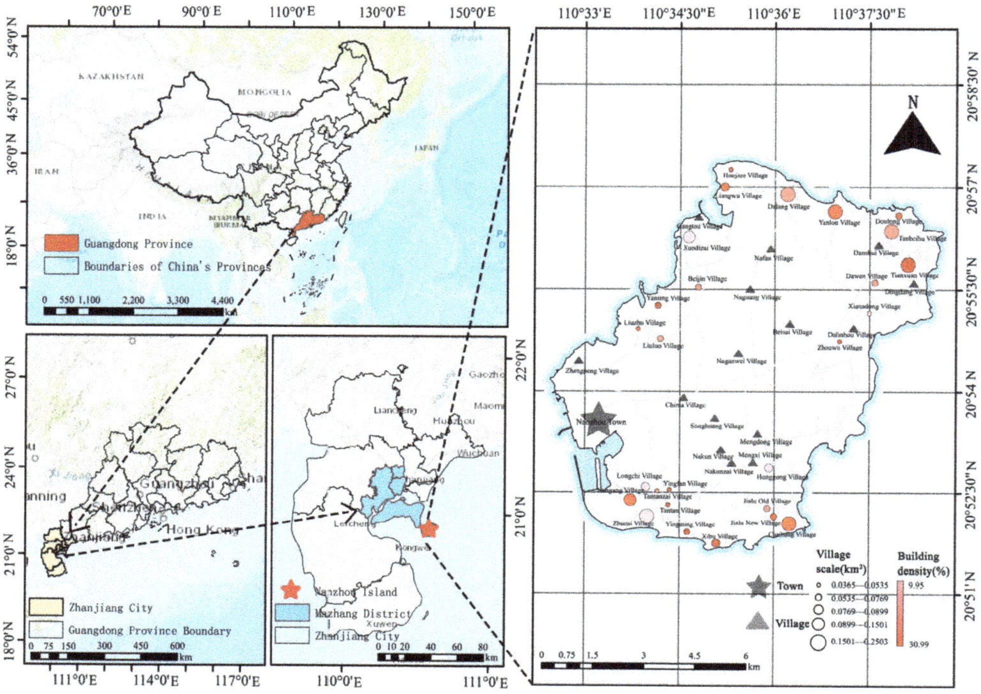

Figure 1. Location map of Naozhou Island.

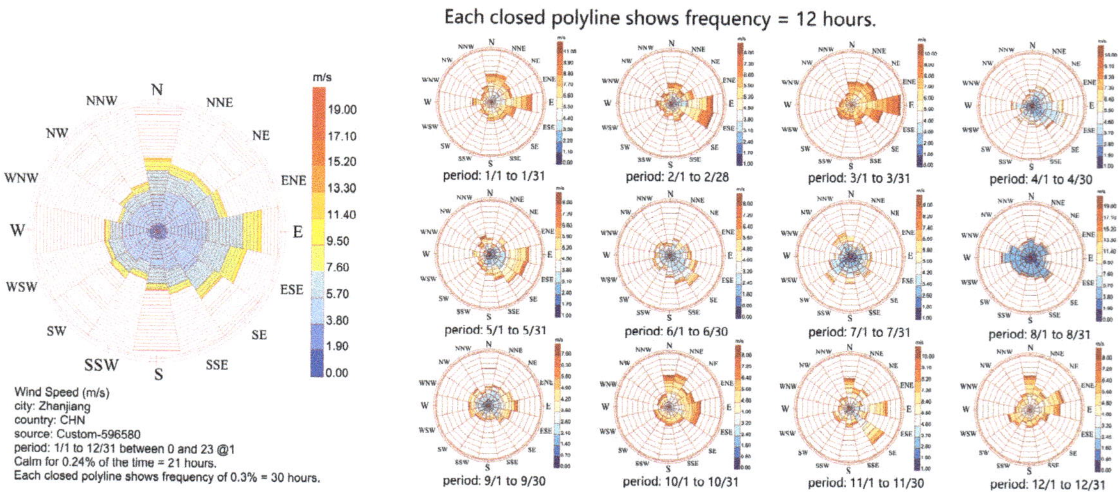

Figure 2. Climate map of Naozhou Island.

The research object of this paper focuses on 42 coastal natural villages on Naozhou Island that have a certain scale (with areas ranging from 0.01 km^2 to 0.26 km^2). The research content includes the spatial morphology of coastal villages on Naozhou Island, including village plan morphology, village scale, village orientation, and building density, providing a foundational basis for summarizing the regularity of spatial morphology indicators of coastal villages on Naozhou Island.

Characteristics of Village Spatial Morphology

- Village Spatial Morphology

Most villages on Naozhou Island exhibit a certain pattern in their planar forms. The coastal villages can be categorized into two types based on their planar layouts: compactly arranged collective villages and loosely arranged dispersed villages (Table 1).

Table 1. The planar morphology of the villages on Naozhou Island.

Village Morphology		Simplified Schematic Diagram	Typical Village Planar Morphology	Feature Description
Set class	Checkerboard layout		 CunLiang Village	The buildings are arranged neatly like chess pieces; the architectural layout is rigorous, with a strong sense of order, and the overall arrangement is compact.
	Clustered layout	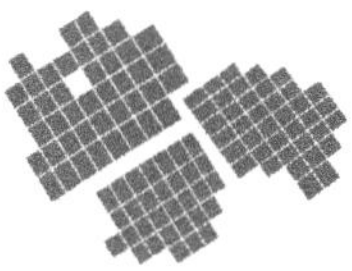	 ZhuCai Village	Several clusters of habitation are separated by roads, woods, and ponds, yet are connected as a whole.
Dispersed Category	Strip layout	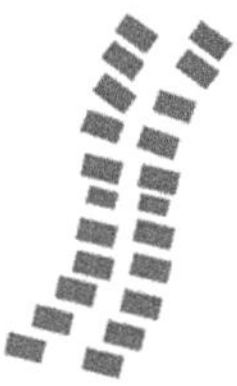	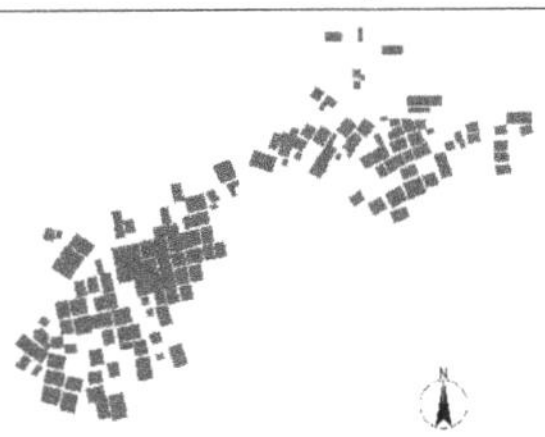 LongChi Village	Buildings are scattered along both sides of straight or curved roads, resulting in poor overall coherence.
	Scatter layout	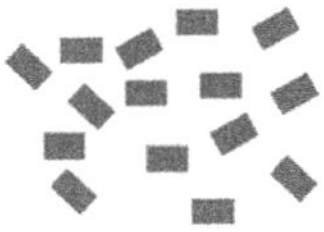	 GangTou Village	The buildings have no fixed orientation and are flexibly distributed within the village at certain distances, resulting in a dispersed and unordered overall layout.

According to statistics (Table A1), the majority of coastal villages on Naozhou Island are clustered layout villages, with both grid layout villages and group layout villages numbering 16 each, accounting for 38.1%; in contrast, the dispersed layout includes strip layout villages and scattered layout villages, which represent a smaller proportion at 21.4%

and 2.3%, respectively. The compact spatial layout demonstrates better wind resistance compared with the loose spatial layout, and grid layout villages and group layout villages are more compact than linear and scattered layouts, with a combined proportion of 76.2%. This is somewhat related to the frequent typhoon attacks on Naozhou Island, as a compact village spatial form can reduce the impact of typhoons to some extent.

- Village Scale

Based on field research and measurement data from Google Earth aerial images, the village scale is divided into five categories (Table 2). A statistical analysis of the village scales of 42 coastal villages on Naozhou Island (Table A1) reveals significant variations in scales among these villages. The smallest village is Dingdang Village, located on the eastern side of Naozhou Island, with an area of only 0.029 km^2; the largest village is Tianxuan Village, also on the eastern side of Naozhou Island, with a total area of 0.26 km^2. In terms of the proportional distribution of village scales, micro-scale villages (0.01–0.04 km^2) and super large-scale villages (greater than 0.2 km^2) account for a small proportion, both at 7.1%; large-scale villages (0.1–0.2 km^2) account for 16.7%; small-scale villages (0.04–0.1 km^2); and medium-scale villages (0.07–0.1 km^2) are the most common on Naozhou Island, each accounting for nearly one-third, at 38.0% and 31.0%, respectively.

Table 2. Statistical table of coastal villages of different scales on Naozhou Island.

Category	Aerial Photograph of a Typical Village	Typical Village Names
Micro-villages (0.01∼0.04 km^2)		Yingfan Village, Tantan Village, Dingdang Village.
Small villages (0.04∼0.07 km^2)		Zhouwu Village, Yingming Village, Dawen Village, Danshui Village, Houjiao Village, Yanting Village, Liuzhu Village, Mengxi Village, Naganwei Village, Tantanzai Village, Jialiu Old Village, Gangtou Village, Zhengpeng Village, Nakunzai Village, Xianadong Village, Liuluo Village.
Medium-sized village (0.07∼0.1 km^2)		Jialixin Village, Honggong Village, Doulong Village, Beijin Village, Chima Village, Songhuang Village, Nakun Village, Beizhai Village, Longchi Village, Xibu Village, Liangwu Village, Nafan Village, Naguang Village.

Table 2. *Cont.*

Category	Aerial Photograph of a Typical Village	Typical Village Names
Large villages (0.1~0.2 km^2)		Cunliang Village, Dalang Village, Nangang Village, Xundizai Village, Tanbeihu Village, Yanlou Village, Dalinhou Village.
Super large village (greater than 0.2 km^2)		Zhucai Village, Tianxuan Village, Mengdong Village.

- Building Density

The overall terrain of Naozhou Island is flat, and the populations in the villages are relatively small, providing ample space for construction and public areas within the villages. At the same time, the lower building density in the villages not only contributes to good ventilation but also serves as an intuitive reflection of the relatively dispersed spatial layout of the various villages on Naozhou Island. According to statistical results (Table A1), the building density of coastal villages on Naozhou Island is low, generally not exceeding 50%. Villages with building densities of 10% to 20% and 20% to 30% account for the largest proportions, comprising 38% and 50% of all villages, respectively; there are 2 villages each with building densities of 10% and 30% to 40%, each accounting for 5%; and among the 42 villages, only 1 village has a building density exceeding 40%. These factors collectively contribute to the characteristic low building density of the villages on Naozhou Island.

- Orientation of the Village

Combining field research with Google Earth aerial images, a statistical analysis of the orientations of 42 coastal villages on Naozhou Island was conducted (Table A1). The results show a clear pattern in village orientations: the most common orientations are north–south and northwest–southeast, each accounting for 28.6%, together comprising nearly two-thirds of the villages. Other orientations have a lower proportion, and, additionally, 4.7% of the villages have unclear orientations. The formation of this orientation distribution is closely related to the unique natural environment and climatic conditions of Naozhou Island. The island experiences prevailing south and southeast winds throughout the year; to adapt to this wind pattern, most villages have chosen north–south or southeast orientations to maximize the use of natural wind power and enhance internal ventilation. At the same time, Naozhou Island enjoys a warm and humid climate with ample sunlight. These climatic conditions allow village orientations to be flexible rather than strictly adhering to the traditional north–south principle, enabling adjustments based on actual needs and topographical features, thus resulting in diverse orientation characteristics. Further analysis reveals that villages close to the coastline, such as Yingming Village and Xibu Village in the south of the island, tend to face the sea, while villages in the north, like Yanlou Village, choose a northern orientation based on topography and wind direction. The orientation of these villages towards the sea serves both ventilation purposes and is related to local customs, such as the construction of Mazu temples facing the sea to pray for the safe return of family members who go out to sea.

2.2. Construction of the Wind Environment Evaluation System for Naozhou Island

Due to Naozhou Island's location in a subtropical monsoon climate region, the effects of humidity and heat are quite pronounced, along with 3 to 4 typhoons occurring each year, which pose a serious threat to the safety of the island's residents. Therefore, this study primarily evaluates the comfort and safety at a height of 1.5 m, corresponding to pedestrian level.

2.2.1. Classification of Wind Speed Levels on Naozhou Island

The existing literature [28] has proposed a classification standard for horizontal wind speed levels at a height of 1.5 m for outdoor pedestrian comfort in Hong Kong during the summer, specifically as follows: stagnant wind ($V < 0.3$ m/s), poor (0.3 m/s $\leq V <$ 0.6 m/s), low (0.6 m/s $\leq V <$ 1.0 m/s), generally acceptable (1.0 m/s $\leq V <$ 1.3 m/s), and good ($V \geq$ 1.3 m/s). Given that both Zhanjiang and Hong Kong belong to the subtropical climate zone, the two regions exhibit a high degree of similarity in climatic characteristics during the summer, both showing high temperatures and humidity. Additionally, according to historical meteorological data, the average summer temperatures in Zhanjiang and Hong Kong are 28.3 °C and 28.2 °C, respectively, with summer relative humidity at 82.2% and 81.9%. Therefore, this wind speed classification standard has a high degree of applicability for evaluating the wind speed values in the outdoor wind environment of Naozhou Island in Zhanjiang during the summer.

The summer gust wind speed on Naozhou Island ranges from 4.0 to 6.0 m/s, and the unpaved land in the villages is prone to dust raising. According to the "Technical Specifications for the Prevention and Control of Urban Dust Pollution"[29], wind speeds above level 4 (with a surface roughness $\alpha = 0.12$ converted to 4.4~6.3 m/s) require dust prevention measures. This paper sets 4.4 m/s as the dust threshold. Additionally, combined with related studies, 5 m/s is defined as the critical value for wind comfort. Beaufort scale level 6 (approximately 8.6 m/s at pedestrian height) and level 7 (approximately 11.0 m/s) significantly affect outdoor activities. This paper refines the assessment of wind speed levels on Naozhou Island in summer and establishes a wind speed evaluation standard for the wind environment at pedestrian height (1.5 m), categorizing wind comfort into five levels (Table 3).

Table 3. Evaluation standards for wind speed values in the wind environment at 1.5 m above ground level on Naozhou Island in summer.

Wind Comfort Level	Wind Speed Grade	Pedestrian Height (1.5 m) Wind Speed	Pedestrian Perception	Remarks
Level 1: unsatisfied, the environment is stuffy and hot	1	$V < 0.3$ m/s	Calm wind, poor thermal comfort	Measures such as adding ventilation facilities or adjusting the landscaping layout should be taken to improve thermal comfort.
	2	0.3 m/s $\leq V <$ 0.6 m/s	Poor thermal comfort	
	3	0.6 m/s $\leq V <$ 1.0 m/s	Lower thermal comfort	Under the shelter of tree shade or building shadows, solar radiation is reduced, making this discomfort somewhat acceptable.
Level 2: satisfied, can engage in outdoor activities freely	4	1.0 m/s $\leq V <$ 1.3 m/s	Basic satisfaction with thermal comfort	Wind energy effectively alleviates heat and enhances human comfort.
	5	1.3 m/s $\leq V <$ 4.4 m/s	Better thermal comfort, without generating dust	

Table 3. *Cont.*

Wind Comfort Level	Wind Speed Grade	Pedestrian Height (1.5 m) Wind Speed	Pedestrian Perception	Remarks
Level 3: discomfort, outdoor activities slightly affected	6	$4.4\ \text{m/s} \leq V < 5\ \text{m/s}$	Normal activities are affected, resulting in dust generation	The comfort of pedestrians decreases, and outdoor activities may be somewhat restricted; at this time, effective dust control measures should be implemented.
	7	$5.0\ \text{m/s} \leq V < 8.6\ \text{m/s}$	Can be tolerated, normal activities are affected, resulting in dust generation	
Level 4: danger, outdoor activities severely affected	8	$8.6\ \text{m/s} \leq V < 11.0\ \text{m/s}$	Discomfort severely affects normal activities and may pose a danger	Windproof reinforcement measures must be taken to ensure pedestrian safety and prevent accidents.
	9	$11.0\ \text{m/s} \leq V < 13.6\ \text{m/s}$	Extremely uncomfortable, severely affecting normal activities, and potentially causing danger	
Level 5: dangerous, not suitable for outdoor activities	10	$13.6\ \text{m/s} \leq V$	The wind environment is severe and unbearable, which may pose a danger	The area is considered extremely unsafe and pedestrian access should be avoided as much as possible, with warning signs and protective measures in place.

2.2.2. Evaluation System for the Wind Environment of Naozhou Island

For the evaluation of the outdoor wind environment, common indicators include average wind speed, wind speed amplification factor, ratio of comfortable wind area, ratio of calm wind area, and ratio of strong wind area, among others. Definitions are provided below:

1. Average wind speed: the average wind speed at a height of 1.5 m for pedestrians at various points within the assessment area, used to evaluate the overall wind speed magnitude in the region.
2. Wind speed amplification factor: The wind speed amplification factor reflects the amplification effect of a building on wind speed, typically referring to the ratio of the maximum wind speed at a height of 1.5 m above the ground around the building to the wind speed at the same height in an open area. The calculation of the wind speed amplification factor can be performed using the following exponential function that describes the variation of average wind speed with height, this is example 1 of an equation [30,31]:

$$\begin{cases} v' = \frac{v_{1.5B}}{v_{1.5f}} \\ v_{1.5f} = v_{10f}\left(\frac{1.5}{10}\right)^{\alpha} \end{cases} \quad (1)$$

   v'—wind speed amplification factor. $v_{1.5B}$ —maximum wind speed at a height of 1.5 m above the ground around the building, $v_{1.5f}$ —wind speed at a height of 1.5 m above the ground in an open area away from the building, v_{10f}—wind speed at a height of 10 m above the ground in an open area away from the building, α—ground roughness index.
3. Comfortable wind zone area ratio: This assesses the ratio of the area of the comfortable wind zone at a height of 1.5 m for pedestrians to the total area of the outdoor open space. A larger comfortable wind zone area ratio (not exceeding 100%) indicates better wind environment quality of the site. In this paper, the wind speed value for the comfortable wind zone is taken as 1.0 m/s to 4.4 m/s.

4. Static wind area ratio: The ratio of the area of the static wind zone at a height of 1.5 m for pedestrians to the total area of outdoor open space within the assessment area. In this paper, the area where wind speed is less than 1 m/s at a height of 1.5 m for pedestrians is defined as the static wind zone. Under summer humid and hot climate conditions, the static wind zone can create a feeling of stuffiness, reducing comfort. A larger static wind area ratio indicates poorer ventilation in the assessment site and lower comfort levels.
5. Strong wind area ratio: The ratio of the area occupied by strong wind zones at a height of 1.5 m for pedestrians to the total area of open space within the assessment area. In this paper, areas where wind speeds exceed 5 m/s at a height of 1.5 m for pedestrians are defined as strong wind zones. Under summer humid and hot climatic conditions, while strong wind zones can dissipate heat, they also generate dust, making outdoor activities uncomfortable and potentially leading to safety issues. The larger the area of strong wind zones, the poorer the wind comfort.

In assessing the wind environment quality of coastal villages on Naozhou Island, this paper employs various indicators that have been widely used in this field [5,32,33]. However, to more comprehensively reflect the characteristics of the village's wind environment under different wind conditions, this article will utilize two wind field coefficients, K1 and K2.

6 The setting of K1 is designed to assess the ventilation efficiency and comfort of the site by comparing the ratio of the area of comfortable wind zones to that of stagnant wind zones under normal wind conditions. This can intuitively reflect the proportion of comfortable wind environments in the village and the proportion of poorly ventilated areas under normal wind conditions, thereby providing us with a basis for improving and optimizing the wind environment.

7 The setting of K2 is designed to address challenges under extreme wind conditions. In extreme wind conditions, the emergence of strong wind zones can significantly reduce the wind comfort of the site and may even pose a threat to the lives and safety of residents. Therefore, by comparing the ratio of the area of comfortable wind zones to the area of strong wind zones, we introduce the coefficient K2 to assess the wind resistance capability and the stability of the wind environment of the site under extreme wind conditions. A larger value of K2 indicates that the site can still maintain good wind comfort under extreme wind conditions, which is of great significance for improving the quality of life and safety of residents.

From the above, it can be seen that the average wind speed, as a fundamental and intuitive indicator, can reflect the overall wind speed level within the assessment area, helping to understand the basic characteristics of the wind environment in the village. The wind speed amplification factor is used to quantify the amplification effect of wind speed in specific areas, which is significant for assessing regions within the village that may experience excessively high local wind speeds. The K1 and K2 coefficients combine three indicators—the area ratios of comfortable wind zones, calm wind zones, and strong wind zones—to obtain ventilation efficiency and comfort under different wind conditions (normal and extreme). Together, they reflect the basic characteristics of the wind environment, local amplification effects, and comfort and stability under varying wind conditions. In summary, this paper conducts a comprehensive evaluation of the wind environment in coastal villages on Naozhou Island using four indicators: average wind speed, wind speed amplification factor, wind field coefficient K1, and wind field coefficient K2.

2.3. Research Methods

This study comprehensively employs both field measurements and numerical simulations to assess the wind environment quality of the villages on Naozhou Island. First, field measurements are conducted in selected typical villages to understand their wind environment characteristics. These measured data are used not only to evaluate the current wind environment quality but also as a key basis for verifying the accuracy of PHOENICS simulations. Subsequently, the PHOENICS software (2019) is used to simulate the wind field at pedestrian height in the village under different spatial morphology indicators, calculating relevant parameters of a single spatial index variation model to conduct an in-depth analysis of the village's wind conditions, and providing a scientific basis for improving wind environment quality.

2.3.1. Measured Wind Environment

To clarify the wind environment characteristics of the coastal villages on Naozhou Island, this paper selected Yingming Village, which has the highest proportion of the four morphological indicators previously mentioned, for field research. Wind environment-related measurements and analyses were conducted (Figure 3). During the measurement process, a total of 10 measurement points were arranged to comprehensively and accurately capture the overall wind environment of Yingming Village, ensuring the completeness and accuracy of the data. The Kestrel 500 weather station was used to measure the wind environment in the streets and alleys of Yingming Village, along with the wind environment at the village boundary. The measurement time selected was the period from 9:00 to 20:00 on a typical summer day (15 July 2022).

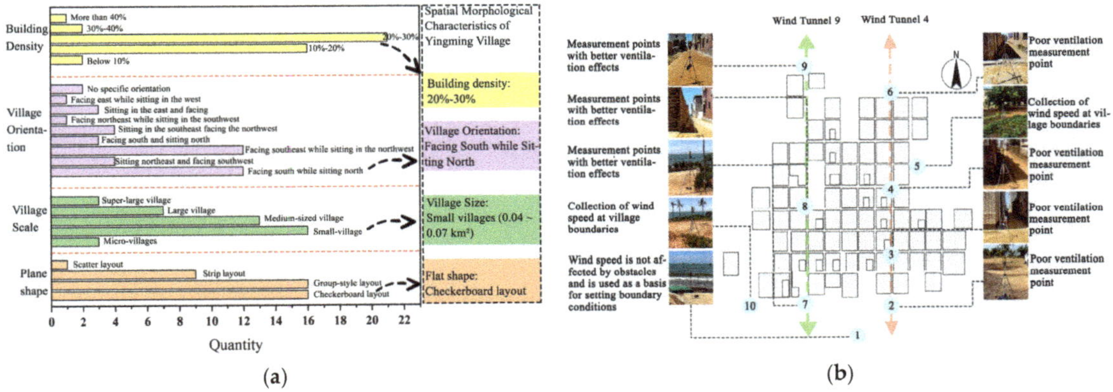

Figure 3. Typical selection and measurement point layout of Yingming Village. (**a**) Typical village selection; (**b**) measurement point distribution map.

2.3.2. Analysis of Measured Results

- Analysis of Wind Sources

From Figure 4, it can be seen that between 8:00 and 20:00, the wind field in Yingming Village exhibits multiple regular fluctuations of "rise/fall." Taking measurement points 1 and 10 as examples, both points are direct wind sources from the southern seaside, and their wind speed and direction throughout the day are relatively close. Measurement point 1 is closest to the seaside, with no plants or structures obstructing it. The average wind speed at measurement point 1 throughout the day is 3.3 m/s, with a maximum wind speed of 4.6 m/s, and the predominant wind direction is southeast. In the morning, the wind speed is relatively stable, averaging 3.0 m/s, while, in the afternoon, the wind speed increases and varies significantly. The average wind speed at measurement point 10 throughout the day is 2.8 m/s, with the main wind direction being south and a maximum wind speed of 4.5 m/s. The wind speed in the morning is relatively stable, but in the afternoon, influenced by surrounding vegetation, the fluctuations in wind speed and direction at measurement point 10 are more pronounced compared with measurement point 1.

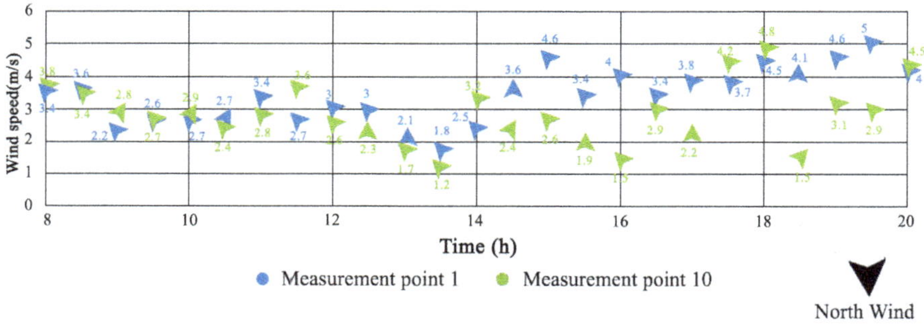

Figure 4. Measurement point 1 and measurement point 10 wind speed and wind direction.

- Wind Channel Analysis

According to the measured results, the wind environment speed in Yingming Village is influenced by internal and external obstacles within the village and the wind pressure of the wind field. The wind speed gradually decreases from south to north, while the wind direction remains relatively constant. Taking the measured results at 8:00 as an example, the wind speed in wind channel 4 (with measurement points 2, 3, 4, and 6) decreases from 2.6 m/s at measurement point 1 to 0.0 m/s at measurement point 6 (Figure 5). In wind channel 9 (with measurement points 7, 8, and 9), the wind speed decreases from 3.1 m/s to 1.3 m/s, and the trend of decreasing wind speed from south to north in wind channel 9 is more pronounced. It can be observed that from 8:00 to 20:00, the wind speed at measurement point 7 is greater than that at measurement point 9 for most of the time.

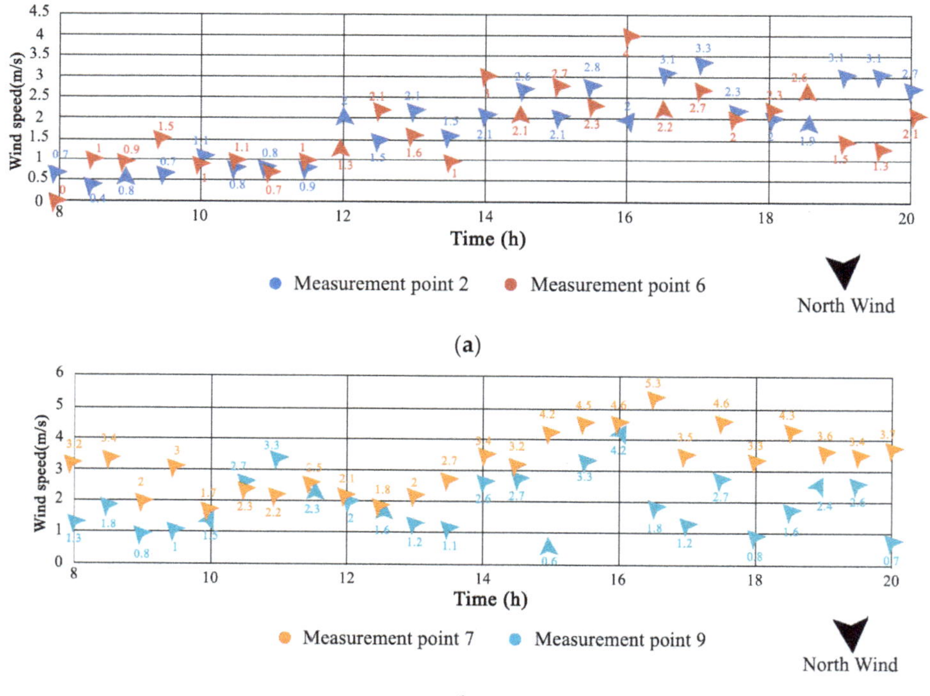

Figure 5. Wind speed and direction at measurement points of two wind channels: (**a**) wind channel 4; (**b**) wind channel 9.

Through the measurement of the wind environment on typical summer meteorological days in Yingming Village, the following characteristics can be observed: the overall wind speed in Yingming Village on typical summer meteorological days is relatively low and varies little, with an average wind speed below 3.0 m/s. The wind direction is also quite stable, predominantly from the southeast, with occasional south winds; the incoming wind is weakened by the village, and, after passing through the village, there is a noticeable attenuation in wind speed.

2.3.3. Software Simulation and Verification

- Simulation Verification

Numerical simulation based on computational fluid dynamics (CFD) is one of the important tools for improving urban wind environments [34,35]. The most common CFD methods include two types: Reynolds-averaged Navier–Stokes (RANS) and large eddy simulation (LES). RANS is known for its cost-effectiveness and reasonable accuracy in various turbulent flows, as it directly solves the mean flow and simulates turbulence across all scales [36]. RANS-based CFD software includes PHOENICS, Fluent, COMSOL, etc. Due to PHOENICS' strong openness, ease of use, and suitability for industrial applications and scientific research, it has been widely used [37]. Whether for small-scale village buildings or large-scale residential areas, the error between measurement results and simulation results is within an acceptable range (less than 10%). Therefore, the use of PHOENICS software for wind environment simulation has good reliability [4,38]. This paper uses PHOENICS for the wind environment simulation of Niao Island (Figure 6), with the initial parameters and boundary conditions for the simulation as follows:

- Grid accuracy setting: the manual setting is used to ensure that each grid cell is less than 4 m, with a grid gradient rate of 1.2 applied.
- Wind environmental boundary conditions: based on the survey, the normal wind condition is set as a southeast wind with a wind speed of 2.57 m/s; the extreme wind condition is set as an east wind with a wind speed of 21.70 m/s according to meteorological station data.
- Number of iterations: 3000 iterations are completed to ensure the stability of the results.
- Surface roughness index is set at 0.12, reflecting the impact of surface characteristics on the wind field.

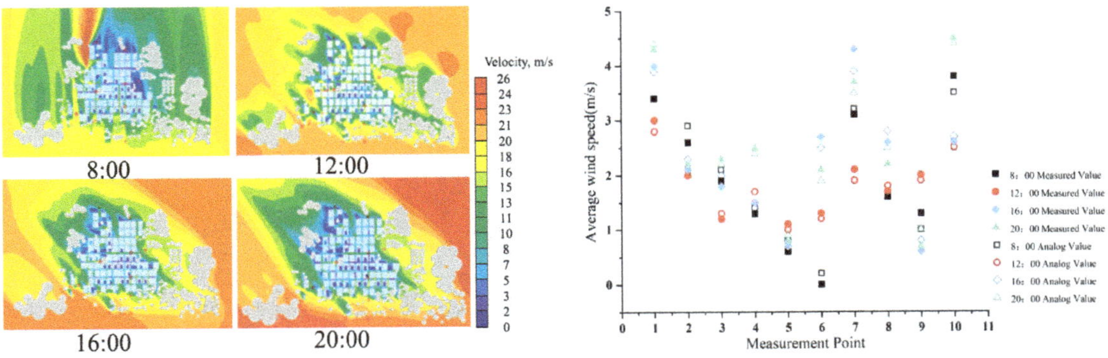

(a) (b)

Figure 6. Measured and simulated analysis of Yingming Village. (**a**) Simulated wind speed cloud map at four moments; (**b**) comparison of simulated wind environment values and measured values.

Through the comparative analysis of the simulated values and the measured values at 10 measurement points (Figure 6), it can be observed that the trends in wind speed and direction for the 10 measurement points are consistent between the simulated and measured values. During the same time period, the trends of rising or falling wind speed are essentially the same for both measured and simulated data. Overall, the measured values and simulated values at the 10 measurement points at 4 different times do not differ significantly, with a difference of 0 to 0.3 m/s, and the measured wind direction is generally consistent with the simulated wind direction.

- Goodness-of-Fit Analysis

The goodness of fit refers to the degree of agreement between the model's predicted results and the actual values. Here, it is used to examine the degree of agreement between the simulated wind speed results and the measured values. Common testing methods include residual sum of squares test, chi-squared test, linear regression test, etc.

This paper uses the residual sum of squares for testing, This is example 2 of an equation:

$$Q = \sum(y - y^*)^2 \tag{2}$$

where y represents the measured value; y* represents the simulated value.

The goodness-of-fit index equation is given in (3):

$$R^2 = 1 - Q \tag{3}$$

The average fitting degree of the 10 measurement points at 8:00 is R^2 (8:00) = 0.67, at 12:00 is R^2 (12:00) = 0.78, at 16:00 is R^2 (16:00) = 0.67, and at 20:00 is R^2 (20:00) = 0.62.

The goodness of fit is averaged for the four time points as follows (4):

$$R^2_{Average} = \left(R^2_{(8:00)} + R^2_{(12:00)} + R^2_{(16:00)} + R^2_{(20:00)}\right)/4 = 0.685 \tag{4}$$

A goodness-of-fit value greater than 0.5 indicates that the fit is reliable. The fitting values for the 10 measurement points across 4 time periods are all greater than 0.6, with an average of 0.685, which is close to 0.7, indicating a high degree of agreement between the simulated and measured values. This suggests that the software can accurately describe the wind environment of a Naozhou Island village.

2.3.4. Model Establishment Based on Single Spatial Index Variation

To accurately analyze the wind environment characteristics of the villages on Naozhou Island, 27 villages with clear planning and uniform building orientations were selected from a total of 42 villages for in-depth discussion (Table A2). In contrast, scattered layout villages, due to their lack of fixed orientations and flexible distribution within a certain distance, exhibited a disordered and discrete overall pattern, which significantly increased the complexity and difficulty of wind environment simulation. Therefore, in this study of the wind environment simulation of coastal villages on Naozhou Island, scattered layout villages were not included in the analysis. Additionally, based on the results of field investigations of the spatial forms of the selected villages, we constructed four sets of models in the software simulation. In the model settings, the height of the buildings and the spacing between buildings were kept constant at 10 m (Figure 7). By adjusting the spatial form index values of each model group, we analyzed the changes in the village wind environment under different spatial form index values and two wind conditions (the normal wind condition being a southeast wind at a speed of 2.57 m/s; the extreme wind condition being an east wind at a speed of 21.70 m/s).

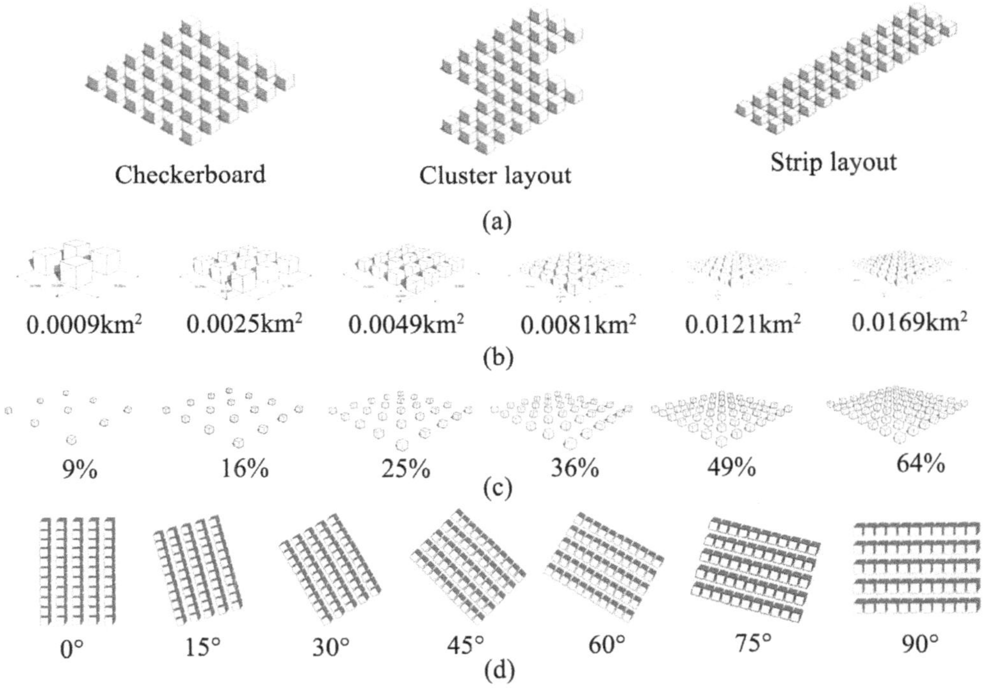

Figure 7. Comparison of spatial morphological indicators of villages. (**a**) Differences in planar morphology; (**b**) differences in scale; (**c**) differences in building density; (**d**) differences in village orientation.

3. Results and Discussion

The previous text has established a wind environment evaluation system for coastal villages on Naozhou Island and verified the feasibility of the software simulation method. The average wind speed, wind speed amplification factor, and wind field coefficients K1 and K2 have been identified as key evaluation indicators. Four sets of models have been established to explore the relationship between different village spatial forms and the wind environment. On this basis, this chapter will discuss the simulation experiment results, analyzing the specific impacts of different spatial form variables (plan form, village scale, building density, village orientation) on the village wind field under normal and extreme wind conditions, aiming to reveal the intrinsic connection between village morphology and wind field changes, providing a scientific basis for subsequent optimization.

3.1. The Impact of Village Spatial Morphology on Wind Environment

Figure 8 shows the wind speed cloud maps of different village forms under two wind conditions, including the simulation results of average wind speed, wind speed amplification factor, and wind field coefficient K1.

From Figure 8, it can be seen that under normal wind conditions (a southeast wind, with a wind speed of 2.57 m/s), the average wind speed in the village is influenced by its layout, with the checkerboard layout having an average wind speed of 2.34 m/s, followed by the strip layout at 2.19 m/s, and the clustered layout at 2.02 m/s. The wind speed amplification factor, however, is relatively unaffected by the village's planar morphology, with all three models having a wind speed amplification factor around 1. As the village morphology changes from a checkerboard layout to a clustered layout, the wind field coefficient decreases from 10.8 to 9.24, demonstrating that the checkerboard layout is optimal for wind field comfort. A comparison of the three indicators indicates that the checkerboard layout allows people to feel more comfortable. In extreme wind conditions (an east wind, with a wind speed of 21.70 m/s), the wind speed amplification factor for the strip layout is lower than that of the other two layouts, indicating it is optimal for wind field stability, while, in terms of wind field safety, the order is strip layout > clustered layout > checkerboard layout. In summary, it can be concluded that, under normal summer wind conditions, the checkerboard layout creates a better wind environment, making it easier for individuals to feel comfortable, and demonstrates good adaptability to the wind environment on the island. In extreme wind conditions, if the construction of the village aims for wind field safety, the strip layout should be chosen.

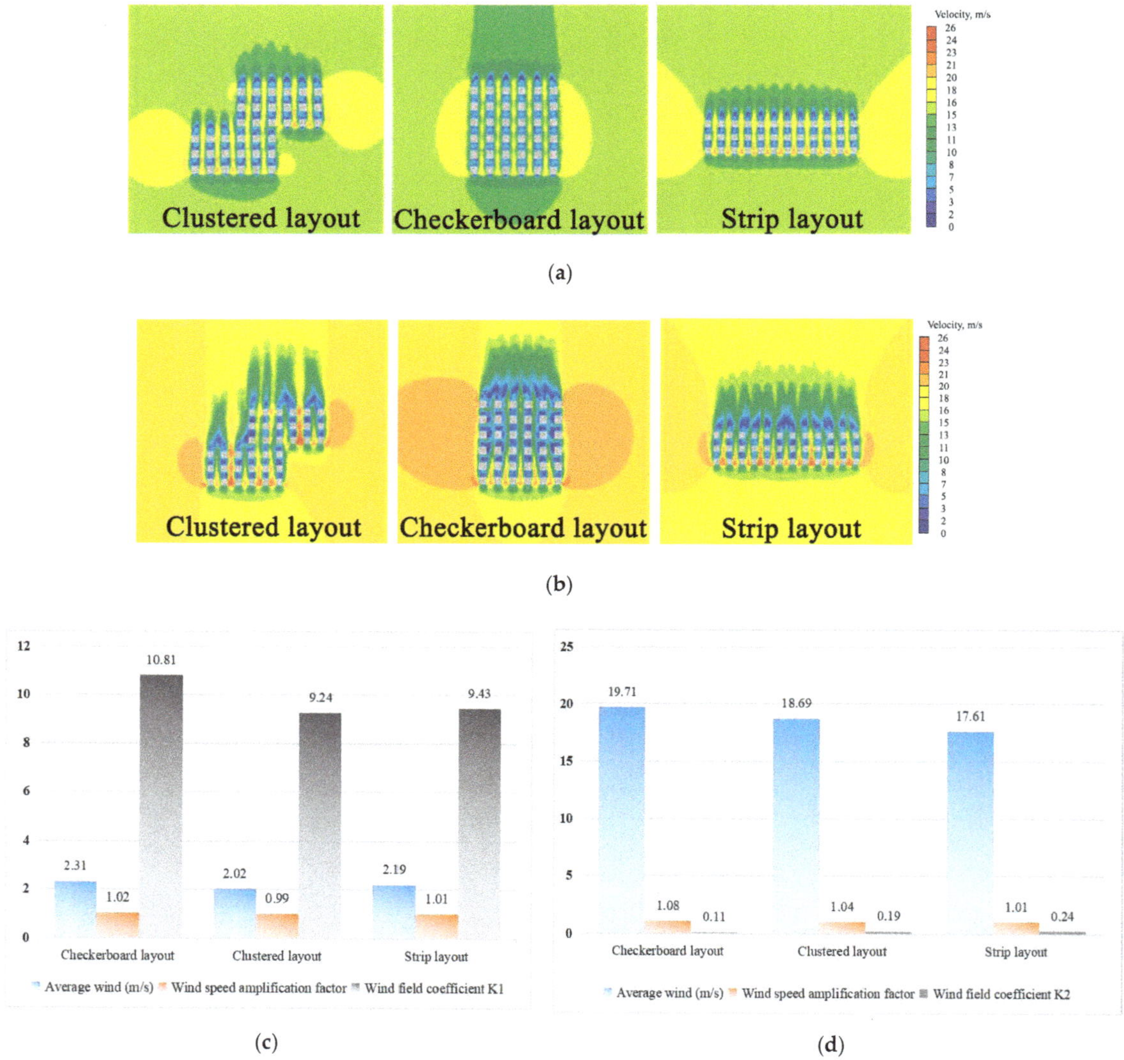

Figure 8. Wind speed cloud maps and simulation values of different village forms under two wind conditions: (**a**,**c**) normal wind conditions; (**b**,**d**) extreme wind conditions.

3.2. Impact of Village Orientation on the Wind Environment

The orientation of the village primarily affects the angle of incidence of the wind. In this subsection of the experiment, under a south wind condition with a wind speed of 10 m/s, the wind environment of villages with different orientations is simulated. In the model setup, the village size is maintained at 11,700 square meters, with the number of buildings kept constant at 55, arranged in 5 columns and 11 rows, with a spacing of 10 m between columns and 2 m between rows, only changing the orientation of the model (Figure 9).

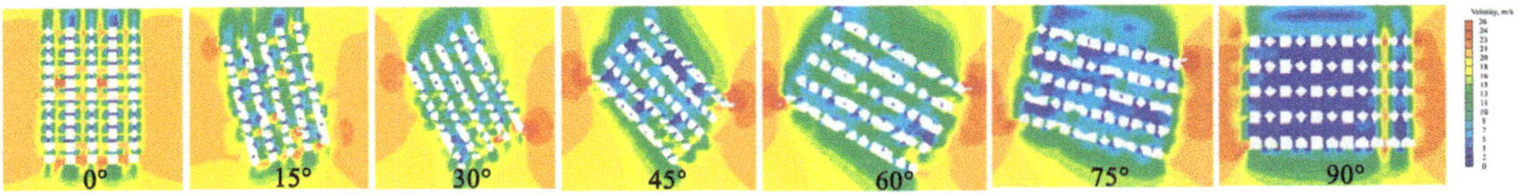

(**a**)

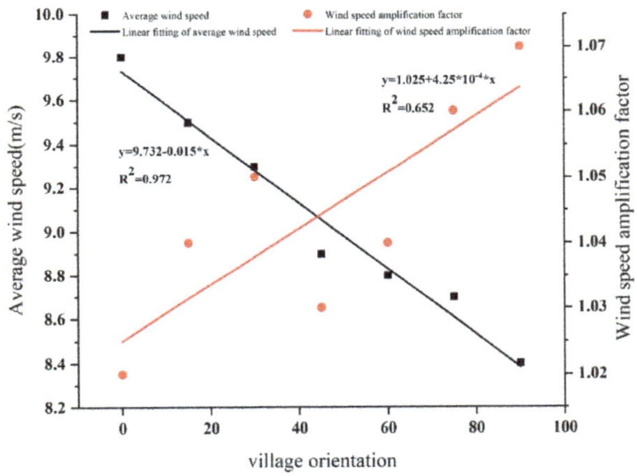

(**b**)

Figure 9. Numerical simulation of different building orientations. (**a**) Village wind speed cloud map; (**b**) simulation data.

By simulating the wind environment of villages with different orientations (Figure 9), and statistically analyzing the simulation results (Table A2), it can be observed that the orientation of the village primarily affects the angle at which wind enters the village. Between 0° and 90°, the angle between the village and the wind direction is inversely proportional to the average wind speed (regression coefficient of −0.015), with a correlation coefficient R^2 of 0.95. This indicates that the smaller the angle between the village and the wind direction, the greater the wind speed. Under normal wind conditions, the wind field is more comfortable, while under extreme wind conditions, the wind field is more dangerous. The stability of the wind field is not significantly correlated with changes in angle (regression coefficient of only 0.0064). In summary, the authors suggest that under normal wind conditions, the orientation of the village should be aligned as closely as possible with the prevailing wind direction to achieve better ventilation benefits. Under extreme wind conditions, it is advisable to avoid alignment with the prevailing wind direction and to form a certain angle to create a larger windward area, thereby resisting strong winds and enhancing the safety of the village's wind field.

3.3. The Impact of Village Scale on Wind Environment

Figure 10 shows the wind speed cloud map for different village sizes, providing the trend and fitting degree of changes in average wind speed, wind speed amplification factor, and wind field coefficient with variations in village size.

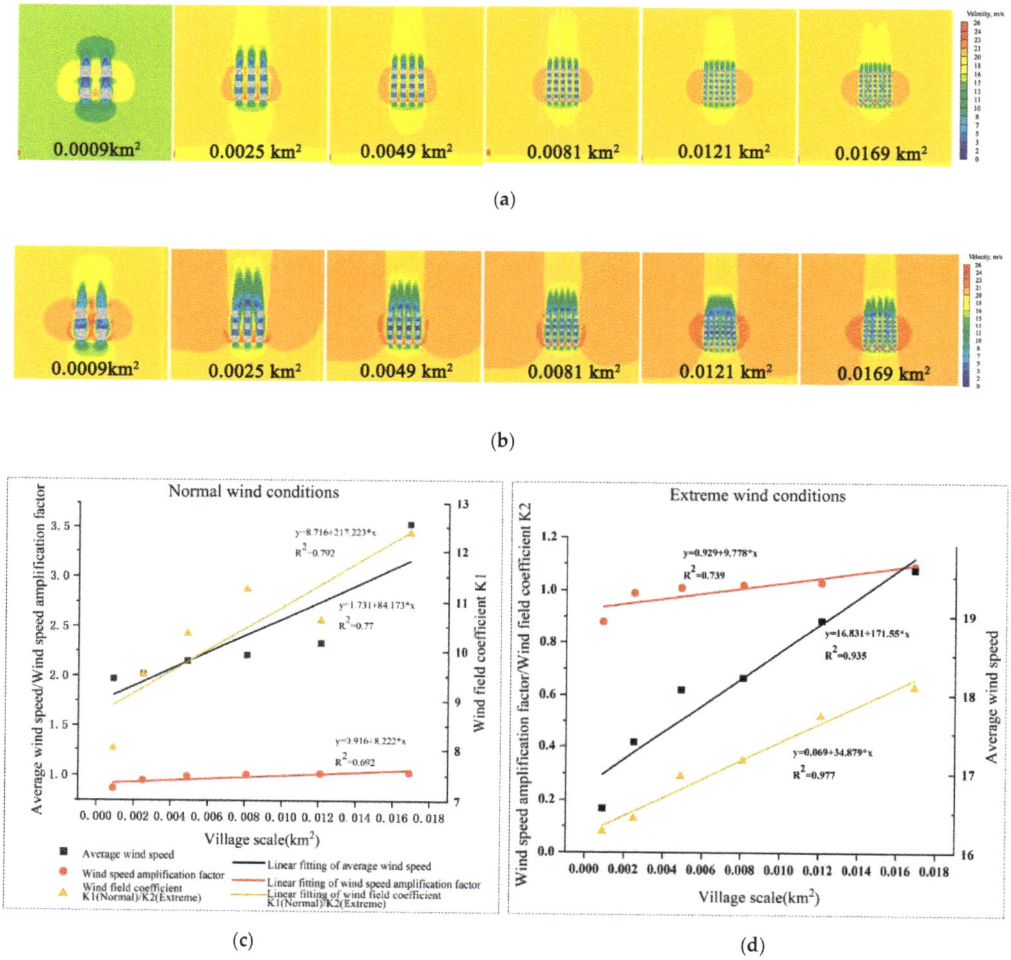

Figure 10. Wind speed cloud maps and simulation values for different village scales under two wind conditions: (**a**,**c**) normal wind conditions; (**b**,**d**) extreme wind conditions.

By simulating the wind environment of villages of different scales, and statistically analyzing the simulation results (Table A2), it can be observed that under both normal wind conditions (a southeast wind, with a wind speed of 2.57 m/s) and extreme wind conditions (an east wind, with a wind speed of 21.70 m/s), the three evaluation indicators are proportional to the village scale within the range of 0.03 km^2 to 0.26 km^2, with correlation coefficients R^2 greater than 0.67. The research results indicate that the village scale has significant explanatory power for the three indicators, particularly for the average wind speed, where the R^2 values for both wind conditions are significantly higher than those for the other two indicators. This clearly indicates a strong correlation between the village scale and the average wind speed, meaning that, as the village scale increases, the average wind speed also increases. Although the village scale also shows a positive correlation with the wind speed amplification factor, its impact is relatively limited. As the village scale continues to expand, the maximum wind speed at a height of 1.5 m above the ground around buildings will also increase. The changes in the wind field coefficients indicate that under normal wind conditions, as the village scale expands, the average wind speed rises, and the comfortable zone is significantly larger than the calm wind zone. Under extreme wind conditions, as the scale increases, the wind field becomes relatively safer. In summary, it can be inferred that the impact of the village scale on the wind environment is significant. Under normal wind conditions, larger village scales correspond to higher average wind speeds and a more comfortable wind field, resulting in a higher overall comfort level of the village's wind field. Under extreme wind conditions, larger villages exhibit a stronger resistance to strong winds.

3.4. The Impact of Building Density on Wind Environment

In the model setup, the village size remains constant at 10,000 square meters, with only the building density being altered. Figure 11 shows the wind speed cloud maps for different village densities, along with the trends and fitting degrees of changes in average wind speed, wind speed amplification factor, and wind field coefficient with variations in village density.

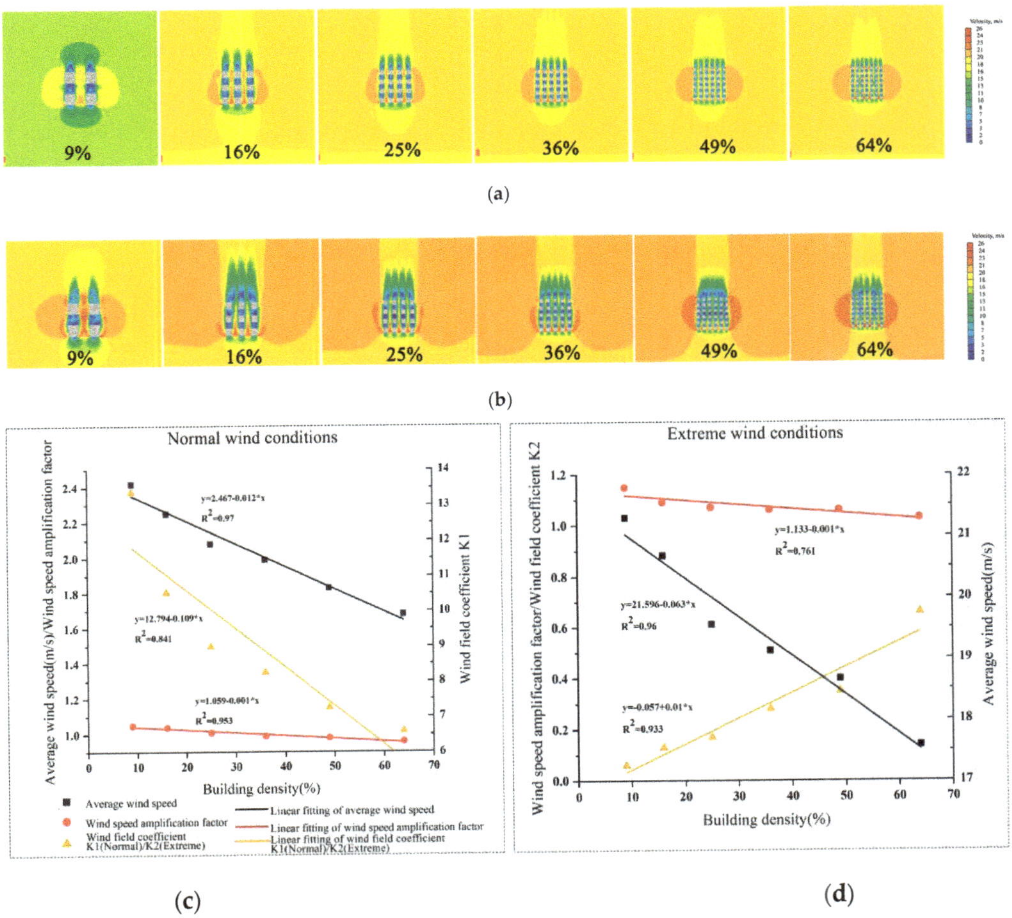

Figure 11. Wind speed cloud maps and simulation values for different building densities under two wind conditions: (**a**,**c**) normal wind conditions; (**b**,**d**) extreme wind conditions.

From the simulation data of two wind conditions (Table A2), it can be observed that the variation in building density has a high explanatory power for the variation of the three dependent variables under both wind conditions. The difference lies in that, under normal wind conditions, building density shows an inverse relationship with average wind speed, wind speed amplification factor, and wind constant K1 (regression coefficients of −0.012, −0.001, and −0.109, respectively). In contrast, under extreme wind conditions, building density exhibits a direct relationship with the wind field coefficient (regression coefficient of 0.01), meaning that the greater the village building density, the safer the wind field. From the wind speed cloud map, it can be seen that when a typhoon approaches, the outdoor wind far exceeds the comfortable range perceived by the human body, and an increase in density helps to block more wind, thereby creating more wind shadow areas. If people stay in this area, they will be relatively safer. Therefore, it can be inferred that, whether under normal or extreme wind conditions, changes in village building density can effectively alter the wind environment.

The above preliminary discussion on the correlation between spatial morphological indicators describing villages and the wind environment at pedestrian height indicates that different spatial forms have varying impacts on the comfort level of pedestrians in villages. Under normal wind conditions, the wind speed in a grid layout can reach 2.34 m/s, which is more suitable for island environments compared with the other two layouts. The density of the village is inversely related to the overall wind environment, while the scale of the village is directly proportional to the overall wind environment. Villages oriented towards the prevailing wind direction will experience higher wind speeds, thereby enhancing comfort within the village. In extreme cases, a strip layout (with a wind speed amplification factor of 1.01) can significantly reduce the hazards posed by typhoons, and villages with high density and large scale can better withstand extreme conditions, creating a larger safe area. In terms of village orientation, it is advisable to avoid alignment with the prevailing wind direction. However, these analyses only consider single factors; in reality, the changes in the wind environment of a village are influenced by multiple factors. Therefore, the following sections will conduct a coupled analysis of various selected spatial forms and wind environments, exploring the correlation and impact degree of each morphological indicator on the wind environment.

3.5. Coupling of Wind Environment and Spatial Form in Village Units

The previous text discussed the changes in a single indicator and environmental indicators, but the wind adaptability of coastal villages on Naozhou Island may not be influenced by a single variable but rather by the combined effects of multiple variables. This section conducts wind environment simulations for the 27 selected villages based on field research data, using SPSS (27) software to analyze the wind environment parameters at pedestrian height in coastal villages and the spatial morphology indicators, distinguishing between those spatial morphologies that have a significant impact on wind environment indicators and those that do not. The planar morphology is a categorical variable (discrete classified category), while the other three morphological indicators are continuous variables (capable of taking continuous values), which are fundamentally different in measurement scale. The mixed use of these may lead to results that are difficult to interpret or even misleading. To ensure the accuracy and interpretability of the analysis results, the following text will only conduct correlation and regression analysis on the three continuous morphological indicators and the four wind environment indicators.

The Pearson matrix in Figure 12 shows the correlation between different morphological indicators and wind environment indicators.

- In two wind conditions, the building density has a strong negative correlation with the average wind speed (extreme wind conditions) and wind speed amplification factor (normal), and there is a weak relationship with the average wind speed (normal) and wind constant coefficient K2, while there is almost no linear relationship with the wind speed amplification factor (extreme).

- In both wind conditions, the village scale has a strong positive correlation with other wind environment indicators, other than having a weak relationship with the average wind speed (extreme).
- In this study, there is no significant relationship between village orientation and wind environment indicators.

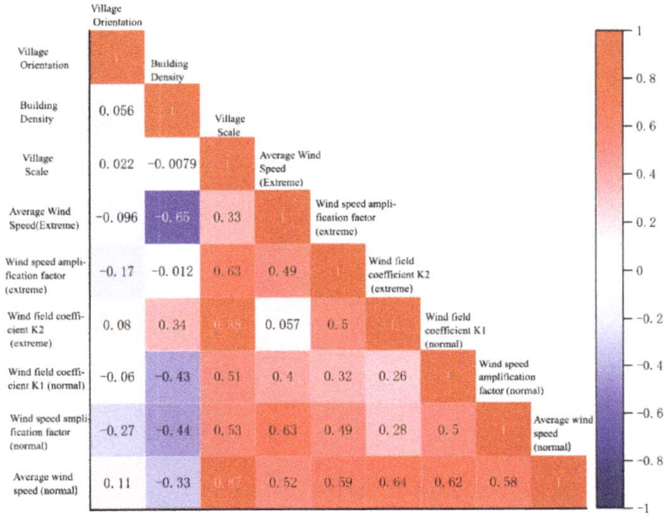

Figure 12. Pearson analysis of morphological and environmental indicators.

According to the results of the Pearson correlation analysis, both building density and village scale have a significant correlation with wind environment indicators, while village orientation shows no obvious relationship with wind environment indicators. This may be due to the complex process of wind field formation, which requires the consideration of multiple linear regression modeling in future research. The following text uses building density, village scale, and village orientation as independent variables in the model to predict the corresponding wind environment effects of a certain pattern.

The fitting of the multiple linear regression is presented in Table A3. A larger R^2 value indicates a better model fit. The constant in the table represents the value of the dependent variable when the independent variable is zero. The regression coefficients are also subjected to hypothesis testing, with corresponding significance provided; it is generally considered significant if the significance is less than or equal to 0.05. Tolerance > 0.1, VIF < 5. There is no multicollinearity in this study.

The regression equation is as follows:

Normal wind conditions:

$$Y_1 = 1.903 + 9.819A_1 - 0.043B_1 + 0.001C_1 \tag{5}$$

$$Y_2 = 1.118 + 0.858A_2 - 0.008B_2 - 3.19 \times 10^{-4}C_2 \tag{6}$$

$$Y_3 = 12.047 + 18.341A_3 - 0.171B_3 - 0.001C_3 \tag{7}$$

Extreme wind conditions:

$$Y_1 = 16.895 + 5.045A_1 - 0.112B_1 + 0.001C_1 \tag{8}$$

$$Y_2 = 1.053 + 0.814A_2 - 0.008B_2 + 4.63 \times 10^{-5}C_2 \tag{9}$$

$$Y_3 = -0.061 + 1.328A_3 + 0.006B_3 - 4.80 \times 10^{-5}C_3 \tag{10}$$

In the formula: Y_1, Y_2, Y_3—represent the average wind speed, wind speed amplification factor, and wind field coefficients K1 (normal)/K2 (extreme); A, B, C represent village scale, village density, and village orientation, respectively.

Through Figure 13, the changes in the goodness of fit and the significance of influencing factors for the two wind condition models can be visually observed. Under normal wind conditions, the R^2 values of the mathematical models are 0.867, 0.545, and 0.449, respectively. Among the three models, the R^2 value for average wind speed is the highest, indicating a high explanatory power for the independent variable, effectively explaining the positive correlation between village scale and average wind speed (regression coefficient of 9.819), meaning that, as the village scale increases, the overall wind comfort of the village also increases, and the negative correlation with building density (regression coefficient of −0.043), indicating that as the village density increases, the overall wind comfort of the village correspondingly decreases. The wind speed amplification coefficient and wind field coefficient K1 have a medium to high explanatory power for the independent variable. In terms of significance, the significance of village orientation is higher than that of other morphological indicators, thus the impact of village orientation on the wind environment is relatively small, and the wind environment indicators can be accurately described by Equations (5)–(7). Under extreme wind conditions, the R^2 values of the mathematical models are 0.538, 0.428, and 0.891, respectively. The R^2 value of the wind field coefficient K2 is the highest, which can better explain its correlation with the village scale and the building density (regression coefficients of −0.061 and 0.006). Meanwhile, the average wind speed and wind speed amplification coefficient have a medium to high explanatory power for the independent variable. From a significance perspective, in the average wind speed and wind field coefficient K2 model, village orientation shows higher significance compared with other morphological indicators. In the wind speed amplification coefficient model, the significance of village orientation and building density exceeds that of the village scale, indicating a relatively strong correlation between the village scale and the wind speed amplification coefficient, the wind environment indicators can be accurately described by Equations (8)–(10).

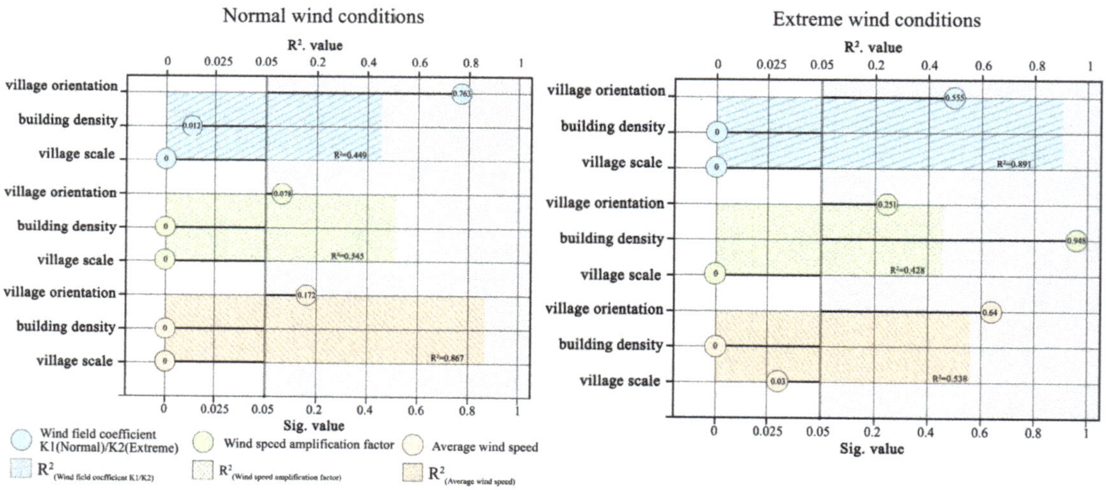

Figure 13. Significance test of village spatial morphology and wind environment indicators under two wind conditions.

In the model, the orientation of the village and the dependent variable showed no significant correlation ($p > 0.05$). Although it can have some impact on wind speed, this variable cannot yield accurate predictive values in the formula. The reason for this result may be the irregular spatial distribution of island villages, where the main orientation is usually closely linked to the direction of the streets. Due to the complexity of the island terrain and the possible lack of unified planning in early construction, the street system is relatively chaotic, characterized by winding paths and variable directions. This irregular street layout further affects the construction of houses within the village, leading to an uneven arrangement of houses, which exacerbates the discontinuity and disorder of the street system. Therefore, in terms of the impact on the wind environment, the orientation of the village does not demonstrate a significant correlation.

Due to the differences in the magnitude and units of the original variables, the regression coefficients of each variable in the formula cannot reflect the differences in the degree of influence. Instead, judgments should be made based on the standardized coefficients, where the sign of the standardized coefficient (positive or negative) indicates the vector relationship between the independent and dependent variables. As shown in Table 4, under normal wind conditions, the negative impact of the building density on the dependent variable is less than the positive impact of the village scale, which is most evident in the variable of average wind speed. In extreme wind conditions, the highest contribution to average wind speed comes from the building density (contribution rate: −62.12%), while the highest contributions to the wind speed amplification factor and wind field coefficient K2 come from the village scale (contribution rates: 76.97%, 69.73%).

Table 4. Table of relative contribution rates.

Wind Conditions	Dependent Variable / Independent Variable	Average Wind Speed	Wind Speed Amplification Factor	Wind Field Coefficient K1	Wind Field Coefficient K2
Normal wind condition	Village Scale	66.10%	43.93%	51.98%	-
	Building Density	−25.67%	−34.75%	−43.23%	-
	Village Orientation	8.22%	−21.31%	−4.78%	-
Extreme wind conditions	Village Scale	31.44%	76.97%	-	69.73%
	Building Density	−62.12%	0.36%	-	27.02%
	Village Orientation	−6.42%	−22.65%	-	3.24%

Relative contribution rate = standardized coefficient of the independent variable/sum of the absolute values of all standardized coefficients of the independent variables.

These three models consist of indicator coefficients and equation constants. Villages with different spatial morphological characteristics will exhibit variations in the types and degrees of wind environment influencing factors, leading to different corresponding regression models. This model is applicable to villages with a scale ranging from 36,500 m^2 to 260,300 m^2, a village density between 9.95% and 30.99%, and an orientation within the range of 70° to 248°. It can serve as an effective tool for assessing the wind environment under normal wind conditions in coastal villages, combined with outdoor wind environment evaluation standards, for wind environment assessment or optimization, aiming to enhance outdoor wind environment comfort. The minimum standard for critical wind speed to meet comfort requirements is established, and through the model, suitable values for the spatial morphological parameters of coastal villages can be inferred, assisting in the formulation of guidelines for village design or renovation and overall morphological optimization.

4. Conclusions

This study takes the coastal village of Naozhou Island in western Guangdong as a case study. First, spatial morphological data were obtained through field surveys. Secondly, the existing policies and standards related to outdoor wind environment evaluation, along with commonly used outdoor wind environment evaluation methods, were analyzed and studied. Based on the research object of this paper, an outdoor wind environment evaluation system for island villages was constructed. Subsequently, numerical simulation software was used to simulate and compare the results with measured data, exploring the relationship between the spatial morphology of island villages and wind adaptability.

The research results indicate that, under normal wind conditions, the layout patterns of villages show a trend of optimal chessboard arrangement, followed by banded arrangement, and relatively weaker group arrangement in terms of adaptability to the wind environment; under extreme wind conditions, the banded layout demonstrates better stability and safety of the wind field. Additionally, we found that spatial morphological indicators such as village scale, building density, and village orientation significantly affect the wind environment. Under normal wind conditions, the contribution rates of village scale to average wind speed, wind speed amplification factor, and wind field coefficient (66.10%, 43.93%, 51.98%) significantly exceed those of village density (−25.67%, −34.75%, −43.23%), highlighting the dominant role of the village scale in optimizing the wind environment.

Based on an in-depth exploration of the relationship between spatial morphology and wind environment, this study reveals the optimization strategies for village layout under different wind conditions. Under normal wind conditions, appropriately expanding the spatial scale of the village helps to improve the quality of the wind environment. Under extreme wind conditions, moderately increasing building density and adopting relatively small-scale village layouts become effective means to enhance the safety of the wind environment. Among these, increasing building density is the more effective method. These principles not only provide a solid scientific basis for the planning and wind environment optimization of island villages but also lay a strong foundation for guiding the sustainable development of island villages and improving the quality of life and safety of the villagers.

However, despite this study employing quantitative analysis methods to explore the climate adaptability patterns of the coastal village layout on Naozhou Island, there are still some limitations:

- The building materials of coastal villages (such as shell waste, granite blocks, and adobe) differ from those in inland areas, which is crucial for wind adaptability and structural resilience. However, this study focuses solely on the impact of spatial morphology on the wind environment and does not delve into the role of building materials. This limits our understanding of the overall adaptability of the village. Future research could explore the performance of building materials in coastal villages within wind environments, revealing their relationship with wind adaptability. Additionally, by enhancing the monitoring of coastal village buildings [39] and employing image processing technology [40], we can gain a clearer understanding of the damage caused by extreme weather and timely develop corresponding solutions.
- In the research process, we focused on investigating whether various independent spatial morphological factors (such as building density, orientation, village orientation, and planar morphology) have a significant effect on the wind environment. This allowed us to deeply analyze the unique contributions of each factor in influencing wind behavior. However, it also led to our failure to explore the potential interactions between spatial morphological indicators and their comprehensive impact on the wind environment. Such interactions may produce wind behavior patterns that differ from expectations, but incorporating them into the research scope would complicate the analysis process, potentially hindering our ability to clearly identify the significant effects of each independent factor on the wind environment. Therefore, in future research, it is necessary to design more complex experiments and simulations to explore the actual impact of these variable combinations on the wind environment.

Author Contributions: Conceptualization, P.X.; methodology, P.X.; software, Z.L. and P.X.; validation, Z.L. and P.X.; investigation, Y.P.; resources, Y.P.; data curation, Z.L.; writing—original draft preparation, Z.L. and L.L.; visualization, Z.L. All authors have read and agreed to the published version of the manuscript.

Funding: This research received no external funding.

Data Availability Statement: The experimental data used to support the findings of this study are included in the article.

Conflicts of Interest: Author Peisheng Xie is employed by the Guangdong Provincial Architectural Design Research Institute Group Co., Ltd. The remaining authors declare that the research was conducted in the absence of any commercial or financial relationships that could be construed as a potential conflict of interest.

Appendix A

Table A1. Table of research on Naozhou Island villages.

	Categorization	Quantities	Proportions	Village Name
Village pattern	Checkerboard layout	16	38.1%	Yingming Village Cunliang Village Jialixin Village Dawen Village Danshui Village Doulong Village Houjiao Village Chima Village Songhuang Village Mengxi Village Naganwei Village Nafan Village Tantanzai Village Tantan Village Yanlou Village Nakunzai Village.
	Cluster Layout	16	38.1%	Zhouwu Village, Zhucai Village, Yanting Village, Liuzhu Village, Nakun Village, Beizhai Village, Yingfan Village, Jialiu Old Village, Tianxuan Village, Tanbeihu Village, Zhengpeng Village, Naguang Village, Mengdong Village, Dalinhou Village, Xianadong Village, Dingdang Village.
	Strip layout	9	21.4%	Longchi Village, Nangang Village, Xibu Village, Honggong Village, Dalang Village, Beijin Village, Xundizai Village, Liangwu Village, Liuluo Village
	Distributed layout	1	2.3%	Gangtou Village
	(grand) Total	42	100%	
Village size	Micro-village	3	7.1%	Yingfan Village, Tantan Village, Dingdang Village.
	Small village	16	38.0%	Zhouwu Village, Yingming Village, Dawen Village, Danshui Village, Houjiao Village, Yanting Village, Liuzhu Village, Mengxi Village, Naganwei Village, Tantanzai Village, Jialiu Old Village, Gangtou Village, Zhengpeng Village, Nakunzai Village, Xianadong Village, Liuluo Village.
	Medium-sized villages	13	31.0%	Jialixin Village, Honggong Village, Doulong Village, Beijin Village, Chima Village, Songhuang Village, Nakun Village, Beizhai Village, Longchi Village, Xibu Village, Liangwu Village, Nafan Village, Naguang Village.
	Large villages	7	16.7%	Cunliang Village, Dalang Village, Nangang Village, Xundizai Village, Tanbeihu Village, Yanlou Village, Dalinhou Village.
	Mega-village	3	7.1%	Zhucai Village, Tianxuan Village, Mengdong Village.
	(grand) total	42	100%	
Village orientation	lit. sit north, face south (idiom); figure facing south	12	28.6%	Yingming Village, Xibu Village, Honggong Village, Danshui Village, Chima Village, Mengxi Village, Nafan Village, Tantanzai Village, Nakunzai Village, Liuzhu Village, Mengdong Village, Dingdang Village, Nakun Village, Beizhai Village, Yanting Village, Liuluo Village.
	sit northeast to face southwest	4	9.5%	NaKun Village, BeiZhai Village, YanTing Village, LiuLuo Village,
	lit. sitting northwest, facing southeast	12	28.6%	Jialixin Village, Cunliang Village, Dawen Village, Songhuang Village, Zhouwu Village, Zhucai Village, Tanbeihu Village, Yingfan Village, Jialiu Old Village, Tianxuan Village, Xianadong Village, Nangang Village.

Table A1. *Cont.*

	Categorization	Quantities	Proportions	Village Name
Village orientation	sit south and face north	3	7.1%	Doulong Village, YanLou Village, NaGuang Village
	sit southeast, face northwest	4	9.5%	Dalang Village, Houjiao Village, Tantan Village, Gangtou Village.
	sit southwest to face northeast	1	2.4%	LiangWu Village
	sit east and face west	3	7.1%	Beijin Village, Naganwei Village, Xundizai Village.
	sit west and face east	1	2.4%	Longchi Village
	no specific orientation	2	4.7%	Zhengpeng Village, Dalinhou Village.
	(grand) Total	42	100%	
Village density	less than 10%	2	5%	Mengdong Village, Xianadong Village
	10–20%	16	38%	Honggong Village, Dawen Village, Dalang Village, Beijin Village, Xundizai Village, Nafan Village, Tantanzai Village, Tantan Village, Zhucai Village, Yingfan Village, Jialiu Old Village, Naguang Village, Dalinhou Village, Liuluo Village, Longchi Village, Gangtou Village
	20–30%	21	50%	Yingming Village, Cunliang Village, Jialixin Village, Danshui Village, Doulong Village, Houjiao Village, Songhuang Village, Mengxi Village, Naganwei Village, Yanlou Village, Nakunzai Village, Yanting Village, Liuzhu Village, Nakun Village, Beizhai Village, Liangwu Village, Tianxuan Village, Tanbeihu Village, Zengpeng Village, Nangang Village, Xibu Village
	30–40%	2	5%	Chima Village, Zhouwu Village
	40% or more	1	2%	Dingdang Village
	(grand) total	42	100%	

Table A2. Measured and software simulated data for villages under normal wind conditions.

Village Name	Village Orientation	Village Size	Building Density	Wind Field Coefficient K1/K2		Wind Speed Amplification Factor		Average Wind Speed(m/s)	
				N	E	N	E	N	E
Liuzhu Village	175°	4.94	20.15	7	0.153	0.913	1.006	1.64	15.03
Yanting Village	230°	6.8	26.77	8.17	0.19	0.86	0.984	2.25	14.31
Zhouwu Village	140°	4.64	30.99	8.35	0.196	0.874	1.068	1.22	13.16
Zhucai Village	135°	22.79	13.71	13.29	0.362	1.237	1.189	4.07	16.54
Yingfan Village	143°	3.65	25.4	7.18	0.168	0.906	1.0887	1.49	14.03
Jialiu Old Village	150°	6.5	15.99	8.17	0.1192	1.067	1.129	1.81	15.12
Tianxuan Village	146°	26.03	27.78	14.19	0.42	1.085	1.38	3.79	14.88
Tanbeihu Village	152°	15.01	20.9	8.93	0.23	0.944	1.148	2.34	15.23
Xianadong Village	120°	4.44	9.95	12.62	0.06	1.008	0.978	2.22	14.68
Yingming Village	185°	6.98	27.24	7.45	0.194	0.877	1.018	1.51	13.16
Tantanzai Village	180°	5.35	16.23	12.23	0.04	1.022	1.105	2.01	15.22
Houjiao Village	300°	5.16	26.61	5.17	0.188	0.76	1.148	1.86	14.3
Tantan Village	310°	3.67	26	10.45	0.175	1.023	1.056	1.37	14.35
Doulong Village	352°	7.69	21.83	12.32	0.186	0.83	0.988	1.9	14.32
Yanlou Village	356°	19.34	21.95	10.25	0.33	1.053	1.138	3.3	15.48

Table A2. *Cont.*

Village Name	Village Orientation	Village Size	Building Density	Wind Field Coefficient K1/K2		Wind Speed Amplification Factor		Average Wind Speed(m/s)	
				N	E	N	E	N	E
Cunliang Village	129°	18.97	27.51	9.4	0.321	1.114	1.068	2.37	14.44
Jialixin Village	127°	7.45	24.25	8.27	0.202	1.027	1.0881	1.6	14.04
Dawen Village	133°	5.81	17.97	11.37	0.138	0.953	1.077	1.75	14.86
Xibu Village	180°	8.92	23.97	9.85	0.21	0.972	1.096	1.49	14.49
Honggong Village	178°	8.35	14.32	9.45	0.15	1.146	1.218	2.23	18.04
Liuluo Village	222°	5.94	17.82	9.74	0.068	0.955	1.0881	1.76	15.27
Beijin Village	273°	7.14	18.36	9.7	0.1142	1.005	1.067	2.22	14.77
Xundizai Village	271°	12.16	10.15	10.09	0.22	1.018	1.088	2.79	15.06
Dalang Village	313°	19.99	18.22	13.49	0.352	0.973	1.128	3.2	15.37
Nangang Village	132°	14.62	20.55	11.28	0.3	0.964	1.106	2.12	15.26
Liangwu Village	42°	8.94	22.8	10.27	0.23	0.965	1.118	1.82	15.32
Longchi Village	90°	8.99	14.09	13.36	0.1082	1.028	1.147	2.65	16.13

N: Normal wind conditions, **E:** Extreme wind conditions.

Table A3. Regression coefficients and their significance tests and multiple covariance tests.

	Mold		Unstandardized Coefficient		Standardized Coefficient	T	Significance	Covariance Statistics		R^2	F-Value
			B	Standard Error				Tolerances	VIF		
Normal wind condition	Average wind speed	(Constant)	1.903	0.254		7.482	<0.001			0.867	50.158 ***
		Village size	9.819	0.867	0.86	11.326	<0.001	0.999	1.001		
		Building density	−0.043	0.01	−0.334	−4.389	<0.001	0.997	1.003		
		Village orientation	0.001	0.001	0.107	1.411	0.172	0.996	1.004		
	Wind speed amplification factor	(Constant)	1.118	0.066		16.931	<0.001			0.545	50.158 ***
		Village size	0.858	0.225	0.536	3.81	0.001	0.999	1.001		
		Building density	−0.008	0.003	−0.424	−3.008	0.006	0.997	1.003		
		Village orientation	-3.19×10^{-4}	1.73×10^{-4}	−0.26	−1.841	0.078	0.996	1.004		
	K1	(Constant)	12.047	1.629		7.398	<0.001			0.449	6.252 **
		Village size	18.341	5.552	0.511	3.304	0.003	0.999	1.001		
		Building density	−0.171	0.062	−0.425	−2.743	0.012	0.997	1.003		
		Village orientation	−0.001	0.004	−0.047	−0.305	0.763	0.996	1.004		
Extreme wind conditions	Average wind speed	(Constant)	16.895	0.64		26.392	<0.001			0.538	8.943 ***
		Village size	5.045	2.182	0.328	2.312	0.03	0.999	1.001		
		Building density	−0.112	0.025	−0.648	−4.564	<0.001	0.997	1.003		
		Village orientation	−0.001	0.002	−0.067	−0.473	0.64	0.996	1.004		
	Wind speed amplification factor	(Constant)	1.053	0.06		17.657	<0.001			0.428	5.740 ***
		Village size	0.814	0.203	0.632	4.004	0.001	0.999	1.001		
		Building density	4.63×10^{-5}	0.002	0.003	0.02	0.984	0.997	1.003		
		Village orientation	−0.0001	0.0001	−0.186	−1.177	0.251	0.996	1.004		
	K2	(Constant)	−0.061	0.03		−2.006	0.057			0.891	62.610 ***
		Village size	1.328	0.104	0.88	12.777	<0.001	0.999	1.001		
		Building density	0.006	0.001	0.341	4.937	<0.001	0.997	1.003		
		Village orientation	4.80×10^{-5}	0.00008	0.041	0.599	0.555	0.996	1.004		

** $p < 0.01$, *** $p < 0.001$.

References

1. Yuan, C.; Ng, E.; Norford, L.K. Improving Air Quality in High-Density Cities by Understanding the Relationship between Air Pollutant Dispersion and Urban Morphologies. *Build. Environ.* **2014**, *71*, 245–258. [CrossRef] [PubMed]
2. Guo, A.; Yang, J.; Xiao, X.; Xia, J.; Jin, C.; Li, X. Influences of Urban Spatial Form on Urban Heat Island Effects at the Community Level in China. *Sustain. Cities Soc.* **2020**, *53*, 101972. [CrossRef]

3. Scislo, L.; Szczepanik-Scislo, N. Influence of Mechanical Ventilation and Cooling Systems on Vibrations of High Precision Machines. *E3S Web Conf.* **2019**, *100*, 00080. [CrossRef]
4. Miao, Y.; Chiou, S.-C. Study on the Wind Environment of the Architecture Communities: Traditional Typical Min Nan Human Settlements' Case. *Math. Probl. Eng.* **2013**, *2013*, 467076. [CrossRef]
5. Asfour, O.S. Prediction of Wind Environment in Different Grouping Patterns of Housing Blocks. *Energy Build.* **2010**, *42*, 2061–2069. [CrossRef]
6. Chen, Y. Application of a Wind Environment Simulation Generation Technology in Urban Planning. *Wirel. Commun. Mob. Comput.* **2022**, *2022*, 7939777. [CrossRef]
7. Ji, H.; Li, Y.; Li, J.; Ding, W. A Novel Quantitative Approach to the Spatial Configuration of Urban Streets Based on Local Wind Environment. *Land* **2023**, *12*, 2102. [CrossRef]
8. Kubota, T.; Miura, M.; Tominaga, Y.; Mochida, A. Wind Tunnel Tests on the Relationship between Building Density and Pedestrian-Level Wind Velocity: Development of Guidelines for Realizing Acceptable Wind Environment in Residential Neighborhoods. *Build. Environ.* **2008**, *43*, 1699–1708. [CrossRef]
9. Ma, T.; Chen, T. Outdoor Ventilation Evaluation and Optimization Based on Spatial Morphology Analysis in Macau. *Urban Clim.* **2022**, *46*, 101335. [CrossRef]
10. Takemi, T.; Yoshida, T.; Horiguchi, M.; Vanderbauwhede, W. Large-Eddy-Simulation Analysis of Airflows and Strong Wind Hazards in Urban Areas. *Urban Clim.* **2020**, *32*, 100625. [CrossRef]
11. Zhao, H.; Jiang, X.; Cao, Y.; Zhang, H.; Zhen, S.; Jia, R.; Zhang, S. Wind Environment Simulation and Optimisation Strategies for Block Spatial Forms in Cold Low Mountainous Areas—a Case Study of Changchun, China. *Sustainability* **2022**, *14*, 6643. [CrossRef]
12. Hsieh, C.-M.; Yu, C.-Y.; Shao, L.-Y. Improving the Local Wind Environment through Urban Design Strategies in an Urban Renewal Process to Mitigate Urban Heat Island Effects. *J. Urban Plan. Dev.* **2023**, *149*. [CrossRef]
13. Qi, L.; Liu, R.; Cui, Y.; Zhou, M.; Bonenberg, W.; Song, Z. Study of the Landscape Pattern of Shuiyu Village in Beijing, China: A Comprehensive Analysis of Adaptation to Local Microclimate. *Sustainability* **2021**, *14*, 375. [CrossRef]
14. Shawei, Z.; Zhong, Y.; Dinghang, W.; Jin, T. Landscape Pattern in Traditional Village--Thinking over the Zhonglou Village, Conghua. *Iop Conf. Ser. Earth Environ. Sci.* **2018**, *153*, 052014. [CrossRef]
15. Xu, X.; Genovese, P.V.; Zhao, Y.; Liu, Y.; Woldesemayat, E.M.; Zoure, A.N. Geographical Distribution Characteristics of Ethnic-Minority Villages in Fujian and Their Relationship with Topographic Factors. *Sustainability* **2022**, *14*, 7727. [CrossRef]
16. Chen, J.; Lin, W.; Zhang, Y.; Dai, Y.; Chen, B. Village Fengshui Forests as Forms of Cultural and Ecological Heritage: Interpretations and Conservation Policy Implications from Southern China. *Forests* **2020**, *11*, 1286. [CrossRef]
17. Zhang, D.; Gao, W.; Lv, Y. The Triple Logic and Choice Strategy of Rural Revitalization in the 70 Years since the Founding of the People's Republic of China, Based on the Perspective of Historical Evolution. *Agriculture* **2020**, *10*, 125. [CrossRef]
18. Xu, F.; Gao, Z.; Xing, Y.; Wu, Z.; Zhang, J.; Liao, Y.; Hu, Y. The Effect of Village Morphological Variation Caused by Economic Development on Residents' Health and Rural Ventilation in Tianjin. *Buildings* **2022**, *12*, 1393. [CrossRef]
19. Zang, H.Y.; Hsieh, C.-M.; Cang, Y.; Gu, M. Microclimate Adaptability Analysis of Waterfront Traditional Villages: A Case Study of Mingyuewan Village in Suzhou. *J. Water Clim. Change* **2023**, *14*, 3329–3344. [CrossRef]
20. Xiong, Y.; Zhang, J.; Yan, Y.; Sun, S.; Xu, X.; Higueras, E. Effect of the Spatial Form of Jiangnan Traditional Villages on Microclimate and Human Comfort. *Sustain. Cities Soc.* **2022**, *87*, 104136. [CrossRef]
21. Sun, R.; Dong, J.; Zhao, H.; Wang, L.; Qin, D. Typical Village and Town Houses in the Cold Region of Northeast China: Simulation Analysis of Courtyard Layout and Winter Wind Environment Habitability. *Buildings* **2023**, *14*, 109. [CrossRef]
22. Gou, S.; Li, Z.; Zhao, Q.; Nik, V.M.; Scartezzini, J.-L. Climate Responsive Strategies of Traditional Dwellings Located in an Ancient Village in Hot Summer and Cold Winter Region of China. *Build. Environ.* **2015**, *86*, 151–165. [CrossRef]
23. Guo, J.; Han, G.; Xie, Y.; Cai, Z.; Zhao, Y. Exploring the Relationships between Urban Spatial Form Factors and Land Surface Temperature in Mountainous Area: A Case Study in Chongqing City, China. *Sustain. Cities Soc.* **2020**, *61*, 102286. [CrossRef]
24. He, J.; Chen, Y.; Zheng, L.; Zheng, J. Research on Wind Environment and Morphological Effects of High-Rise Buildings in Macau: An Example from the New Reclamation Area around Areia Preta. *Int. J. Environ. Res. Public Health* **2023**, *20*, 4143. [CrossRef] [PubMed]
25. Du, Z.; Guo, W.; Li, W.; Gao, X. A Study on the Optimization of Wind Environment of Existing Villa Buildings in Lingnan Area: A Case Study of Jiangmen's "Yunshan Poetic" Moon Island Houses. *Buildings* **2022**, *12*, 1304. [CrossRef]
26. Meir, I.A.; Pearlmutter, D.; Etzion, Y. On the Microclimatic Behavior of Two Semi-Enclosed Attached Courtyards in a Hot Dry Region. *Build. Environ.* **1995**, *30*, 563–572. [CrossRef]
27. Liu, K.S.; Chan, J.C.L. Climatological Characteristics and Seasonal Forecasting of Tropical Cyclones Making Landfall along the South China Coast. *Mon. Wea. Rev.* **2003**, *131*, 1650–1662. [CrossRef]
28. Cheng, V.; Ng, E. Thermal Comfort in Urban Open Spaces for Hong Kong. *Archit. Sci. Rev.* **2011**, *49*, 236–242. [CrossRef]
29. *HJ/T 393-2007*; Technical Specifications for Urban Fugitive Dust Pollution Prevention and Control. China Environmental Science Press: Beijing, China, 2007.
30. *ASCE/SEI 49-21*; Standard for Wind Tunnel Test of Buildings and Structures. China Architecture & Building Press: Beijing, China, 2015.

31. *JGJ/T 449-2018;* Standard for Green Performance Calculation of Civil Buildings. Ministry of Housing and Urban-Rural Development of the People's Republic of China: Beijing, China, 2018.
32. Tsang, C.W.; Kwok, K.C.S.; Hitchcock, P.A. Wind Tunnel Study of Pedestrian Level Wind Environment around Tall Buildings: Effects of Building Dimensions, Separation and Podium. *Build. Environ.* **2012**, *49*, 167–181. [CrossRef]
33. Ma, T.; Chen, T. Classification and Pedestrian-Level Wind Environment Assessment among Tianjin's Residential Area Based on Numerical Simulation. *Urban Clim.* **2020**, *34*, 100702. [CrossRef]
34. Toparlar, Y.; Blocken, B.; Maiheu, B.; Van Heijst, G.J.F. A Review on the CFD Analysis of Urban Microclimate. *Renew. Sustain. Energy Rev.* **2017**, *80*, 1613–1640. [CrossRef]
35. Blocken, B. Computational Fluid Dynamics for Urban Physics: Importance, Scales, Possibilities, Limitations and Ten Tips and Tricks towards Accurate and Reliable Simulations. *Build. Environ.* **2015**, *91*, 219–245. [CrossRef]
36. Blocken, B. LES over RANS in Building Simulation for Outdoor and Indoor Applications: A Foregone Conclusion? *Build. Simul.* **2018**, *11*, 821–870. [CrossRef]
37. Yang, L.; Li, Y. Thermal Conditions and Ventilation in an Ideal City Model of Hong Kong. *Energy Build.* **2011**, *43*, 1139–1148. [CrossRef]
38. Wu, S.; Zhang, L.; Han, Z.; Hu, C.; An, D. Research on Microclimate Optimization of Traditional Residential Buildings in Central Anhui Based on Humid and Hot Climate Characteristics and Regional Architectural Features. *Buildings* **2024**, *14*, 2323. [CrossRef]
39. Sun, H.; Burton, H.V.; Huang, H. Machine Learning Applications for Building Structural Design and Performance Assessment: State-of-the-Art Review. *J. Build. Eng.* **2021**, *33*, 101816. [CrossRef]
40. Pimenov, D.Y.; Da Silva, L.R.R.; Ercetin, A.; Der, O.; Mikolajczyk, T.; Giasin, K. State-of-the-Art Review of Applications of Image Processing Techniques for Tool Condition Monitoring on Conventional Machining Processes. *Int. J. Adv. Manuf. Technol.* **2024**, *130*, 57–85. [CrossRef]

buildings

MDPI

Article

An Operational Carbon Emission Prediction Model Based on Machine Learning Methods for Urban Residential Buildings in Guangzhou

Lintao Zheng [1], Kang Luo [2] and Lihua Zhao [2,*]

1 School of Built Environment and Design, Guangdong Polytechnic of Water Resources and Electric Engineering, Guangzhou 510925, China; zhenglintao@gdsdxy.cn

2 State Key Laboratory of Subtropical Building Science, School of Architecture, South China University of Technology, Guangzhou 510641, China

* Correspondence: lhzhao@scut.edu.cn; Tel.: +86-18819826874

Abstract: The carbon emissions of urban residential buildings are substantial. However, the standard operating conditions specified in current energy-saving standards significantly differ from the actual energy consumption under real operating conditions. Therefore, it is essential to consider the impact of residents' actual energy consumption behavior in carbon emission forecasts. To improve the accuracy of carbon emission predictions for urban residential buildings, this paper focuses on residential buildings in Guangzhou. Taking into account the energy consumption behavior of residents, parameterized modeling is carried out in the R language, and simulation is carried out using EnergyPlus software. Analysis revealed that the higher the comfort level of residential energy consumption behavior, the more it is necessary to encourage residents to adopt energy-saving behaviors. Combining carbon emission factors, air-conditioning energy efficiency, and the power consumption models of lighting and electrical equipment, a comprehensive operational carbon emission prediction model for urban residential operations in Guangzhou was developed. By comparing the prediction model with an actual case, it was found that the prediction deviation was only 4%, indicating high accuracy. The proposed operational carbon emission model can quickly assist designers in evaluating the carbon emissions of urban residential buildings in the early stages of design, providing an accurate basis for decision-making.

Keywords: operational carbon emission prediction; urban residential buildings; sensitivity analysis; energy consumption behavior; machine learning method

1. Introduction

With the rapid advancement of global urbanization, fuel consumption has significantly increased, leading to serious environmental issues and social concerns. Among all energy-consuming activities, the construction and maintenance of buildings account for over 40% of the world's total energy consumption and one-third of total carbon dioxide emissions [1]. Many countries have established carbon peaking and carbon neutrality goals based on their domestic conditions [2]. China has also steadily promoted various initiatives with carbon reduction as a key strategic direction, proposing a gradual shift from dual control of energy consumption to dual control of total carbon emissions and intensity. Urban and rural construction is one of the main sources of China's carbon emissions, with building energy consumption exceeding 25% of the country's total energy consumption [3]. Statistics show that in 2022, the energy consumption of urban residential buildings in China during the operation phase was 410 million tons of coal equivalent, accounting for 39% of the total energy consumption of national buildings during the operation phase. Corresponding carbon emissions were 900 million tons of CO_2, representing 42% of the total carbon emissions during the operation phase of national buildings [4]. Moreover, as living standards continuously improve, people's requirements for living comfort are also increasing, leading to a continuous rise in total carbon emissions from building operations. Clearly, the energy-saving and carbon-reduction efforts for urban residential buildings have a long way to go.

Currently, carbon peaking and carbon neutrality in the construction sector have garnered global attention. However, most existing domestic forecasts for building carbon peaks are from a macro perspective, lacking in-depth consideration of individual energy consumption behaviors. Consequently, research results are often limited to an analysis of the overall construction sector [5–7]. This is particularly problematic given the significant variations in local carbon emissions and the heterogeneity of individual energy consumption behaviors. Existing research has yet to provide adequate support for the development of regional carbon peak and carbon neutrality pathways. By considering complex factors such as the physical characteristics of the building, climate models, and occupancy behavior, simulation can provide a comprehensive performance evaluation of the building automation and control system, which helps to predict energy efficiency savings more accurately [8]. Recent studies highlight the growing impact of individual energy consumption behavior on building energy use and carbon emissions [9–11], yet the focus remains on appliance usage, overlooking user behavior in the setting and operation of air conditioners. As a result, personal energy consumption patterns have not been fully integrated into existing carbon peak implementation plans.

When analyzing the energy-saving and emission-reducing potential of buildings, it is essential to conduct a detailed comparison of energy consumption and carbon emissions under different energy-saving measures. By establishing a building carbon emission prediction model, researchers can quickly obtain energy consumption and carbon emission data for various energy-saving programs, significantly reducing research costs and time consumption. Currently, various prediction methods such as multivariate linear regression, neural networks, support vector machines, and time series analysis are widely used. For example, multivariate linear regression can effectively predict long-term energy consumption trends by establishing a linear relationship model between multiple influencing factors and building energy consumption, making it suitable for large-scale building energy consumption analysis and prediction [12]. In contrast, neural networks are better at handling complex nonlinear problems, extracting patterns from historical data through adaptive learning, and can make short-term and high-precision energy consumption predictions [13,14]. Support vector machines perform well in minimizing both training error and the upper limit of the generalization error, making them suitable for high-precision building energy consumption predictions, such as predicting the cooling load of commercial buildings [15,16]. Time series methods can effectively predict short-term fluctuations in building energy consumption by analyzing the temporal patterns of historical data, particularly suitable for predicting and optimizing the energy consumption of single buildings [17,18]. Meanwhile, supervised machine learning models offer more nuanced predictions by accounting for multifactorial influences such as weather, production rates, and equipment operations [19]. Transfer learning, a relatively new approach, enhances model performance by applying knowledge from related fields [20]. These techniques collectively improve forecasting accuracy, providing essential insights for developing targeted strategies to enhance energy efficiency and reduce emissions in the building sector.

At present, although existing research has provided substantial theoretical and technical support for building energy conservation and emission reduction, there is still a need for in-depth studies on the actual operation of different regions and building types. This is particularly important given the significant variations in local carbon emissions and the heterogeneity in individual energy consumption behaviors. Such research is essential to formulate more targeted and effective carbon peaking and carbon neutrality pathways. Research on the factors influencing carbon emissions from building operations can be divided into macro-level and micro-level studies. At the macro level, key factors include residents' average wage, population density, electricity intensity, educational attainment, and per capita household liquefied petroleum gas consumption in urban areas [21]. At the micro level, considerations include operation time, start-up time of the HVAC system, setpoint temperature level, geometry characterization, and building envelopes properties [22]. However, there are relatively few studies on the influence of indoor occupants' energy consumption behavior on residential energy consumption. Most building energy consumption predictions focus on the objective conditions of the buildings themselves, with fewer studies comprehensively considering subjective factors such as occupants' energy consumption behavior. This paper addresses this gap by establishing a carbon emission prediction model for urban residential buildings in Guangzhou based on the different energy-using behaviors of residents. In developing this model, careful selection of input parameters is essential due to the many influencing factors. Common approaches, such as parameterized methods or probability distribution functions combined with Monte Carlo or Latin hypercube sampling, generate extensive datasets for simulation. The Standardized Regression Coefficient (SRC) and Partial Correlation Coefficient (PCC) can then assess each parameter's influence, helping to isolate the most critical factors for prediction accuracy [22]. Recently, feature selection methods like Elastic Net have proven effective for optimizing prediction models by reducing redundant features and enhancing accuracy, particularly in building cooling load predictions [23,24]. This study uses SRC to filter key features [25], subsequently applying them as inputs to a machine learning model that predicts urban residential carbon emissions under varying energy-saving scenarios [26]. This model also incorporates changes in residents' energy-use behavior, providing insights into potential carbon reduction pathways, including carbon peak scenarios and strategies for achieving carbon neutrality.

The structure of this article is arranged as follows: first, it introduces the influencing factors and current situation of carbon emissions in the construction field, followed by an analysis of the shortcomings and challenges in current research. It then details the research methods and data sources, showcasing the process and technical details of model establishment. Finally, the experimental results are summarized and analyzed, and suggestions and policy recommendations for future research are put forward, aiming to contribute new ideas and methods to the planning of the carbon peaking and carbon neutrality pathways for urban residential buildings in China.

2. Materials and Methods

As the capital of Guangdong Province, Guangzhou is a central city in the Guangdong–Hong Kong–Macao Greater Bay Area. It has a high urbanization rate and a dense population, resulting in a large urban residential construction area. Consequently, its operational carbon emissions are significant and cannot be overlooked. Moreover, Guangzhou is located in a region with hot summers and mild winters, characterized by long summers and no real winters. As people's demands for indoor thermal comfort increase, energy consumption for comfort is becoming more prevalent. Therefore, the task of building energy conservation and emission reduction in Guangzhou is challenging, necessitating urgent action to achieve peak carbon emissions from building operations.

The purpose of this paper is to develop carbon emission prediction models for typical urban residential operations with different energy consumption behaviors. The aim is to enable rapid and accurate predictions of carbon emissions from urban residential operations and to assist policymakers and designers in achieving precise carbon emission forecasts at early stages. This will facilitate quick evaluations of the carbon emission characteristics of urban residential operations in Guangzhou in response to changes in residents' energy consumption behaviors in the future. To achieve these goals, data on energy-use behavior was collected through online questionnaires and on-site interviews. A total of 300 questionnaires were recovered, including 254 online questionnaires and 46 interview questionnaires, with 289 valid responses, resulting in an efficiency rate of 96.3%. The typical building model was based on data from 200 residential settlements across 11 administrative districts in Guangzhou. Basic data such as building layout, unit form, number of buildings, floor area ratio, and outdoor site layout were collected and analyzed. The research process of this study is shown in Figure 1, and the process can be divided into four steps:

1. Analyzing typical energy consumption behavior: through questionnaires, typical energy consumption behaviors of urban residents in Guangzhou were analyzed, including indoor stay and rest, air conditioning use, lighting, and electrical appliance use. And typical energy consumption behavior patterns were extracted;
2. Parameterized modeling: using the R language [27], typical urban dwellings in Guangzhou were parameterized, taking into account geometric characteristics, internal heat sources, and envelope structures affecting energy consumption. The Latin hypercube sampling method was used to generate multiple sets of parameter combinations [28];
3. Energy consumption simulation: EnergyPlus software was used to simulate and calculate the energy consumption of these parameterized models, obtaining residential energy consumption data under different scenarios [29]. Key influencing factors were determined through SRC sensitivity analysis [30];
4. Carbon emission intensity prediction model: based on the typical established energy consumption behavior model and energy consumption model, a carbon emission intensity prediction model for typical urban residential operations in Guangzhou was developed. Different machine learning methods were screened to establish the most effective model.

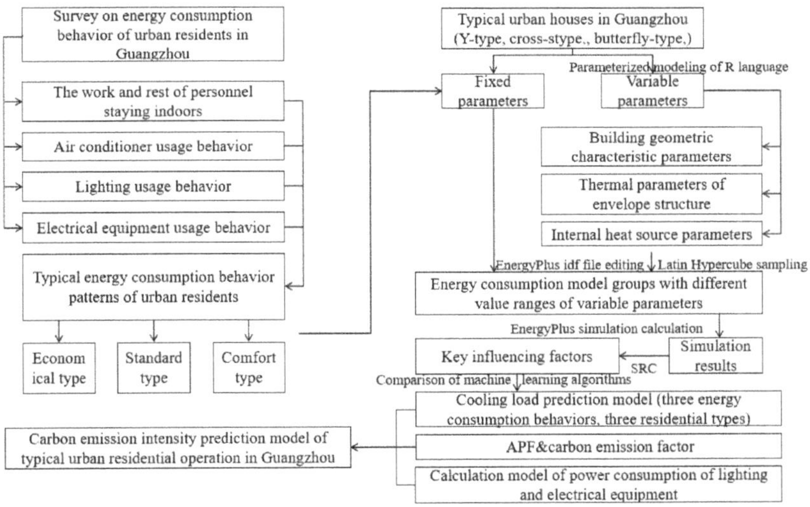

Figure 1. Technology roadmap.

This structured approach aims to provide a scientific basis for achieving carbon peaking and carbon neutrality in urban residential buildings in Guangzhou, offering new ideas and methods for the planning and implementation of energy-saving and emission-reducing strategies. The tools used in each process are described below:

1. Data collection tool—questionnaire survey: through a questionnaire survey, the energy consumption behavior of urban residents in Guangzhou was studied. The survey results were used to refine and summarize the typical energy consumption behavior patterns of these residents;
2. Drawing analysis tool—statistical method: resident information, building information, and especially the lifestyle of residents in the collected samples were analyzed. The urban residents of Guangzhou were divided into two groups: intermittent users and continuous users. The characteristics and differences in residential indoor stay, rest,

and air-conditioning use behavior of these two groups were analyzed separately. Their air-conditioning use behavior was clustered and analyzed, and typical air-conditioning use behavior patterns were refined and summarized. Combining the typical indoor stay, rest, and typical lighting and electrical equipment use behaviors obtained from the survey, three typical energy-use behavior patterns for intermittent users and two for continuous users were identified;

3. Modeling tool—parameter modeling: the LHS package in the R language was used for Latin hypercube sampling. Generally, the number of samples should be more than 10 times the number of variables to ensure the rationality of the sampling results. To obtain enough samples, this study set the number of samples to 100 times the number of variables. The sampling results were output to the edited idf file through the code, resulting in the final residential energy consumption calculation model;
4. Sensitivity analysis tool—SRC: by calculating the value of SRC, the key influencing variable parameters of residential energy consumption for the two groups of intermittent users and continuous users under different energy consumption behavior patterns were determined;
5. Simulation calculation tool—EnergyPlus: this study used the R language for parameterized modeling and EnergyPlus to simulate and calculate the energy consumption model. This approach provided a large number of energy consumption values, creating conditions for sensitivity analysis and the establishment of energy consumption prediction models;
6. Data analysis tool—generalized linear model: building energy consumption prediction is usually a linear problem. Therefore, different machine learning methods were introduced and compared to determine the best algorithm for the building carbon emission prediction model. This model was established to predict carbon emissions from urban residential operations in Guangzhou, accommodating the research of multiple variables affecting residential energy consumption in this study.

3. Results

3.1. Typical Energy Consumption Behavior Patterns of Urban Residential Residents in Guangzhou

The energy consumption and operational carbon emissions of residential buildings are influenced by various factors such as the enclosure structure, internal heat sources, occupant energy consumption behavior, and building geometric characteristics. Among these, the energy consumption methods and work–rest habits of indoor occupants are critical factors leading to differences in energy consumption and carbon emissions. To improve the accuracy and reliability of predictions regarding energy consumption and carbon emissions in residential buildings, it is essential to analyze residents' energy consumption behaviors and summarize their typical patterns. Therefore, this study investigated the energy consumption behavior of urban residential residents in Guangzhou.

First, a questionnaire on energy use behavior was designed according to the research objectives, and data were collected through online platforms and on-site interviews. A total of 300 questionnaires were distributed. The survey content primarily included residents' indoor stay and rest, air-conditioning usage and settings, lighting usage, and electrical equipment usage, as shown in Table 1.

Table 1. Questionnaire on energy consumption behavior of urban residents in Guangzhou.

Survey Items	Main Content
Basic resident information	Gender, age, number of permanent residents in the family, number of elderly people living together
Basic building information	Administrative district, residential type, apartment type
Lifestyle of residents	Indoor air conditioning work and rest, air conditioning work and rest and temperature setting, electrical equipment and lighting work and rest

The survey results indicate that the types of urban residential buildings in Guangzhou are predominantly middle- and high-rise (7–10 floors) and high-rise residential buildings (11–30 floors). The residential apartment types are mainly two-bedroom and three-bedroom units, with three-bedroom units being more common. Households of urban residential residents in Guangzhou typically consist of three-person and four-person households. The proportion of households without elderly members is close to that of households with elderly members, and in those with elderly members, there are generally one or two elderly individuals. Typical indoor stay and rest patterns for urban residents in Guangzhou are as follows: on weekdays, residents usually leave home around 8 a.m., return home around 7 p.m., and go to bed around 11 p.m. On holidays, residents typically get up around 9 a.m., stay at home for most of the day, and go to bed around 11 pm.

Analysis of the air-conditioning usage behavior of urban residents reveals significant differences in usage patterns and a diversity of air-conditioning schedules and temperature settings within the group. These differences impact the overall building energy consumption and operational carbon emissions. To provide representative input conditions for the prediction model, it is necessary to cluster and analyze these air-conditioning usage patterns and temperature settings and summarize the typical air-conditioning behavior patterns.

In this study, second-order clustering analysis is performed using log likelihood as the distance measure and the Schwarz Bayesian Criterion as the clustering criterion. After clustering, the most common air-conditioning usage pattern for each type of schedule is selected as the typical air-conditioning behavior. The typical air-conditioning usage characteristics and cooling temperatures are summarized, as shown in Table 2. The air-conditioning usage behavior patterns are categorized into three types: economical, standard, and comfortable.

Table 2. Three types of typical air-conditioning work and rest of personnel and air-conditioning set temperature.

Type	Is the Bedroom Heated?	Is the Living Room Heated?	Is the Bedroom Cooling/ Temperature?	Is the Living Room Cooling/ Temperature?
Economical	no	no	often/28 °C	no
Standard	no	no	often/26 °C	occasionally/26 °C
Comfort	occasionally	occasionally	often/24 °C	often/24 °C

Combining the above-mentioned typical indoor stay and rest patterns with the provisions of the latest energy-saving standards [31] regarding the hourly occupancy rate of personnel in each room of a residential building, we derived the typical hourly occupancy rate, typical lighting usage time, and typical electrical equipment usage rate and indoor air conditioning temperature for urban residential residents in Guangzhou, as shown in Figure 2. There are three types of air-conditioning usage and cooling temperature settings: economical, standard, and comfortable.

According to the research and analysis results, the cooling time in bedrooms generally lasts all night, while air-conditioning in living rooms is used predominantly when people are present and it feels hot. Most bedrooms are heated for 1 to 3 h before bedtime, averaging a 2 h heating duration. The preferred heating temperature is between 23 °C and 25 °C, with 24 °C being the most desirable. For the vast majority of elderly residents, bedroom air-conditioning is set for up to 3 h, typically around 2 h, with the consensus being that it is only needed when it feels hot. The air conditioner in the living room is turned on only when it is occupied. In summary, the typical air-conditioning usage behavior patterns for the available personnel are shown in Figure 2g,h. It can be seen that according to the research results, this study does not consider heating energy consumption.

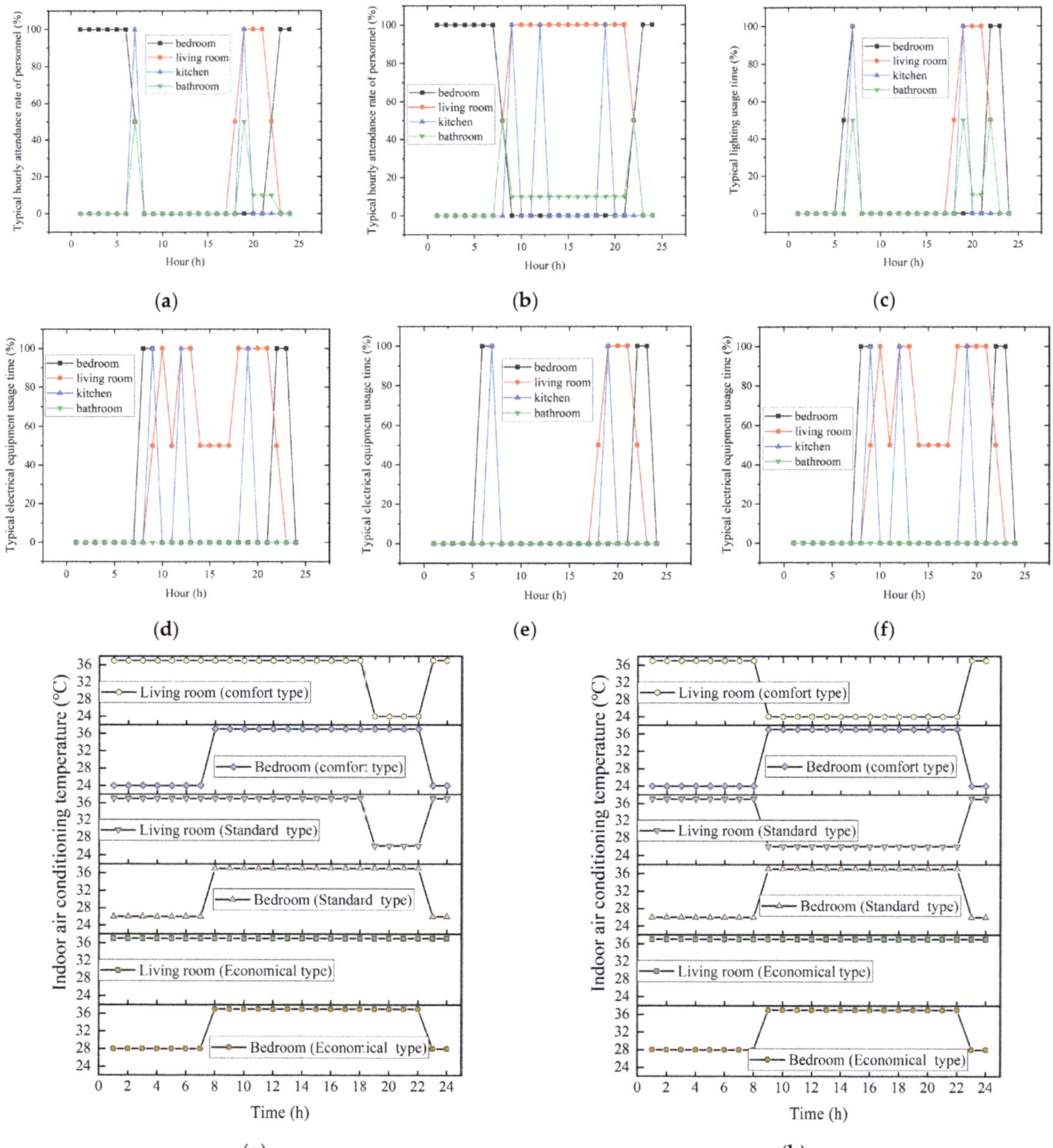

Figure 2. Input condition settings for three typical energy consumption modes, including the following: (**a**) weekday staff attendance hours, (**b**) holiday staff attendance hours, (**c**) weekday lighting usage duration, (**d**) holiday lighting usage duration, (**e**) weekday electrical equipment operation time, (**f**) holiday electrical equipment operation time, (**g**) weekday indoor air conditioning temperature settings, and (**h**) holiday indoor air conditioning temperature settings.

3.2. Carbon Emission Prediction Model for Typical Urban Residential Operations in Guangzhou

3.2.1. Typical Building Types of Urban Dwellings in Guangzhou

Based on the research results of 200 residential districts, urban residential buildings in Guangzhou can be categorized into three main types: Y-type, cross-type, and butterfly-type. This study uses these three types as the standard floor plan layouts, depicted in Figure 3. In the figure, yellow represents the living room, green represents the bedroom, blue represents the bathroom, red represents the kitchen, and black represents the elevator hall. The Y-type standard floor covers an area of 515.92 m^2, the cross-type standard floor covers 453.7 m^2, and the butterfly-type standard floor covers 658.4 m^2.

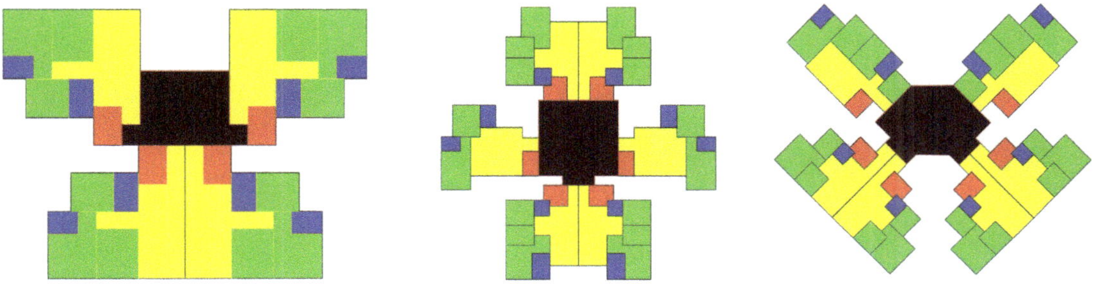

(a) (b) (c)

Figure 3. Typical plan types of urban residential buildings in Guangzhou, including the following: Y-type (**a**), cross-type (**b**), and butterfly-type (**c**).

3.2.2. Model Parameter Setting

The energy consumption of buildings is influenced by four main types of factors: the geometric characteristics of the building, the thermal properties of the building envelope, the internal heat sources, and the indoor air design parameters. For the Guangzhou area, the indoor air design parameters primarily refer to the set temperature for air conditioning used by residents. According to Table 2, these parameters are categorized into three types: economical, standard, and comfortable.

Regarding the geometric characteristics, the standard floor plan layout of the building is fixed, and the typical residential building height is approximately 3 m. Therefore, the geometric factors affecting the building load include the number of floors, the building orientation, and the window-to-wall area ratio in air-conditioned rooms such as bedrooms and living rooms.

The heat dissipation from occupants, a component of the internal heat source parameters, is set based on the literature [32]. The range of variability for these influencing factors is determined based on relevant standards [31–33] and research data, as shown in Table 3.

Table 3. Parameter setting of typical urban residential building model.

Category	Symbol	Value
the geometric characteristics of the building	NF	18~33
	OR	−180~180
	BR/LR	The minimum value is calculated and determined according to the window-to-wall area ratio of each room not being less than 1/7, and the maximum value is determined according to the limit of the window-to-wall area ratio of each room in the literature [31].

Table 3. *Cont.*

Category	Symbol	Value
the thermal properties of the building envelope	WK	0.3~1.5
	RK	0.25~0.4
	CK	2.5~3.5
	SHGC	0.15~0.35
	WA	0.7~0.8
	RA	0.7~0.8
the internal heat sources	OC	1 or 2
	LI	5~6
	EQ	3.8~6
	PH	Bedroom: 108 Living room: 134 Kitchen: 235
the indoor air design parameters	T	According to Table 2
	V	1.0

3.2.3. Model Group Generation and Load Simulation Results

The fixed parameters needed for the simulation model are initially configured in EnergyPlus to create the idf file. Using the Latin hypercube sampling method in the R language, random combinations of simulation models for Y-type, cross-type, and butterfly-type houses are generated. These combinations are created within the range of variable parameter values and account for three different energy-use behaviors: economical, standard, and comfortable.

This process produces 2100 Y-type residential simulation models, totaling 6300 models, and 3600 cross-type and butterfly-type residential simulation models, totaling 10,800 models. The results of the load simulations conducted using EnergyPlus 24.1.0 for these generated models are shown in Figure 4.

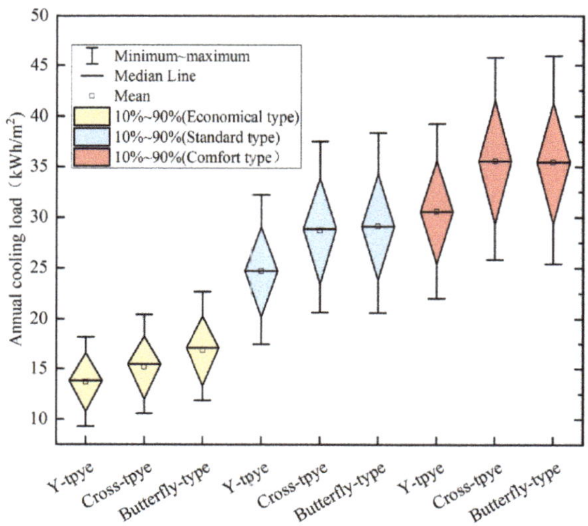

Figure 4. Simulation results of annual cooling load of three typical urban residential buildings under three different energy consumption behaviors.

The abscissa represents three different typical urban residential types in Guangzhou, and the ordinate is the annual cooling load value of each model under different energy consumption behaviors. The results show that the energy consumption behavior of residents has a greater impact on the annual cooling load of residential buildings. The annual cooling load of economical residential buildings is the lowest, with an average value of around 15 kWh/m^2, followed by the standard type between 25 and 30 kWh/m^2. The annual cooling load of residential buildings under the comfortable energy consumption behavior is the highest, with an average value of 30 kWh/m^2 and above. The economical cooling load is very different from the standard type and the comfortable type. The standard type is 1.6 times that of the economical type, and the comfortable type is 2 times that of the economical type. Economical users only turn on air conditioning and cooling in the bedroom in summer, while standard and comfortable users also cool living rooms in summer. Energy behavior has a huge impact on operating energy consumption, and ignoring the prediction of carbon emissions by energy consumption behavior may cause large errors.

Under the economical energy consumption behavior, the variable parameters of the model have the greatest impact on its annual cooling load, reaching about 65%, about 60% for the standard type, and 56% for the comfortable type. It can be seen that if you want to reduce residential energy consumption by optimizing the thermal performance of the envelope, the geometric characteristic parameters of the building, and other factors (in addition to energy consumption behavior), the effect on economical energy-using residents is relatively better. In addition, under the same energy consumption behavior model, the Y-type residential buildings have the lowest annual cooling load, and the cross-type and butterfly-type residential buildings have the same level. This is related to the type of residential floor layout. The standard floor area of Y-type residential buildings ranks second among these three types, but it can only accommodate four households. Compared with the six-family cross-type and butterfly-type residential buildings, the total energy demand is less, so the annual cooling load level is the lowest.

3.3. Sensitivity Analysis of Parameters Affecting Residential Load

To further clarify the key factors influencing the annual cooling load of urban dwellings, a global sensitivity analysis was conducted on the above load simulation results and corresponding variable parameters using the SRC value for evaluation. Figure 5 presents the sensitivity analysis results of the annual cooling load and variables for three typical urban residential models under different energy consumption behavior patterns. The ordinate represents the variable parameters of the input parameterized model, while the abscissa displays the SRC sensitivity analysis result value for each variable parameter.

Overall, the results indicate that BR and LR have minimal impact on the annual cooling load of residential buildings. This is because their variable range is relatively small. To meet the natural lighting requirements of residential buildings, regulations stipulate that the area ratio of the room to the window shall not be less than 1/7. The current energy-saving design standards in Guangzhou limit the window–wall area ratio to a maximum of 0.4 in each direction, with only one room per residential unit allowed with a window–wall area ratio not exceeding 0.6. Therefore, for newly built urban houses in Guangzhou, under the current energy-saving standards, the energy-saving potential of the room window–wall area ratio is very limited.

According to the sensitivity ranking in Figure 5, the primary factors impacting the annual cooling load of residences under the three energy consumption behaviors are OC, LI, EQ, and the thermal properties of the exterior walls and windows (including WK, WA, and SHGC). Among these, the number of people in the room has the greatest impact on the cooling load. This is because human metabolism generates 108 W of heat even when at rest. Consequently, the more people in a room, the greater the cooling load needed to reach the set temperature during summer.

Economical users typically do not use air conditioning in the living room during summer. As a result, the SHGC sensitivity of living room windows ranks higher in the economical type compared to the standard and comfort types, though still higher than that of the bedroom. This is due to the large area of living room window walls. During summer days, a significant amount of heat enters the living room through these windows, which then transfers through partition walls into adjacent bedrooms, affecting the bedroom's cooling load at night.

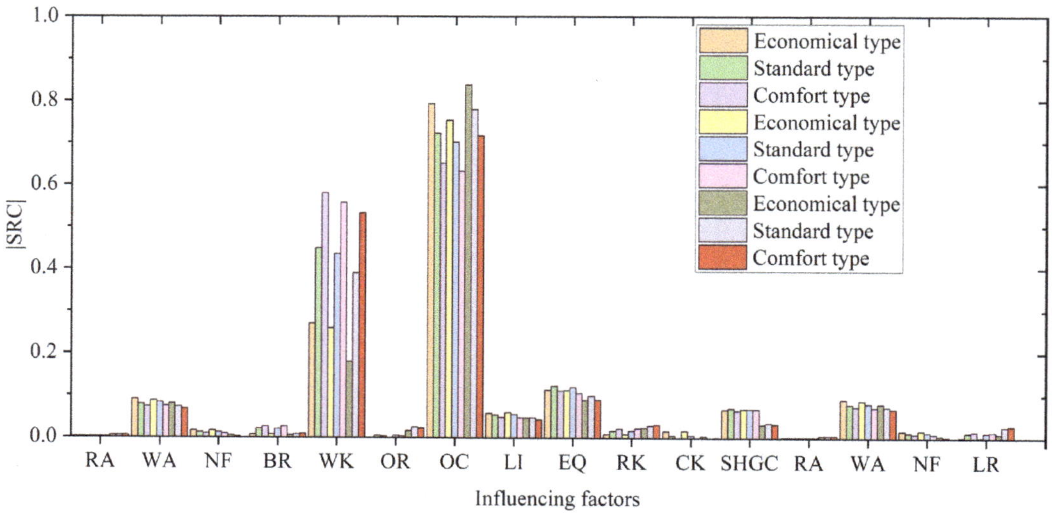

Figure 5. Sensitivity analysis of annual cooling load and variables of residential buildings under different energy consumption behaviors.

Cross-type houses differ from the other two types of houses. The SHGC of bedroom and living room windows has a minimal impact on the annual cooling load. Even under standard and comfort energy consumption behaviors, only some living room windows show greater sensitivity to SHGC. This indicates that for cross-type houses, the primary influencing factors for the annual cooling load are the disturbances within the room and the thermal performance of the exterior walls, with the energy-saving potential of windows being relatively low.

In summary, the main factors affecting the annual cooling load of urban dwellings in Guangzhou are internal disturbances, the thermal performance of exterior walls, and SHGC. Among these, OC is the most significant factor, followed by WK. For Y-type and butterfly-type homes, special attention should be given to the SHGC of all bedroom and living room windows.

3.4. Establishment of Carbon Emission Prediction Model for Urban Residential Operation

The carbon emissions of urban residential operations are mainly composed of emissions from air-conditioning systems, lighting, and electrical equipment. The carbon emissions from lighting and electrical equipment can be directly calculated based on power density, corresponding usage patterns, and carbon emission factors. Therefore, determining the carbon emissions from air-conditioning operation is the most critical step in predicting the overall carbon emissions from residential operations.

Based on the aforementioned cooling load sensitivity analysis, the key variables for the cooling load prediction model are selected. Appropriate algorithms are then employed to predict the cooling load of air conditioners. Given the common use of split-type air conditioners in residential buildings, their energy consumption is calculated using the Annual Performance Factor (APF) to obtain the carbon emissions of the air conditioners.

In this study, nine common machine learning models were selected. Datasets consisting of 2100 groups, 6300 groups, and 3600 groups generated by Y-type, cross-type, and butterfly-type houses, respectively, were used as input data. The k-fold cross-validation method (k = 10) was utilized to calculate and predict cooling load. By comparing the generalization capabilities of different algorithms, the best prediction method for air-conditioning carbon emissions was determined. As shown in Table 4, the analysis found that the Gradient Boosting Regression Tree, Gradient Boosting Nearest Neighbors, and gradient boosting algorithms provided the best prediction results, with R^2 values higher than 0.98. This is significantly better than the prediction accuracy of linear regression and support vector machines. Gradient boosting algorithms typically outperform others in energy consumption simulation tasks due to their ability to handle complex nonlinear relationships, robustness to missing data and noise, and faster training speed, offering better interpretability and generalization. While neural networks and support vector machines excel in specific tasks, tree-based models often deliver superior efficiency and performance in practical applications like energy simulation, requiring less feature engineering and parameter tuning, and being less sensitive to noise. The prediction results for the three building layouts under three different energy-use patterns were consistent. Therefore, any of these three algorithms can be selected as a prediction model for the carbon emissions of air-conditioning operations.

Table 4. R^2 for predicting total energy consumption per unit area.

	Economical			Standard			Comfort		
	Y-Type	**Cross-Type**	**Butterfly-Type**	**Y-Type**	**Cross-Type**	**Butterfly-Type**	**Y-Type**	**Cross-Type**	**Butterfly-Type**
Linear Regression	0.779	0.759	0.765	0.818	0.789	0.793	0.851	0.820	0.828
Decision Tree	0.950	0.946	0.955	0.925	0.954	0.964	0.941	0.950	0.961
Random Forest	0.974	0.978	0.976	0.970	0.977	0.978	0.972	0.977	0.980
Support Vector Machine	0.731	0.734	0.747	0.746	0.751	0.769	0.774	0.781	0.804
Ridge Regression	0.775	0.758	0.762	0.814	0.788	0.791	0.848	0.819	0.826
Gradient Boosting	0.980	0.983	0.981	0.980	0.983	0.981	0.982	0.985	0.984
Extreme Random Trees	0.972	0.978	0.978	0.971	0.976	0.979	0.973	0.975	0.981
Gradient Boosting Nearest Neighbors	0.980	0.983	0.981	0.980	0.983	0.981	0.982	0.985	0.984
Gradient Boosting Regression Tree	0.980	0.983	0.981	0.980	0.983	0.981	0.982	0.985	0.984

Combined with the cooling load prediction model established above, and considering the power consumption of lighting and electrical equipment, a final model of carbon emissions from urban residential operations can be established, as shown in Equations (1)–(4).

$$CE = (Q_c \ / \ \mathrm{APF} + Q_l + Q_e) \times N \quad (1)$$

$$Q_c = f\ (OC, WK, EQ, WA, SHGC, LI) \quad (2)$$

$$Q_l = LI \times 1.045 \quad (3)$$

$$Q_e = LI \times 1.460 \quad (4)$$

where CE is the carbon emission from building operation, in $\mathrm{kgCO_2/(m^2{\cdot}a)}$. APF is the annual operating energy efficiency of air-conditioning equipment. N represents the carbon emission factor of the power grid, in $\mathrm{kgCO_2/kWh}$. Q_c is the annual cooling load of the building, in $\mathrm{kW/(m^2{\cdot}a)}$. Q_l is the annual power consumption of building lighting, in $\mathrm{kW/(m^2{\cdot}a)}$. Q_e is the annual power consumption of building electrical equipment, in $\mathrm{kW/(m^2{\cdot}a)}$.

To verify the accuracy of the prediction model, the measured data of the carbon emissions from the operation of a residential building in Guangzhou are compared with the predicted values of the model. The information indicators of the residential building are shown in Table 5. Judging from the family structure and air-conditioning usage behavior, the energy consumption behavior pattern of the residents is close to the standard energy consumption behavior pattern. Therefore, the prediction model under the standard energy consumption behavior pattern for Y-type residential buildings is used for calculation.

Table 5. Residential information.

Information Indicators	Indicator Value
Household population	3 people are the main
Family structure	Mainly urban white-collar workers
WK	1.38 W/(m^2·K)
LI	7.0 W/m^2
EQ	3.8 W/m^2
WA	0.48
SHGC	0.45
Air conditioner cooling set temperature	26 °C
Bedroom air conditioning on time	22:00~6:00
Living room air conditioning on time	17:00~22:00
Annual cooling load	29.40 kWh/m^2

The average annual air-conditioning cooling load of the residential building is 28.54 kWh/m^2, which is lower compared with the measured data of 29.40 kWh/m^2 in the table. The error is 4.0%, which is less than 10% and thus within the acceptable range. This demonstrates that the accuracy of the prediction model is good and also proves that the carbon emission prediction model of building operation proposed in this study has sufficient accuracy to assist designers and policymakers in program planning and decision-making.

4. Discussion

At present, rapid assessment tools for the carbon emissions of residential building operations exist, but they are often limited by the varied energy consumption characteristics of different residential buildings. This variability makes it challenging to predict carbon emissions simply and directly. To address this problem, it is essential to first determine the energy consumption characteristics of residential buildings through research and then classify them accordingly. Using machine learning methods, parameterized models, and the energy consumption simulation software EnergyPlus, an annual cooling load prediction model for urban residential buildings can be developed. By integrating air conditioning energy efficiency, lighting, and electrical equipment power consumption, a comprehensive carbon emission prediction model for urban residential operations is established. This model provides rapid evaluation tools for designers and policymakers, aiding in the transformation of old city residential buildings and the construction of new districts to support informed decision-making. However, there are still some limitations that need to be addressed in this model:

1. Building types: this study focuses solely on high-rise tower houses as a typical urban residential building type in Guangzhou. The model's applicability is limited due to this single building type. Future research can expand to include various building types such as low-rise buildings, multi-story buildings, and slab buildings;
2. Survey methodology: the energy consumption behavior survey of urban residents in Guangzhou faced certain challenges, primarily relying on online questionnaires for data collection. Supplementing this with follow-up data from summer field research can provide a more accurate energy consumption behavior pattern for Guangzhou urban residents;
3. Geographical scope: the carbon emission prediction model developed in this study is specific to the Guangzhou area. To make it applicable to other regions, further research can be conducted to include different climates. An in-depth analysis based on local climate conditions can offer relevant guidance for residential building planning in various climate zones.

By addressing these limitations, the model can be refined and its applicability broadened, thereby enhancing its utility for a wider range of residential building types and regions.

5. Conclusions

The establishment of a carbon emission prediction model for urban residential operations in Guangzhou involves the following three steps:

First of all, a questionnaire survey was conducted on the energy consumption behavior of urban residents in Guangzhou City, analyzing the characteristics and differences of people's indoor stay and rest, air-conditioning use behavior, and clustering analysis of their air-conditioning use behavior to obtain typical air-conditioning use behavior patterns. Combined with the investigation of typical indoor stay and rest, typical lighting, and electrical equipment use behavior, three typical energy-use behavior patterns of personnel were refined: economical, standard, and comfortable.

Secondly, the parameterized models of three typical urban dwellings in Guangzhou are established through the R language, and the various variables that affect the energy consumption of residential operations are entered into the parameterized model. Latin hypercube sampling is performed to randomly generate a large number of energy consumption calculation models. EnergyPlus is used to perform batch energy consumption simulation of the residential energy consumption calculation model, and the global sensitivity analysis method is used to determine the key influencing variables of residential energy consumption of personnel under different energy consumption behavior patterns. Regardless of the type of group, residence, and energy consumption behavior pattern, the number of rooms and the heat transfer coefficient of the exterior wall are the two parameters that have the greatest impact on the energy consumption of residential operations. Other more sensitive variables are equipment power density, lighting power density, solar radiation absorption coefficient, and the solar heat gain coefficient of the outer windows of the bedroom and living room. It can be seen that in order to reduce the operating energy consumption and carbon emissions of residential buildings, in addition to the energy consumption behavior of residents, urban residential buildings in Guangzhou also need to focus on indoor thermal disturbance, exterior wall thermal performance, and the sun's heat coefficient of the outer windows of air-conditioned rooms. For the heat transfer coefficient of the outer windows, the thermal performance of the roof and the area ratio of the window and wall, etc., it is enough to meet the mandatory requirements of the current energy-saving standards.

Finally, a database of residential air conditioning load samples of three layout types was established separately, and K-fold cross-verification was used for segmentation. The learning and modeling performance of nine common machine learning algorithms were compared. As a result, three prediction algorithms, Gradient Boosting Regression Tree, Gradient Boosting Nearest Neighbors, and gradient boosting algorithms, can be used as prediction models for the cooling load of urban residential air conditioners in Guangzhou. After combining carbon emission factors, air conditioning energy efficiency APF, lighting energy consumption, and electrical equipment energy consumption, a carbon emission prediction model for urban residential buildings in Guangzhou was finally established and compared with actual residential cases. The deviation is only 4.0%, proving that the carbon emissions of urban residential buildings in Guangzhou can be predicted very accurately.

In addition, through analysis, when other variables are the same, residents' energy consumption behavior changes from comfortable to standard and economical: the carbon emission intensity of Y-type residential operation is reduced by 17.5% and 52.4%, respectively; cross-type residential is reduced by 18.3% and 56.2%, respectively; butterfly-type residential is reduced by 16.7% and 50.9%, respectively. This indicates that in the process of realizing the carbon emission control of urban residential operations, focusing on guiding residents' energy consumption behavior is also a very effective method.

Author Contributions: All authors contributed to the study conception and design. Material preparation, data collection and analysis were performed by L.Z. (Lintao Zheng) and K.L. The first draft of the manuscript was written by L.Z. (Lintao Zheng) and all authors commented on previous versions of the manuscript. All authors have read and agreed to the published version of the manuscript.

Funding: This research was funded by SZ02 Architectural Engineering Technology Specialty Group of Guangdong Polytechnic of Water Resources and Electric Engineering.

Data Availability Statement: Data is available on request due to privacy restrictions.

Conflicts of Interest: The authors declare no conflict of interest.

Nomenclature

SRC	Standardized regression coefficient
PCC	Partial correlation coefficient
NF	Number of floors (floors)
OR	Building orientation (°)
BR/LR	The window–wall area ratio of bedroom/living room
WK	Heat transfer coefficient of exterior wall ($W{\cdot}m^{-2}{\cdot}K^{-1}$)
RK	Heat transfer coefficient of roof ($W{\cdot}m^{-2}{\cdot}K^{-1}$)
CK	Heat transfer coefficient of exterior window ($W{\cdot}m^{-2}{\cdot}K^{-1}$)
SHGC	Solar heat gain coefficient
WA	Solar radiation absorption coefficient of exterior wall
RA	Solar radiation absorption coefficient of roof
OC	Number of people in the room
LI	Lighting power density ($W{\cdot}m^{-2}$)
EQ	Equipment power density ($W{\cdot}m^{-2}$)
PH	Personnel heat dissipation ($person^{-1}$)
T	Air conditioner cooling set temperature (°C)
V	Number of ventilations (h^{-1})
CE	The carbon emission from building operation ($kgCO_2/(m^2{\cdot}a)$)
APF	Annual operating energy efficiency of air-conditioning equipment
N	The carbon emission factor of the power grid ($kgCO_2/kWh$)
Q_c	Annual cooling load of the building ($kW/(m^2{\cdot}a)$)
Q_l	Annual power consumption of building lighting ($kW/(m^2{\cdot}a)$)
Q_e	Annual power consumption of building electrical equipment ($kW/(m^2{\cdot}a)$)

References

1. Liu, Z.; Fan, G.; Sun, D.; Wu, D.; Guo, J.; Zhang, S.; Yang, X.; Lin, X.; Ai, L. An innovative distributed energy system that combines hybrid energy storage with a multi-objective optimization approach for nearly zero-energy communities and buildings. *Energy* **2022**, *239*, 122577. [CrossRef]
2. Wang, S.J. Corporate Green Development Strategy Under the Carbon Peaking and Carbon Neutrality Goals. *Learn. Educ.* **2022**, *10*, 105–106. [CrossRef]
3. Du, X. Thoughts on strategies and paths to achieve carbon peaking and carbon neutrality in China. *Front. Energy* **2023**, *17*, 324–331. [CrossRef]
4. China Building Energy Conservation Association Building Energy Consumption and Carbon Emission Data Professional Committee. A Series of Research Reports on Carbon Emissions in the Field of Urban and Rural Construction in China in 2022. Available online: https://www.cabee.org/upload/file/20230104/1672820934145324.pdf (accessed on 2 January 2024).
5. Tan, X.C.; Lai, H.P.; Gu, B.H.; Zeng, Y.; Li, H. Carbon emission and abatement potential outlook in China's building sector through 2050. *Energy Policy* **2018**, *118*, 429–439. [CrossRef]
6. Zhou, N.; Khanna, N.; Feng, W.; Ke, J.; Levine, M. Scenarios of energy efficiency and CO_2 emissions reduction potential in the buildings sector in China to year 2050. *Nat. Energy* **2018**, *3*, 978–984. [CrossRef]
7. Yang, T.; Pan, Y.Q.; Yang, Y.Q.; Lin, M.; Qin, B.; Xu, P.; Huang, Z. CO_2 emissions in China's building sector through 2050: A scenario analysis based on a bottom-up model. *Energy* **2017**, *128*, 208–223. [CrossRef]
8. Vandenbogaerde, L.; Verbeke, S.; Audenaert, A. Optimizing building energy consumption in office buildings: A review of building automation and control systems and factors influencing energy savings. *J. Build. Eng.* **2023**, *76*, 107233. [CrossRef]
9. Ge, S.; Li, J.; Liu, H.; Wang, Y.; Zhou, H. Domestic Energy Consumption Modeling per Physical Characteristics and Behavioral Factors. *Energy Procedia* **2019**, *158*, 2512–2517. [CrossRef]

10. Ge, S.; Li, J.; Liu, H. Modelling of household energy consumption characteristics considering physical features and behavior factors. *Electr. Power Autom. Equip.* **2019**, *39*, 36–44.
11. Yuan, P.; Gao, K.; Zhao, X.; Liu, X.; Kong, W. Electricity consumption and load prediction method for Chinese rural residences based on the randomness and seasonality in electricity usage behavior. *Energy Build.* **2023**, *279*, 112647. [CrossRef]
12. Catalina, T.; Virgone, J.; Blanco, E. Development and validation of regression models to predict monthly heating demand for residential buildings. *Energy Build.* **2008**, *40*, 1825–1832. [CrossRef]
13. Olinger, M.S.; de Araújo, G.M.; Dutra, M.L.; Silva, H.A.D.; Júnior, L.P.; de Macedo, D.D. Metamodel Development to Predict Thermal Loads for Single-family Residential Buildings. *Mob. Netw. Appl.* **2022**, *27*, 1977–1986. [CrossRef]
14. Chen, Z.; Zhao, Z.; Deng, Q.; Tang, P.; Yang, C.; Li, X.; Gui, W. A knowledge embedded graph neural network-based cooling load prediction method using dynamic data association. *Energy Build.* **2023**, *278*, 112635. [CrossRef]
15. Deb, C.; Eang, L.S.; Yang, J.; Santamouris, M. Forecasting diurnal cooling energy load for institutional buildings using Artificial Neural Networks. *Energy Build.* **2016**, *121*, 284–297. [CrossRef]
16. Senthil Kumar, D.; George Washington, D.; Reshmy, A.K.; Noorunnisha, M. Energy Performance Prediction of Residential Buildings Using Nonlinear Machine Learning Technique. In *International Conference on Innovative Computing and Communications: Proceedings of ICICC 2021*; Springer: Singapore, 2022.
17. Li, G.; Tian, W.; Zhang, H.; Fu, X. A novel method of creating machine learning-based time series meta-models for building energy analysis. *Energy Build.* **2023**, *281*, 112752. [CrossRef]
18. Li, L.; Su, X.; Bi, X.; Lu, Y.; Sun, X. A novel Transformer-based network forecasting method for building cooling loads. *Energy Build.* **2023**, *296*, 113409. [CrossRef]
19. Kapp, S.; Choi, J.K.; Hong, T. Predicting industrial building energy consumption with statistical and machine-learning models informed by physical system parameters. *Renew. Sustain. Energy Rev.* **2023**, *172*, 113045. [CrossRef]
20. Pinto, G.; Wang, Z.; Roy, A.; Hong, T.; Capozzoli, A. Transfer learning for smart buildings: A critical review of algorithms, applications, and future perspectives. *Adv. Appl. Energy* **2022**, *5*, 100084. [CrossRef]
21. Liu, S.; Zhang, Z.; Yang, J.; Hu, W. Exploring Increasing Urban Resident Electricity Consumption: The Spatial Spillover Effect of Resident Income. *Energies* **2022**, *15*, 4249. [CrossRef]
22. Prataviera, E.; Vivian, J.; Lombardo, G.; Zarrella, A. Evaluation of the impact of input uncertainty on urban building energy simulations using uncertainty and sensitivity analysis. *Appl. Energy* **2022**, *311*, 118691. [CrossRef]
23. Yousaf, S.; Bradshaw, C.R.; Kamalapurkar, R.; San, O. Investigating critical model input features for unitary air conditioning equipment. *Energy Build.* **2023**, *284*, 112823. [CrossRef]
24. Fan, C.; Xiao, F.; Zhao, Y. A short-term building cooling load prediction method using deep learning algorithms. *Appl. Energy* **2017**, *195*, 222–233. [CrossRef]
25. Li, X.; Yao, R. Modelling heating and cooling energy demand for building stock using a hybrid approach. *Energy Build.* **2021**, *235*, 110740. [CrossRef]
26. Ma, J.; Cheng, J.C. Estimation of the building energy use intensity in the urban scale by integrating gis and big data technology. *Appl. Energy* **2016**, *183*, 182–192. [CrossRef]
27. R Language. Available online: https://www.rstudio.com/ (accessed on 16 September 2024).
28. Tian, W.; Heo, Y.; De Wilde, P.; Li, Z.; Yan, D.; Park, C.S.; Feng, X.; Augenbroe, G. A review of uncertainty analysis in building energy assessment. *Renew. Sustain. Energy Rev.* **2018**, *93*, 285–301. [CrossRef]
29. EnergyPlus. Available online: https://energyplus.net/ (accessed on 30 September 2024).
30. Saltelli, A.; Ratto, M. *Global Sensitivity Analysis*; John Wiley: Hoboken, NJ, USA, 2008.
31. *GB 55015-2021*; General Code for Energy Efficiency and Renewable Energy Application in Buildings. Energy Efficiency Design Standards for Public Buildings. China Construction Industry Press: Beijing, China, 2021.
32. *JGJ/T449—2018*; Standard for Green Performance Calculation of Civil Buildings. Energy Efficiency Design Standards for Public Buildings. China Construction Industry Press: Beijing, China, 2018.
33. *GB/T 51350-2019*; Technical Standard for Nearly Zero Energy Buildings. China Construction Industry Press: Beijing, China, 2019.

buildings

MDPI

Article

Energy-Efficiency-Oriented Spatial Configuration of VRV Outdoor Units in an Equipment Layer Under Background Wind Conditions

Lin Liu, Haoran Huang and Xiaoyu Tian *

School of Civil and Transportation Engineering, Guangdong University of Technology, Guangzhou 510006, China; liulin@gdut.edu.cn (L.L.); 2112109130@mail2.gdut.edu.cn (H.H.)
* Correspondence: tianxiaoyujian@163.com

Abstract: This study provides a spatial configuration method to improve the cooling efficiency of multiple VRV outdoor units placed on equipment layers with high floors. Relevant factors include wind parameters, the placement of multiple outdoor units, and louver. A total of 96 cases were designed. CFD simulations were used to obtain the inlet air temperature distributions of multiple outdoor units and then calculate their cooling efficiency. The results found that these factors have effects on the average cooling efficiency of outdoor units in a single row to a certain extent. The influencing degrees of these factors, from large to small, were the louver angle, wind parameters, and the placement of multiple outdoor units. When the cooling efficiency of outdoor units was maximum and the louver angle was 15°, the louvers could be oriented perpendicular to the dominant wind direction (90°) when wind speed was ≥6 m/s, and the unit spacing was 600 mm. Based on this, when the number of outdoor units was expanded in the limited space, staggered arrangements with different directions of heat exchange surfaces were a recommended optimization layout. This study provides technical support for improving the working efficiency of VRV outdoor units in an equipment layer.

Keywords: VRV; equipment layer; cooling efficiency; inlet air temperature; influencing factors

1. Introduction

Rapid urbanization, expanding population, and increased air temperature have elevated energy demand [1,2]. In 2019, the International Energy Agency reported that global electricity consumption reached 25,027.3 TWh and showed an increasing trend [3]. Global carbon dioxide emissions increased from 22.28 Gt in 1990 to 35.47 Gt in 2019 [4]. Building operations generate about 10 billion tons of CO_2 emissions, accounting for about 37% of global emissions [5]. However, the energy consumption of air conditioning systems contributed more than 40%, especially for commercial buildings. For example, air conditioning systems accounted for 47.17% of the total energy consumption of office buildings in Guangzhou, China [6], and 50% of the total energy consumption of high-rise public buildings in Chongqing, China [7]. Therefore, how to control air conditioning energy consumption is a crucial factor in reducing anthropogenic carbon emissions.

Variable Refrigerant Volume (VRV) has been widely used to meet public buildings' cooling/heating demands due to its saving space, easy installation, reliable operation, and flexibility [8,9]. A primary factor influencing the higher energy consumption was the heat exchange effects of VRV outdoor units, which then influenced the operating efficiency of outdoor units. The outdoor unit is like an air-cooled heat exchanger; its heat exchange effects depend on external airflow. Especially, urbanization has resulted in the transformation of natural underlying surfaces into high-rise and dense building clusters [10,11]. Usually, the outdoor units are placed on the roofs or on the top floor of the building. However, some outdoor units are inevitably placed on the middle floors of the building. Due to wind speed, wind pressure, louvers, and other factors, indoor airflow could be affected, and then outdoor units' cooling efficiency can be worsened. In severe cases, it may cause the units to shut down. For example, Kabeel et al. found that the supply air temperature was reduced by 21% and the coefficient of performance improved by 71% in a hybrid air conditioning system using an indirect evaporation cooler with an internal baffle as pre-cooling units compared to previous designs of the hybrid air conditioning system [12]. Yau and Pean found that the total cooling capacity of the system is decreased by 2% when the outdoor temperature is increased by 1 °C [13]. Therefore, it is necessary to explore an optimization design method to improve the cooling performance of outdoor units in an equipment layer. Before that, it is first necessary to determine which factors influence the inlet air temperature of outdoor units.

Nowadays, many studies have investigated the factors influencing the inlet air temperature of air conditioning systems' outdoor units [14–17]. These factors mainly include the louver [18–21], placement of air conditioner units [22–26], layout of the outdoor unit [14,23,27,28], horizontal distance between the condenser unit and wall [29–35], unit spacing [35–37], building re-entrant shapes [38], outdoor wind speed [39–41], and directions of exhaust [42]. These studies provided some recommended parameters to obtain a lower inlet air temperature, namely a higher cooling efficiency. For example, in these studies, they observed that the inlet air temperature of outdoor units increases with the increase in louver angle and the recommended louver angle was not higher than 15°. Chow et al. found that a face-to-face layout of outdoor units can be appropriate [14]. Some studies observed that the staggered placement of the front and rear outdoor units effectively reduces the inlet air temperature of the outdoor units [25,42]. A detailed literature review is shown in Table 1. However, three limitations existed in the literature. Firstly, most of these studies only explored the effect of a single factor or some factors without considering the comprehensive impact of these factors, and the influencing degree of these factors was not explored. Secondly, these studies investigated the inlet air temperature of outdoor units placed outside buildings, not considering the scenario of outdoor units inside buildings and placed on high floors of the building. Thirdly, the comprehensive effect between background wind parameters was not explored in these studies.

Table 1. A detailed literature review on the outdoor units of air conditioners.

Reference	Influencing Factors	Results
Chow et al., 2002 [14] Locations: Hongkong, China	condenser unit layouts	• The effect of 10 types of condensing unit layouts in the building re-entrant on the COP of air conditioning systems. The results found that face-to-face layouts of AC units can be an appropriate layout.
Chow et al., 2002 [15] Locations: Hongkong, China	louver, terrace	• The effect of outdoor units in building re-entrant on the condenser units' surrounding temperature and the energy coefficient of the AC units. The results obtained an optimal arrangement.
Chen et al., 2009 [18] Location: Fushan, China	louver	• Operating environments can be improved when the louver angle decreases.
Yang, 2017 [19] Locations: Shandong, China	louver	• The inlet air temperature increases as the inclination angle of the louvers increases.
Liu et al., 2017 [20] Location: Qingdao, China	louver	• Operating environments can be improved when the louver angle decreases. The louver angle was recommended as 15° or less than 15°.

Table 1. *Cont.*

Reference	Influencing Factors	Results
Yang et al. [21] Locations: Chongqing, China	louver	• The recommended parameters when the operating environment was best were that the louver spacing is 80 mm and its angle is 10°.
Chow et al., 2001 [22] Locations: Hongkong, China	placement of air conditioner units	• The effect of airflow characteristics around the condenser units on the AC characteristics. The results show that discharging the outdoor units' air into a light well leads to better unit performance and low power consumption compared to the case of installing the outdoor units in a building re-entrant.
Nada and Said, 2017 [23] Location: Benha, Egyp	different arrangements	• In the case of an open-bottom shaft, it is preferable to distribute all the outdoor units on one wall of the shaft, except that at the top two levels it is preferable that the units be distributed on two adjacent walls in staggered arrangements.
Li 2018 [24] Locations: Harbin, China	placement, louver	• Investigate the relationship between the inlet air temperature of outdoor units and factors including distance between the units and the wall, distance from the louver, louver angle, louver spacing, outdoor wind speed, and outdoor air temperature.
Zhou et al., 2010 [25] Locations: Beijing, China	placement	• The incoming wind speed significantly affects the heat dissipation of outdoor units.
Zhao et al., 2017 [26] Locations: Guangzhou, China	placement of outdoor units	• The important factors influencing the thermal environment of the outdoor units are the distance between the units and the wall.
Yu et al., 2022 [27] Locations: Shandong, China	layout of outdoor unit and placement	• The CFD simulation results found that the outdoor units' performance is optimal when one outdoor unit is installed close to the louver and the other is lifted away from the louvers at the top of the groove.
Huang et al., 2017 [28] Locations: Beijing, China	layout of outdoor units	• The staggered placement of the front and rear outdoor units effectively reduces the inlet air temperature of the outdoor units.
Avara and Daneshgar, 2008 [29] Location: Bushehr, Iran	spacing between two sidewalls, distance	• The effect of the spacing between the two supporting walls of the outdoor units on the on-coil temperature of condenser units. • Recommended various optimum distances between the outdoor AC unit and supporting wall in different spacing between the sidewalls.
Xue et al., 2007 [30] Location: Henan, China Climate: Temperate	horizontal distance between condenser unit and wall, the space per condensing unit	• Exploring the air temperature and wind velocity distribution around the outdoor units. • A distance between two sidewalls of 7 m is needed to keep the temperature raised less than 10 °C. • The space per condensing unit should be at least larger than 18.2 m^2.

Table 1. *Cont.*

Reference	Influencing Factors	Results
Duan et al., 2016 [31] Location: Beijing, China	distance between neighbouring outdoor units, louver spacing, louver angle, and arrangement style	• To reduce operation temperature and save energy, the following design points are recommended: Vertical arrangement for outdoor units. A larger distance between neighboring outdoor units. Smaller louver for the louver angle. 100 degrees for the louver angle.
Harithkhan et al., 2021 [32] Locations: Moratuwa, Sri Lanka	horizontal distance between condenser unit and wall	• The selected building with a re-entrant size of 6 m × 2.5 m distance of 1.5 m from the exterior wall was found to be the optimum position for the condenser.
Jiang 2013 [33] Locations: Tianjin, China	distance between the units and wall, louver	• It is recommended that the distance between the outdoor unit condenser and the wall be greater than 100 mm, the distance between the left side of the outdoor unit and the wall be at least 200 mm, the distance between the fan and the louver be around 100 mm, the louver angle do not exceed 20°, and the direction be downward with a minimum louver spacing of 50 mm.
Bruelisauer et al., 2013 [34] Location: National University of Singapore Singapore	spacing between two sidewalls, horizontal distance between condenser unit and wall	• The air temperature in the recessed space increased continuously to reach up to 12.7 °C higher temperatures than undisturbed conditions at the top of 20 floors when air conditioners were switched on. • The inlet temperature increases by 9.0 °C on the lowest level along with the 50 °C maximum inlet temperature and 59 °C outlet temperature.
Zhang, 2009 [35] Locations: Dalian, China	Unit spacing, the distance between the units and pillars	• The maximum distance between outdoor units is 0.2 m, the maximum distance from the pillar is 0.8 m, and the maximum distance between multiple outdoor unit arrays is 0.8 m
Zhang et al., 2008 [36] Locations: -	Unit spacing	• Outdoor units of VRV have lower inlet air temperature when the unit spacing is greater than 1.65 m.
Wei and Jiang, 2021 [37] Locations: Zhejiang, China	Unit spacing	• The inlet air temperature is close to the background air temperature when unit spacing is greater than 2 m
Chow et al., 2000 [38] Locations: Hongkong, China	building re-entrant shapes	• Explored the effect of different re-entrant shapes (T-, I- and L-shapes) on which the outdoor AC units were placed. The results found that the T-shaped re-entrant showed lower outdoor air temperatures around the buildings compared with the I- and L-shaped re-entrants.
Choi et al., 2005 [39] Location: Seoul, South Korea	Wind speed, wind direction and layout of the outdoor unit	• The performance of the condensers above the 30th floor decreased. • The performance of heat dissipation of the condensers for the higher floor drooped significantly when the frontal- and side-winds were greater than 4 m/s and 8 m/s, respectively.
Li and Liu, 2019 [40] Locations: Shandong, China	Wind speed	• To effectively solve the airflow short circuit, the inlet and exhaust surfaces of the unit are not in the same direction. The airflow velocity of the incoming air should be controlled below 1–1.5 m/s.

Table 1. *Cont.*

Reference	Influencing Factors	Results
Zi and Wu, 2017 [41] Locations: Guangdong, China	Wind speed	• Keeping the exhaust direction of the outdoor unit consistent.
Ge et al., 2007 [42] Locations: Sichuan, China	Directions of exhaust	• Compared with the units with the same side of exhaust air, the units with the opposite side of exhaust air show better performance.

Considering the above limitations, this study investigates an optimization method for the cooling efficiency of outdoor units in the equipment layer placed on the high floors of the buildings. Firstly, the factors influencing outdoor units' cooling efficiency were analyzed. These factors can be divided into three aspects: wind parameters, the placement of multiple outdoor units, and louvers. Wind parameters include wind speed and wind direction. The placement of multiple outdoor units includes unit spacing, the distance between the units and the front wall, and the distance between the unit and the left/right pillar. Secondly, an arrangement-combination method was used to design 96 cases. CFD simulations were used to obtain the overall inlet air temperature of these cases. The optimal parameter configuration of multiple outdoor units in a single row was obtained considering the maximum cooling efficiency. Finally, on this basis, an optimal layout for multiple outdoor units with an expanded dual row was obtained. This study provides a parametric and systematic optimization design method for obtaining a better cooling efficiency from multiple outdoor units placed on an equipment layer with high floors. It also contributes to the sustainable development of future society [43] and the analysis of energy efficiency factors in buildings in smart cities [44].

2. Methods

2.1. Cooling Efficiency Indicator

Figure 1a shows the circulation process and main hardware components of refrigerant in the VRV air conditioning system. In general, VRV systems mainly include indoor units and outdoor units. The outdoor units mainly involve two components: the condenser and compressor, which are used to deliver the outdoor air temperature to the indoors through air conditioning technology. The heat exchange between the refrigerant and the air occurs in the evaporator (indoors) and the condenser (outdoors). Ideally, the maximum cooling efficiency of the refrigeration machine (ε) can be expressed as following formulas according to the vapor compression refrigeration cycle (a T-S diagram of the Carnot cycle was shown in Figure 1b when operating at the specific temperature T_k and T_0 [45].

$$\varepsilon = \frac{1}{T_k/T_0 - 1} \tag{1}$$

where T_0 is the cooled temperature and T_k is the temperature of the heat source, namely the inlet air temperature.

It is assumed that T_z is the overall inlet air temperature of outdoor units, which can be expressed by each outdoor unit's inlet air temperature and its corresponding cooling capacity ratio (Q_i) to the total cooling capacity (Q_z). Its expression was

$$T_Z = \sum_{k=1}^{n} (T_k \frac{Q_k}{Q_z}) \tag{2}$$

Based on formulas (1) and (2), the average cooling efficiency of refrigeration machines ($\overline{\varepsilon}$) can be written as

$$\overline{\varepsilon} = \frac{1}{\sum\limits_{k=1}^{n} (T_k \frac{Q_k}{Q_z})/T_0 - 1} \tag{3}$$

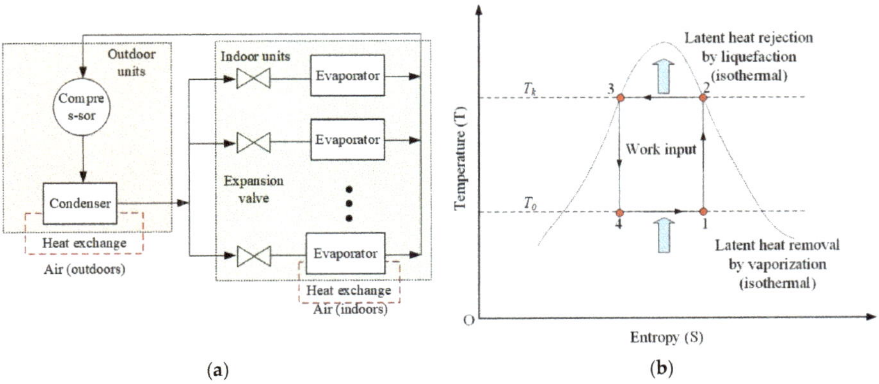

Figure 1. Principle of VRV air conditioning systems. (**a**) Schematic of VRV, (**b**) Carnot cycle.

2.2. Influencing Factors

In this study, the research object is the outdoor units of VRV, which are placed on the equipment layer of high-rise commercial buildings in Guangzhou, China. The equipment layer is located on the 44th floor of the building, which has a height of 4.5 m (see Figure 2) and is higher than 100 m from the ground. During a typical summer in Guangzhou, China, the average wind speed around the building is 1.9 m/s in 10 m of building height, the outdoor temperature is 35 °C, and southeast winds prevail. These air conditioning systems were used to provide the cooling/heating demands on the 43rd and 45th floors. Then, according to the cooling/heating demand, four outdoor units of VRV were arranged in parallel (see Figure 2a). Based on the results of the field investigation, a widely used VRV air conditioning system was selected. The specifications of the outdoor unit are presented in Table 2. The dimension of outdoor units is 1657 × 1240 × 765 mm^3 (L × W × H). The cooling capacity is 61.5 KW, and power consumption is 18.6 KW. The maximum inlet air temperature is 50 °C.

Table 2. The specifications of the outdoor unit.

Model	RUXYQ22AB
Cooling capacity (KW)	61.5
Power consumption (KW)	18.6
Dimension of outdoor unit (L × W × H, mm × mm × mm)	1657 × 1240 × 765
Air volume (m^3/min)	271
Maximum inlet air temperature (°C)	50

According to the results of previous studies (see Table 1), the factors influencing the cooling efficiency of all outdoor units from the perspective of airflow could involve three categories. One is the louver angles. The louvers are always used for ventilation, air exchange, and rainproofing. The second one is the placement of multiple outdoor units in the equipment layer, including unit spacing (D1), distance between the units and the front wall (D2), and distance between the unit and the left/right pillars (D3), as shown in Figure 2b,c.

Besides that, background wind conditions include wind speed and direction. The prevailing wind direction in a season is basically unchanged. When the buildings are built, the windward and leeward sides of the building are already determined (see Figure 2d). The angle is the angle between the background wind direction and the louvers. However, the placement of the outdoor unit on which side of the building can be designed and planned.

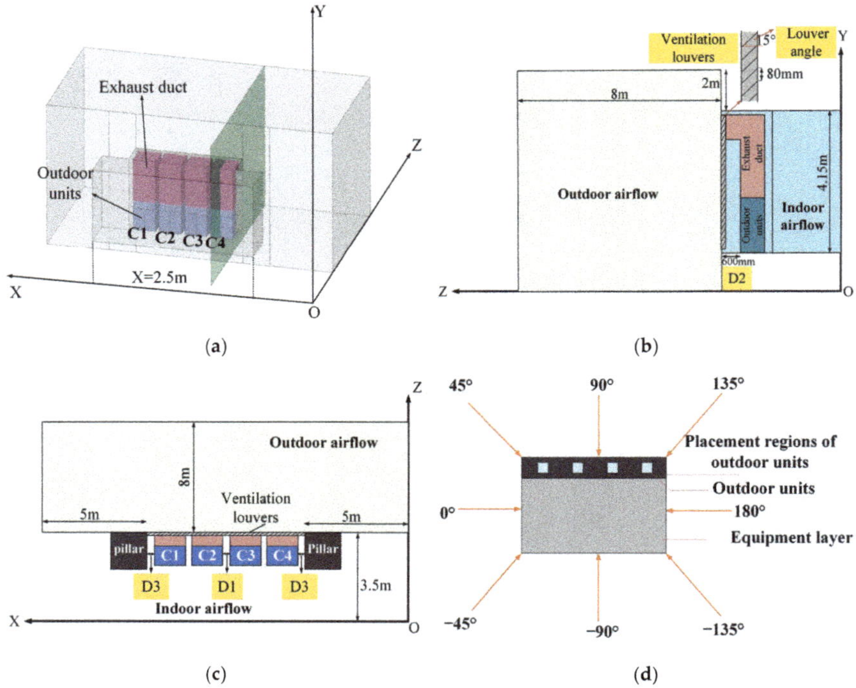

Figure 2. Schematic diagram of a numerical physical model. (**a**) 3D diagram, (**b**) side view, (**c**) vertical view, (**d**) wind directions.

2.3. Case Designs

Based on the results of previous studies and field investigations, a total of 96 cases were designed, including four angles of louver (45°, 30°, 15°, 0°), three unit spacings (200, 400, and 600 mm), three distances between the unit and the left/right columns (200, 400, and 600 mm), three distances between the units and the front wall (400, 600, and 800 mm), three wind speeds (4, 6, and 8 m/s), and eight wind directions (−135°, −90°, −45°, 0°, 45°, 90°, 135°, 180°). Table 3 lists the detailed parameter configurations for the main 12 basic cases. Among them, wind speeds were set as 6 m/s in Cases 1~10 and 4 and 8 m/s in Case 11 and Case 12, respectively. Each case involves eight wind directions.

Table 3. The parameter configurations of cases.

Cases	Louver Angle (°)	D1 (mm)	D2 (mm)	D3 (mm)
1	45	200	600	200
2	30	200	600	200
3	15	200	600	200
4	0	200	600	200
5	15	400	600	200
6	15	600	600	200
7	15	200	400	200
8	15	200	800	200
9	15	200	600	400
10	15	200	600	600
11	15	200	600	200
12	15	200	600	200

2.4. CFD Simulations

2.4.1. Computational Domains

Computational domain was generated by using the SCDM 2024 software built in ANSYS, which was developed by SpaceClaim Corporation [46]. Its dimension was determined and calibrated based on the results of field investigations, including the dimensions of building layers, the number of external air conditioning units, the unit spacing, the distance between the units and the front wall, the distance between the unit and the left/right pillars, and the louver spacing. The domain mainly includes three parts: the building interior, the outdoor space, and the external air conditioning unit (see Figure 2a). The relevant information of the computational domains is shown in Figure 3a,b. In this study, to further reduce the computational cost, the computational domain was simplified through three assumptions:

- In the physical model, all walls were set as insulated walls, such as building exterior walls, louvers, exhaust systems, etc.
- Ignoring the effects of outdoor solar radiation, outdoor temperature was considered and set as 33.2 °C. The indoor temperature was set at 26 °C.
- All outdoor units were set to working at full loads and were regarded as a black-box-like heat source by ignoring their internal structure. The heat dissipation was the sum of the cooling capacity and cooling power consumption of the VRV system. The heat source per unit volume was 37,147 W/m^3.

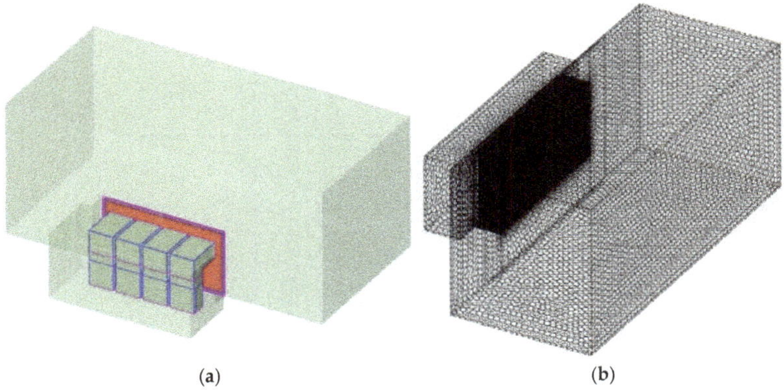

Figure 3. Computational domains and girds. (**a**) computational domains, (**b**) grid cells.

The computational grids were structured and created using a mosaic grid-generation technique to limit the total number of computational cells in the domain and improve the quality of grid topology [47]. The ANSYS Fluent Meshing [48] was used for the generation of computational grids (Figure 3b). All surface grids were spatially discretized with polyhedral cells, and the poly-hexforce cells were used in the external AC unit and automatically combined with the boundary layer grid near the surface walls. The Tecplot 360 software was used to visualize and analyze the results of the CFD simulation [49]. Table 4 lists the physical properties of the air. The specific heat capacity is 1.005 J/(kg·°C). Thermal conductivity is set as 2.715 × e^{-2} W/m·K. The coefficient of thermal expansion is 0.0034.

Table 4. Physical properties of the air.

Parameters	Values
Density (kg/m^3)	1.1465
Specific heat capacity (J/(kg·°C))	1.005
Thermal conductivity (W/m·K)	2.715 × e^{-2}
Viscosity (m·Pa·s)	18.85 × e^{-4}
Coefficient of thermal expansion	0.0034

2.4.2. Boundary Conditions

In this study, the main boundaries include the exterior walls of the equipment layer, casings of outdoor units, air ducts, louvers, fans of outdoor units, the air inlets of outdoor units, and the air inlets and outlets of the external computational domain. In this study, referring to existing relevant literature [15–42], the boundary conditions in ANSYS Fluent software 2024 r1 are set as shown in Table 5. As shown in Table 5, the main boundary conditions include the exterior wall, casing of outdoor units, air ducts, louvers, air inlets of outdoor units, fans of outdoor units, and corresponding computational domain.

Table 5. Boundary condition settings in ANSYS Fluent software 2024 r1.

Parts.	Types	Settings
Exterior walls	Adiabatic wall	Heat flux and heat radiation are 0
Casings of outdoor units	Adiabatic wall	
Air ducts	Adiabatic wall	
Louvers	Wall without thickness	0
Air inlets of outdoor units	Porous jump	$\alpha = 2.13e^{-6}$, C2 = 0.1997 [47]
Fans of outdoor units	Fan	Pressure difference is set to the pressure at the rated air volume of outdoor units
Air inlets of the external computational domain	Velocity Inlet	$V = V_c\left(\frac{h}{h_c}\right)^n$ [48]
Outlets of external computational domain	Static pressure outlet	0 Pa

Note: V is the wind speed at a height. V_c is the wind speed of reference height, which is 1.9 m/s. h is a height. h_c is the reference height, which is 10 m. n is the roughness index, which is set as 0.4.

2.4.3. Solver Settings

The simulations were performed with the commercial ANSYS Fluent 2024r1 [50]. The RANS equation was solved in combination with the Standard k-ε model [29]. Finite volume discretization technology was used to solve the coupled partial differential equation and initial conditions in numerical modes. The SIMPLE algorithm was used for pressure–velocity coupling. Pressure interpolation was a second-order discretization scheme which was used for the pressure terms and the viscous terms of the governing equations. The pressure and momentum relaxation factors were set as 0.3 and 0.7, respectively. The volume and viscosity relaxation factors were 0.6 and 0.7, respectively. The momentum and turbulent energy relaxation factors were 0.6 and 0.9, respectively. Convergence was assumed to be obtained when reaching a minimum of 10^{-6} for momentum.

The governing equations for airflow and heat transfer include the Mass conservation equation, the Momentum conservation equation and the Energy conservation equation. Their expressions are as follows:

(a) Mass conservation equation:

$$\frac{\partial \rho}{\partial t} + \nabla \cdot \rho \vec{v} = 0 \tag{4}$$

where ρ is the pressure, Pa and $\vec{v}$ is the velocity vector, m/s.

(b) Momentum conservation equation:

$$\begin{aligned} &\tfrac{\partial}{\partial t}(\rho \vec{v}) + \nabla \cdot (\rho \vec{v} \vec{v}) = -\nabla \rho + \nabla \cdot (\overline{\overline{\tau}}) + \rho \vec{g} + \vec{F} \\ &\overline{\overline{\tau}} = \mu[(\nabla \vec{v} + \nabla (\vec{v})^T) - \tfrac{2}{3} \nabla \vec{v} I] \end{aligned} \tag{5}$$

where, ρ is the pressure, Pa, $\vec{v}$ is the velocity vector, $\vec{F}$ is the forces acting on micro-elements, and $\overline{\overline{\tau}}$ is the viscous stress.

(c) Energy conservation equation:

$$\frac{\partial(\rho T)}{\partial t} + div(\rho u T) = div(\frac{k}{c_p} gradT) + S_T$$
$$h = \int_{Tref}^{T} cpdT \text{ and } k_t = \frac{c_p \mu_t}{Pr_t} \quad (6)$$

where, T is the air temperature, k is the heat transfer coefficient, W/(m·K), C_p is the specific heat capacity, J/(kg·K), and S_T is a viscous dissipation term.

2.5. Grid Independence Test

The grid independence was used to obtain the optimal number of computational grid cells based on a certain threshold. It indicates that the number of grid cells did not impact the simulation results when the threshold was reached. That is, when $T_Z(m + 1) - T_Z(m) < 0.01$ °C. Where m is the number of grid cells.

Independence tests were performed for five grid resolutions in total, and the numbers of grid cells were 4.57707×10^5, 8.57362×10^5, 12.73952×10^5, 14.05279×10^5, and 16.99537×10^5. The results showed that the overall inlet air temperatures were 40.510, 39.490, 38.753, 38.752, and 38.753 °C, respectively. When the number of grids is not less than 1,273,952, the computational load and speed of computer hardware, and the prediction accuracy of the physical model, the grid size of 12.73952×10^5 is selected for calculations in this study.

The detailed flowchart of this study is shown in Figure 4.

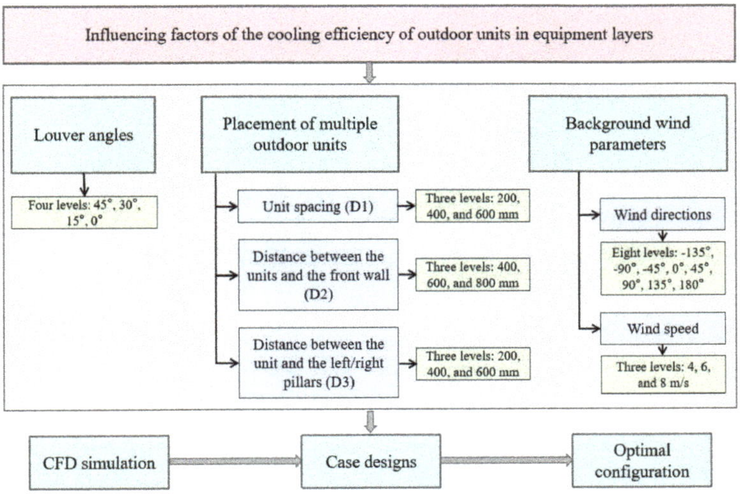

Figure 4. The flowchart of this study.

3. Results and Discussion

3.1. Louver Angle

Figure 5a shows the changes in the average cooling efficiency of outdoor units when the louver angles ranged from 45° to 0° under eight wind directions. With the louver angles decreased (from Case 1 to Case 4), the average cooling efficiency of outdoor units showed an increasing trend. For example, at a wind direction of 90°, the average cooling efficiency was 12.8, 17.9, 23.9, and 24.1 when the louver angles were 45°, 30°, 15°, and 0°, respectively. The most significant difference in average cooling efficiency was 11.4 when the louver angles ranged from 45° to 0°, and the wind directions were 90° and −90°.

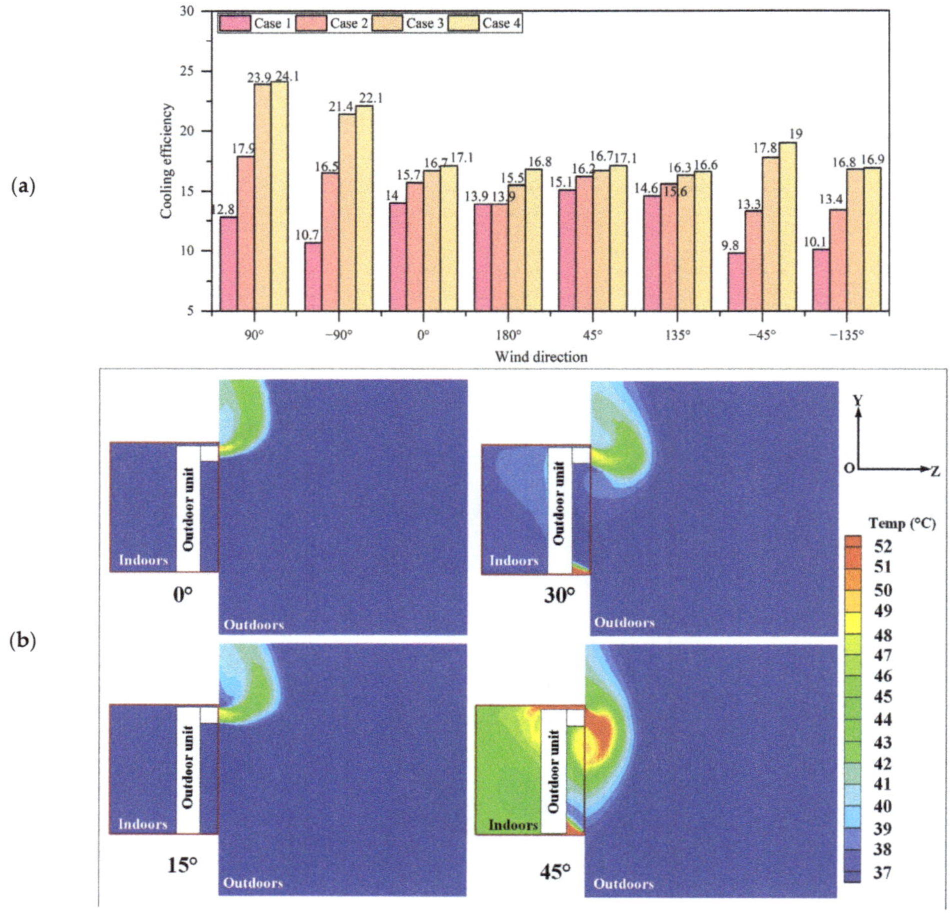

Figure 5. Average cooling efficiency and temperature cloud maps of outdoor units under different louver angles (0°~45°). (**a**) average cooling efficiency, (**b**) temperature cloud maps with 90° wind direction.

Figure 5b shows the air temperature distribution cloud maps when the wind direction was 90°. The distribution range involved in high temperatures is relatively wide when the louver angles are larger. When the louver angles ranged from 45° to 0° under a 90° wind direction, the air temperature difference was 14.9 °C. Additionally, when the louver angle changed from 15° to 0°, the overall air temperature difference was much smaller compared with others. In practical engineering, louvers are always used in ventilation, air exchange, and rainproofing. Therefore, the louver angle was not recommended to be set as 0° for rainproofing. Then, a 15° louver angle was recommended in the letter cases.

3.2. Placement of Outdoor Units

3.2.1. Unit Spacing

Figure 6a shows the changes in the average cooling efficiency of outdoor units when unit spacing ranges from 200 to 600 mm under eight wind directions. The results show that the average cooling efficiency of outdoor units increases with the unit spacing (Case 3, Case 5, and Case 6). For example, at the wind direction of 90°, the air inlet air temperatures were 23.9, 24.5, and 24.7 when the unit spacing was 200, 400, and 600 mm, respectively.

The biggest difference in the average cooling efficiency was 1.5 when the wind direction was 135°.

(**a**)

(**b**)

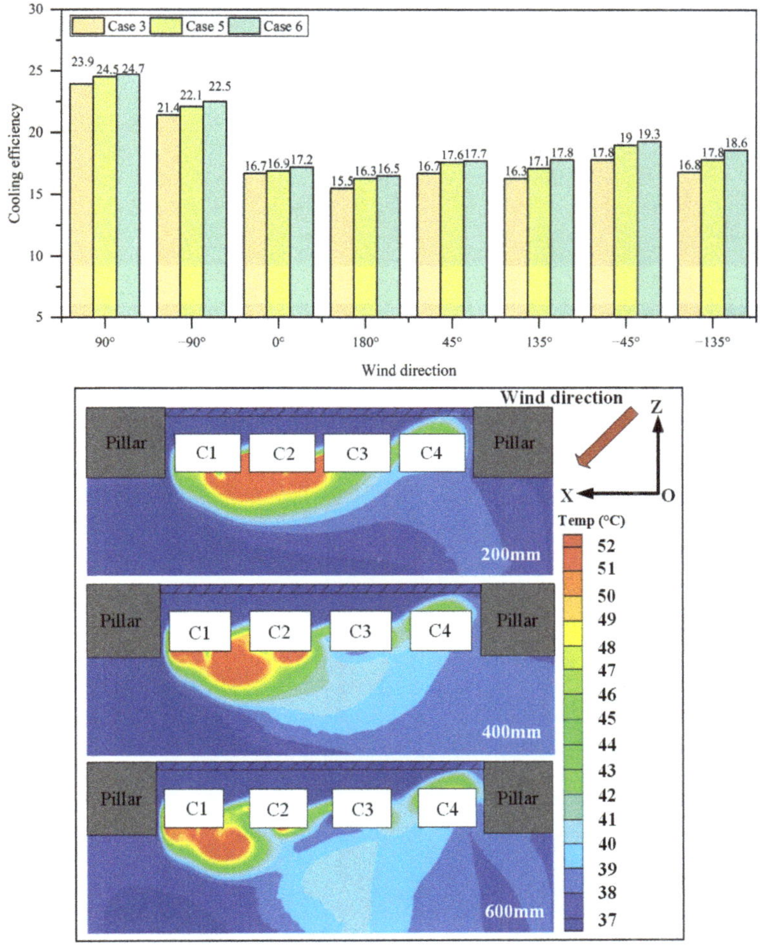

Figure 6. Average cooling efficiency and temperature cloud maps under different unit spacings. (**a**) average cooling efficiency, (**b**) temperature cloud maps with 135° wind direction and Y = 2.65 m.

Figure 6b shows the air temperature distribution cloud maps with 135° wind direction under different unit spacings. The air temperature difference was 1.7 °C as the unit spacing ranged from 200 to 600 mm. Additionally, the larger the unit spacing, the smaller the distribution of high-temperature areas (see red areas). The high-temperature regions moved to the downwind side with increased unit spacing. For example, the red areas were close to units C1~C3 when the unit spacing was 200 mm, while the areas were close to unit C1 when the unit spacing was 600 mm.

3.2.2. Distance Between the Units and the Front Wall

Figure 7a shows the changes in the average cooling efficiency of outdoor units when the distance between the units and the front wall ranges from 400 to 800 mm under eight wind directions. The results show that as the distance between the units and the front wall increases (Case 7, Case 3, and Case 8), the average cooling efficiency of outdoor units displays different changing trends. When wind directions were 90°, 135°, and −135°, the average cooling efficiency of outdoor units first increased and then decreased. When wind directions were −90° and −45°, the average cooling efficiency of outdoor units increased. When wind directions were 0°, 180° and 45°, the average cooling efficiency of outdoor units decreased. For example, at the wind direction of 135°, the average cooling efficiency of outdoor units was 16, 16.3, and 15.7 when the distance between the units and the front wall was 400, 600, and 800 mm, respectively. At a wind direction of −45°, the average cooling efficiency of outdoor units was 16.3, 17.8, and 18.3 when the distance between the units and the front wall was 400, 600, and 800 mm, respectively. At a wind direction of 45°, the average cooling efficiency of outdoor units was 17.9, 16.7, and 16.5 when the distance between the units and the front wall was 400, 600, and 800 mm, respectively. Among them, the biggest average cooling efficiency of outdoor units was 2.

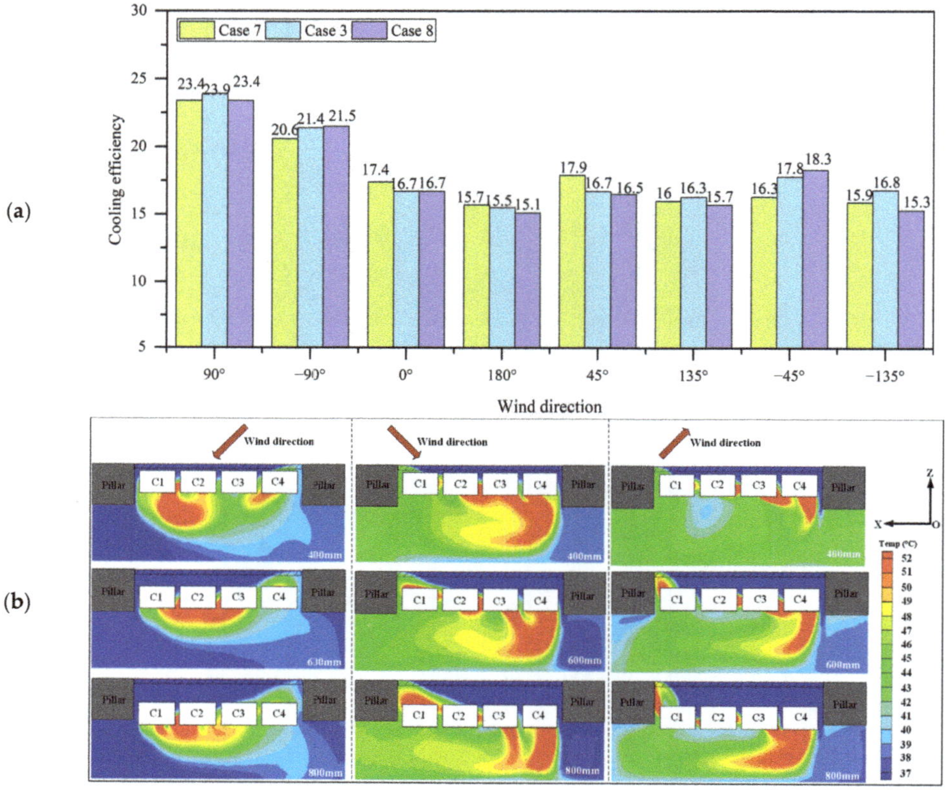

Figure 7. Average cooling efficiency and temperature cloud maps under different distances between the units and the front wall (400~800 mm). (**a**) average cooling efficiency, (**b**) temperature cloud maps with three wind directions (135°, 45°, and −45°) and Y = 2.65 m.

Figure 7b shows the air temperature distributions with three wind directions (135°, 45°, and −45°) under different distances between the units and the front wall. The air temperature difference was 1.7 °C when the distance between the units and the front wall ranged from 400 to 800 mm. Similarly, the higher the inlet air temperature, the larger the distribution of high-temperature areas, namely the red regions. Additionally, high-temperature regions occurred on the downwind side.

3.2.3. Distance Between the Unit and the Left/Right Pillars

Figure 8a shows the changes in the average cooling efficiency of outdoor units when the distance between the unit and the left/right pillars ranged from 200 to 600 mm under eight wind directions. The results show that as the distance between the units and the left/right pillars increases (Case 3, Case 9, and Case 10), the average cooling efficiency of outdoor units displays different changing trends. When wind directions were 90°, −45°, and −135°, the average cooling efficiency of outdoor units first increased and then decreased. When wind directions were 0°, 180°, 45°, 135°, and −90°, the average cooling efficiency of outdoor units increased. For example, when the wind direction was 90°, the average cooling efficiency of outdoor units was 23.9, 24.3, and 22.8 when the distance between the unit and the left/right pillars was 200, 400, and 600 mm, respectively. When the wind direction was 0°, the average cooling efficiency of outdoor units was 16.7, 17.2, and 18.2 when the distance between the unit and the left/right pillars was 200, 400, and 600 mm, respectively. Among them, the biggest difference in the average cooling efficiency of outdoor units was 2.5 as the distance between the unit and the left/right pillars ranged from 200 to 600 mm and wind direction was −90°.

(a)

(b)

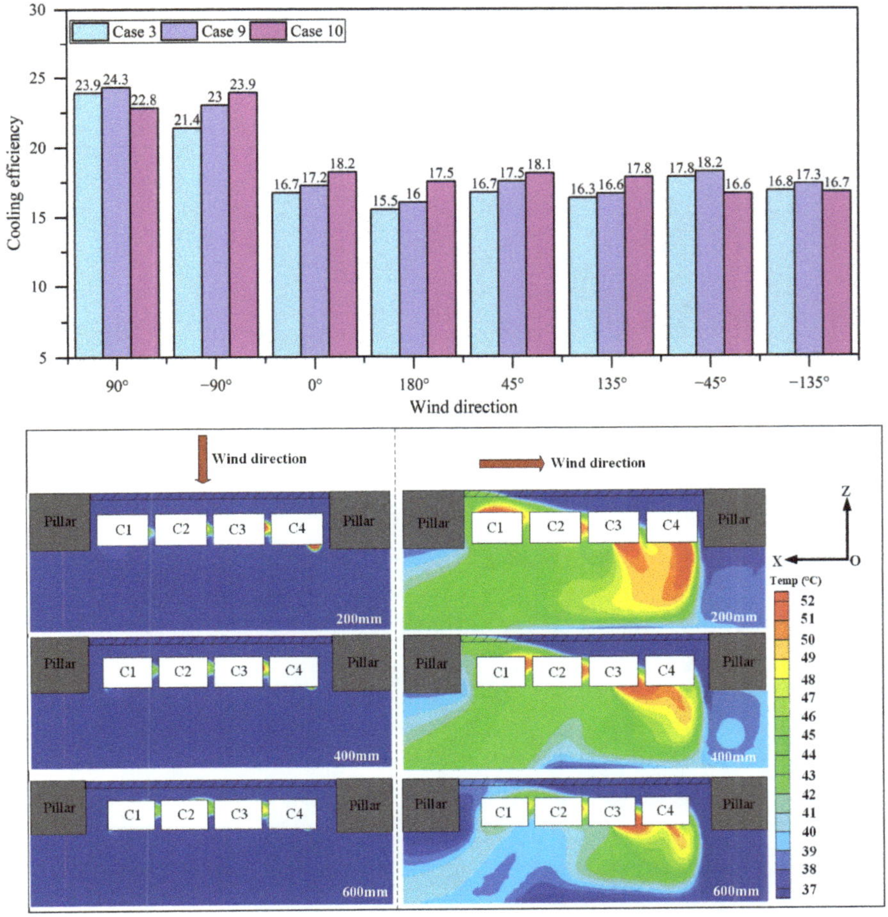

Figure 8. Average cooling efficiency and temperature cloud maps under different distances between the units and the left/right pillars (200~400 mm). (**a**) average cooling efficiency, (**b**) temperature cloud maps with 90° and 0° of wind directions and Y = 2.65 m.

Figure 8b shows the air temperature distributions with 90° and 0° of wind directions under different distances between the units and the front wall. Similarly, the higher the inlet air temperature, the larger the distribution of high-temperature areas, namely the red regions.

3.3. Wind Parameters

Figure 9a shows the changes in the average cooling efficiency of outdoor units when wind speeds ranged from 4 to 8 m/s under eight wind directions. The results show that with wind speed increases (Case 11, Case 3, and Case 12), the average cooling efficiency of outdoor units displayed different changing trends. When wind directions were 90°, 180°, 135° and −135°, the average cooling efficiency of outdoor units first increased and then decreased. When wind directions were 0° and −45°, the average cooling efficiency of outdoor units increased. When wind directions were −90° and 45°, the average cooling efficiency of outdoor units first increased and then remained unchanged. For example, when the wind direction was 90°, the average cooling efficiency of outdoor units was 14.9, 23.9, and 22.5 when wind speeds were 4, 6, and 8 m/s, respectively. When the wind direction was 0°, the average cooling efficiency of outdoor units was 16.6, 16.7, and 16.9 when wind speeds were 4, 6, and 8 m/s, respectively. When the wind direction was −45°, the average cooling efficiency was 17.4, 17.8, and 18 when wind speeds were 4, 6, and 8 m/s, respectively. Among them, the most significant difference in the average cooling efficiency of outdoor units was 9 when the wind direction was 90°.

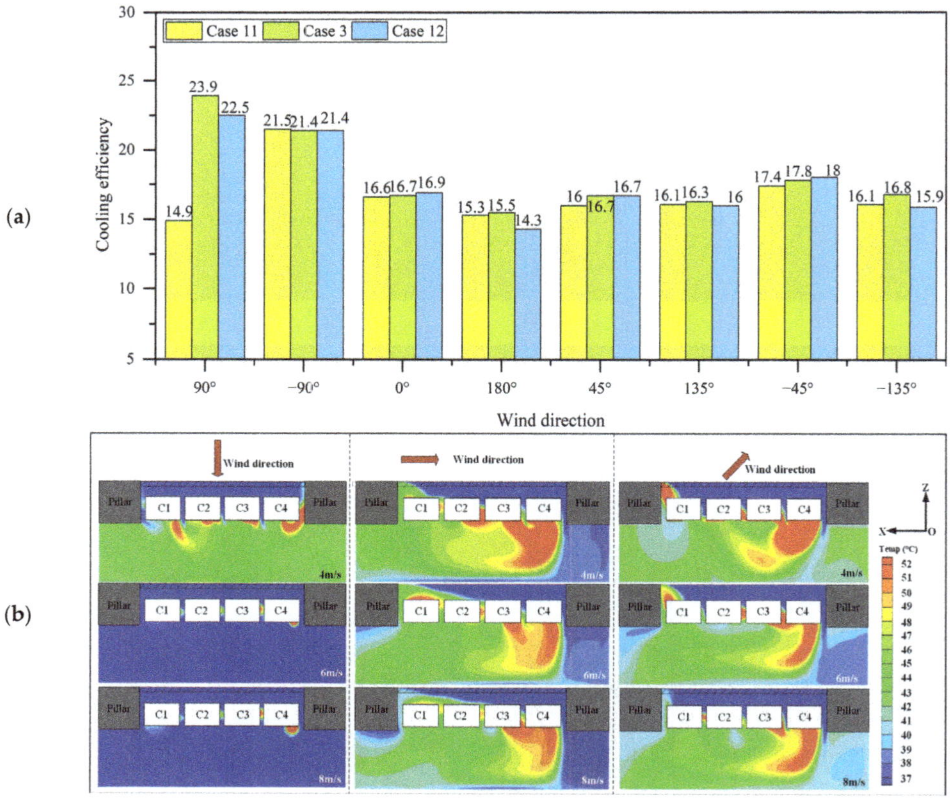

Figure 9. Average cooling efficiency and temperature cloud maps under different wind speeds (4~8 m/s). (**a**) average cooling efficiency, (**b**) temperature cloud maps with 90°, 0°, and −45° wind directions and Y = 2.65 m.

Figure 9b shows the air temperature distributions with 90°, 0°, and −45° wind directions under different wind speeds. The biggest air temperature difference was 7.6 °C as wind speeds ranged from 4 to 8 m/s. Similarly, the higher the inlet air temperature, the larger the distribution of high-temperature areas, namely the red regions.

As shown in Figures 4–8, the average cooling efficiency of outdoor units was 21.8, 20.7, 16.7, 15.5, 16.9, 16.3, 17, and 16 as the wind direction was 90°, −90°, 0°, 180°, 45°, 135°, −45°, −135°, respectively. The corresponding mean overall inlet air temperatures were 40.5, 41.2, 44, 45.3, 43.7, 44.4, 44.3, and 45.3 °C, respectively. The mean overall inlet air temperature at a 90° wind direction was the lowest. The second was a −90° wind direction, and the biggest one occurred at a 180° wind direction. Among them, the biggest difference in air temperature was 15.3 °C at −90° in the direction of the wind.

4. Discussion

4.1. Optimal Configurations

This study adopted a CFD simulation method to explore the effects of multiple factors on the cooling efficiency of VRV systems' outdoor units at the equipment layer in high-rise buildings. These factors include the louver angle, unit spacing, distance between the units and the front wall, distance between the units and the left/right pillars, wind speed, and wind direction. Through CFD simulation, the results showed that these factors have effects on the cooling efficiency of outdoor units to a certain extent. Firstly, it can be seen from Figures 4–8 that the louver angle and wind direction have a relatively significant impact on the cooling efficiency compared with the other factors. For example, the maximum cooling efficiency occurred when the louver angle was 0°and the wind direction was 90° (Figure 5a). The biggest air temperature difference was 14.9 °C as the louver angles ranged from 45° to 0°, and the corresponding cooling efficiency increased by 1.3. Secondly, the factor that had a relatively higher impact on cooling efficiency was wind speed. When wind speeds ranged from 4 to 8 m/s with 90° wind direction, the biggest difference in the average cooling efficiency of outdoor units was 9, and the air temperature difference was 7.6 °C (see Figure 9a). Thirdly, placing multiple outdoor units at the equipment layer has a relatively lower impact on the average cooling efficiency and the overall inlet air temperature. When considering a factor, the maximum cooling efficiency difference could be reduced by 1.5 (see Figures 5a and 6a).

The angle of the louver is linearly correlated with the average cooling efficiency of outdoor units. When the angle of the louver decreases, the average cooling efficiency of outdoor units increases (see Figure 5). Correlation analysis shows a significant positive correlation between the two, with a 0.631 correlation coefficient ($p < 0.01$). Some studies found that the inlet air temperature increases as the inclination angle of the louvers increases [18,19]. Considering the functions of louvers and CFD simulation results, a 15° louver angle can be recommended as the optimal parameter, which ensures a better heat exchange performance and a rain-proof function. Some similar results were found in previous studies. These studies observed that the larger the angle of the louver, the worse the working performance of VRV air conditioning systems. Additionally, the recommended louver angle should not exceed 20°, and the optimal recommended value was 15° [20,33].

As shown in Figures 4–8, when the wind directions were 90° and −90°, the average cooling efficiency was relatively higher compared with other wind directions. In most cases, the average cooling efficiency in the 90° wind direction was highest compared with that in the −90° wind direction. Among them, the temperature difference between 90° and −90° wind directions ranged from 0.5~1.7 °C in most cases. However, when wind speed was 6 m/s and 8 m/s, the average cooling efficiency of outdoor units in the 90° wind direction was significantly higher than in the −90° wind direction (see Figure 9). Consequently, considering the dominant wind direction around the selected building, the louvers on the equipment layer can be oriented perpendicular to the wind direction (90°) when the wind speed is not less than 6 m/s or the wind direction (−90°) when wind speed was less than 6 m/s. According to the calculation formula for wind speed at different building heights (see Table 5), the equipment layer's height was 177.2 m (39~51 floors) when the wind speed was 6 m/s. In this study, the prevailing wind direction around the selected building is southeast. That is, the louver can be installed in the southwest direction of the building when the equipment layer' is higher than 177.2 m. When the equipment layer is less than 177.2 m, the louver can be installed in the northeast direction of the building. Some studies found similar results. For example, Choi et al. observed that the performance of the condensers could decrease when the floors were higher than 30th or the frontal- and side-wind speeds were greater than 4 m/s and 8 m/s, respectively [39].

Subsequently, as shown in Figure 6, the average cooling efficiency of outdoor units increased with the increase in unit spacing. However, for the distance between the units and the front wall and the distance between the units and the left/right pillars, no consistent changing trends were observed in the average cooling efficiency. In general, 600 mm can be recommended as the optimal distance between the units and the front wall vs. the left/right pillars, according to results from Figures 6 and 7. For example, when determining the louver angle, the equipment layer, and their placement directions, the average cooling efficiency of Case 3 was 0.4 higher than Case 7 and Case 8. The recommended values were consistent with other studies. For example, the distance between the outdoor unit condenser and the wall was greater than 100 mm [33], and the distance between the left side of the outdoor unit and the wall was at least 200 mm [33]. Zhang found that the maximum distance from the pillar is 0.8 m [35]. Based on the above results, when the cooling efficiency was maximum, the optimal parameter configuration was a 15° louver angle. The louvers on the equipment layer could be oriented perpendicular to the wind direction (90°) when the wind speed was not less than 6 m/s, or the wind direction (−90°) when wind speed was less than 6 m/s, the unit spacing was 600 mm, the distance between the units and the front wall was 600 mm, and the distance between the units and the left/right pillars was 400 or 600 mm.

Considering that the influencing degree of louvers and wind parameters were relatively higher, the louver angle was designed as 15°, and the louvers on the equipment layer were oriented perpendicular to the wind direction (90°) in practical scenarios. Based on the CFD simulation results, a multiple linear regression formula was established by taking unit spacing, the distance between the units and the front wall, the distance between the unit and the left/right pillars, and wind speed as the input parameters, and the average cooling efficiency and overall inlet air temperature were taken as dependent variables. Then, the regression formulas of the average cooling efficiency and overall inlet air temperature can be expressed as

$$T_z = 36.671 - 0.001 \times D1 + 0.001 \times D2 + 0.0001 \times D3 \; (R^2 = 0.735) \quad (7)$$

$$\bar{\varepsilon} = 27.194 + 0.003 \times D1 - 0.001 \times D2 + 0.0001 \times D3 \left(R^2 = 0.739\right) \quad (8)$$

According to the regression formulas, T_z and $\bar{\varepsilon}$ were 36.731 °C and 28.454, respectively.

4.2. Optimization Schemes

Through the above analysis, the optimal parameter configuration was obtained when outdoor units showed a single-row arrangement at the equipment layer. However, in practical application, considering the increases in cooling/heating demands, the outdoor units in the equipment layer need to be expanded. Due to space limitations and the optimal parameter configuration, six outdoor units can be placed. Considering those possible arrangements for outdoor units in the equipment layer, six layouts were used for further analysis, as shown in Figure 10a. Among them, Layout 1 is that six outdoor units were in a single row. Layout 2 is that six outdoor units are arranged in a double row and a parallel arrangement. Layout 3~6 is that six outdoor units are arranged in a double row and staggered patterns. The heat exchange surfaces of Layout 3~4 were different from that of Layout 5~6. Among them, Layouts 3 and 4 showed staggered arrangements in the same direction as the heat exchange surfaces of all units. Layouts 5 and 6 showed a staggered arrangement of units with the different directions of the heat exchange surfaces of all units. The distances between units in different rows were 600, 800, and 1000 mm, respectively.

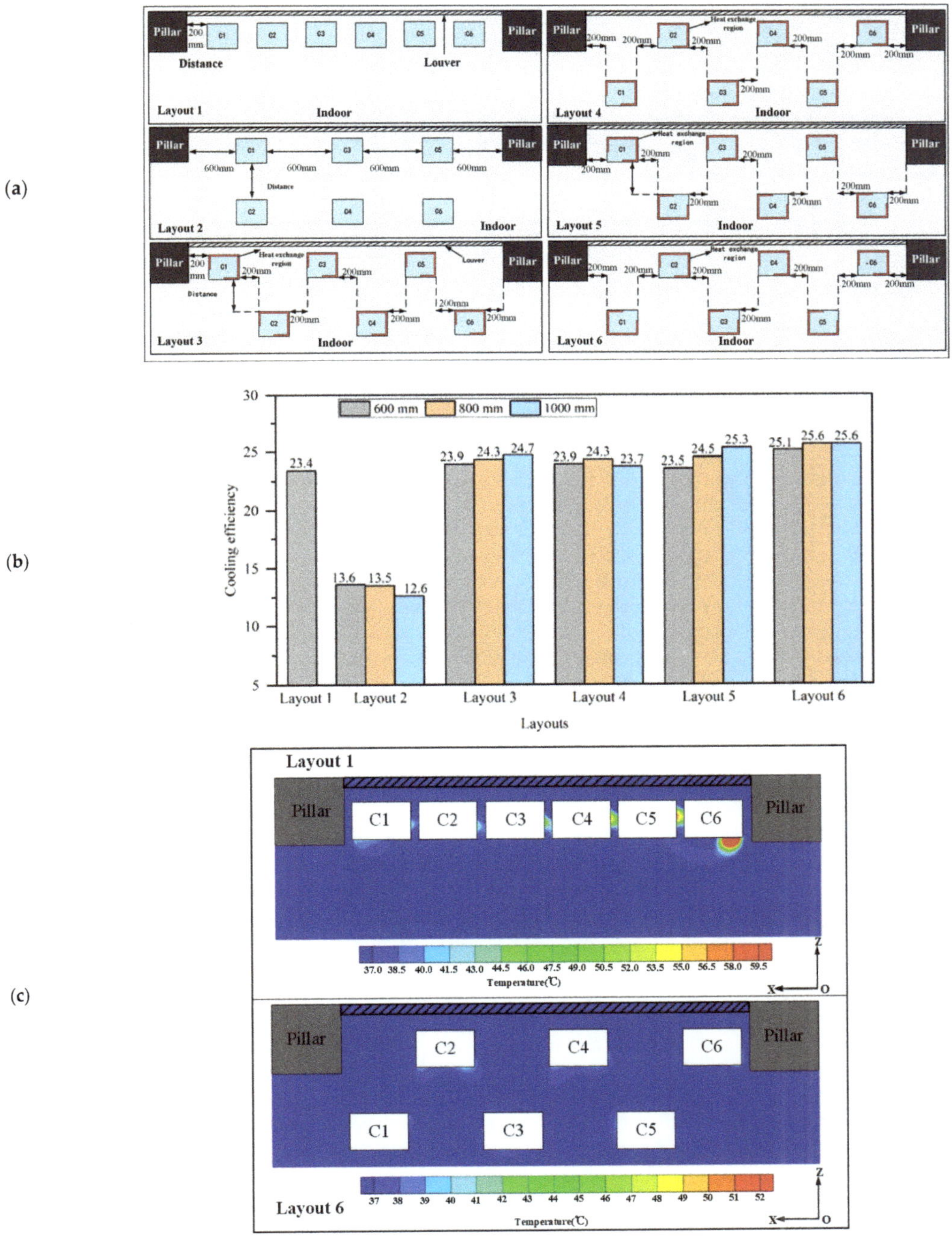

Figure 10. Outdoor units under different layouts. (**a**) Arrangement of outdoor units. (**b**) Average cooling efficiency. (**c**) air temperature distribution cloud maps.

Figure 10b shows the average cooling efficiency of outdoor units with the optimal parameter configurations under different layouts. The average cooling efficiency of outdoor units was lowest when outdoor units were in a parallel arrangement (Layout 2) compared with other layouts. Compared with the single row, the average cooling efficiency of outdoor units showed higher values in double arrangements (see Layouts 3, 4, 5, and 6). Under Layouts 3, 5, and 6, the average cooling efficiency of outdoor units showed an increasing trend with the distances between units in different rows increased. For Layout 4, the average cooling efficiency of outdoor units occurred when the distance was 800 mm. Besides that, when the average cooling efficiency of outdoor units was maximum, the optimal layout was Layout 6. Figure 10c shows the air temperature distribution cloud maps of Layout 1 and Layout 6. Ge et al. found that compared with units showing the same side of exhaust air, the units with the opposite side of exhaust air show a better performance [42].

5. Conclusions and Future Works

5.1. Main Conclusions

This study investigates an optimization parameter configuration that produces a higher heat exchange efficiency of outdoor units placed in the equipment layer on the high floors of buildings. The relevant parameters include three aspects: wind parameters, the placement of outdoor units, and the louver. Among them, wind parameters include outdoor wind speed and wind direction. The placements of outdoor units are unit spacing, the distance between the units and the front, and the distance between the unit and the left/right pillar. A total of 96 cases were designed based on these factors, including four louver angles (45°, 30°, 15°, 0°), three unit spacings (200, 400, and 600 mm), three distances between the unit and the left/right columns (200, 400, and 600 mm), three distances between the units and the front wall (400, 600, and 800 mm), three wind speeds (4, 6, and 8 m/s), and eight wind directions (−135°, −90°, −45°, 0°, 45°, 90°, 135°, 180°). The CFD simulation method was used to obtain the inlet air temperature distributions of multiple outdoor units in the equipment layers, and then the average cooling efficiency of outdoor units was calculated. The main conclusions are as follows:

Firstly, these factors have effects on the average cooling efficiency of outdoor units in a single row to a certain extent. Among them, the angle of the louver and wind direction have a relatively significant impact on the average cooling efficiency and overall inlet air temperature compared with other factors. The average cooling efficiency increased by 11.4 as the angle of the louver ranged from 45° to 0°. When wind speeds ranged from 4 to 8 m/s, the average cooling efficiency increased by 1.5 when the wind direction was 135°. However, placing multiple outdoor units at the equipment layer has a relatively lower impact on the average cooling efficiency and overall inlet air temperature.

Secondly, when the cooling efficiency was maximum, the optimal parameter configuration and the order of configuration were: (1) a louver angle was set as 15°, (2) the louvers on the equipment layer could be oriented perpendicular to the wind direction (90°) when wind speed was not less than 6 m/s, or the wind direction (−90°) when wind speed was less than 6 m/s, (3) the unit spacing was 600 mm, (4) the distance between the units and the front wall was 600 mm, and (5) the distance between the units and the left/right pillars was 400 or 600 mm.

Thirdly, when the number of outdoor units was expanded in the limited equipment layer, the recommended optimization schemes for the layout of outdoor units were that the outdoor units showed a staggered arrangement and the direction of the heat exchange surfaces of all units were different (Layout 6).

5.2. Limitations and Future Works

This study provides a parametric and systematic optimization design method for obtaining a better heat exchange performance of multiple outdoor units placed in an equipment layer of high-rise buildings. The results of this study can effectively guide the designs of practical engineering projects to save building energy consumption and reduce economic costs. Importantly, the proposed optimal configurations can be generalized to other climates. However, for different air conditioning systems, the optimal configurations could be different due to the different working power levels.

During CFD simulations, the wall was assumed to be insulated without considering the effects of outdoor solar radiation. Actually, outdoor solar radiation increases indoor heat gain to a certain degree [51,52]. The increased heat gain exacerbates the rise in the indoor air temperature at the equipment layer. There may be a coupling effect between the increased air temperature and the outdoor unit's heat dissipation, which then influences the temperature distribution, thus leading to the rise in building energy consumption, economic, and carbon saving [53]. Thus, it is necessary to explore the impact of outdoor solar radiation in the future. Additionally, experiments will be conducted to validate the effectiveness of the CFD simulation results of this study, and the placements of outdoor units will be further optimized when the space is limited. This study used only typical wind speeds to explore the impacts on system performance. More experiments would be conducted to examine the effects of seasonal wind variations on long-term system performance and to verify the validation of these results.

Besides that, in this study, the optimization of cooling efficiency focuses on airflow, louver angles, and spatial arrangements. In the future, other factors, including the perovskite smart windows [54], noise, maintenance, and installation costs, will be considered.

Author Contributions: Conceptualization, L.L., H.H. and X.T.; methodology: H.H.; software, H.H.; formal analysis, X.T.; data curation: H.H.; writing—original draft preparation, L.L.; writing—review and editing, X.T.; visualization, X.T.; supervision, L.L. and X.T.; project administration, L.L.; funding acquisition, L.L. All authors have read and agreed to the published version of the manuscript.

Funding: This research was funded by National Natural Science Foundation of China (Grant No. 52208099).

Data Availability Statement: The data presented in this study are available on request from the corresponding author.

Conflicts of Interest: The authors declare no conflicts of interest.

Nomenclature

CFD	Computational Fluid Dynamics	T_0	cooled temperature
CO_2	Carbon dioxide	T_k	temperature of the heat source
Gt	Grand Tourer	$T_Z(m+1)/T_Z(m)$	Mean inlet temperature of (m + 1)th/mth iteration
TWh	Tera Watt Hour	Q_i	cooling capacity ratio
VRV	Variable Refrigerant Volume	Q_z	the total cooling capacity
hc	reference height, which is 10 m	Vc	wind speed of reference height, which is 1.9 m/s.
h	a height.	V	wind speed at a height.
n	Roughness index, which is set as 0.4.	ε	refrigeration machine
T-S diagram	Temperature- Entropy diagram	$\overline{\varepsilon}$	average cooling efficiency of refrigeration machine

References

1. Tian, X.; Zhang, H.; Liu, L.; Huang, J.; Liu, L.; Liu, J. Establishment of LCZ-based urban building energy consumption dataset in hot and humid subtropical regions through a bottom-up method. *Appl. Energy* **2024**, *368*, 123491. [CrossRef]
2. Kang, C.; Zhou, T.; Chen, Q.; Xu, Q.; Xia, Q.; Ji, Z. Carbon emission flow in networks. *Sci. Rep.* **2012**, *2*, 479. [CrossRef] [PubMed]
3. Zheng, S.; Huang, G.; Zhou, X.; Zhu, X. Climate-change impacts on electricity demands at a metropolitan scale: A case study of Guangzhou, China. *Appl. Energy* **2020**, *261*, 114295. [CrossRef]
4. World Bank Group. CO_2 Emissions. 2023. Available online: https://data.worldbank.org/topic/climate-change (accessed on 30 September 2024).

5. UN Environment Programme. *2022 Global Status Report for Buildings and Construction: Towards a Zero-Emission, Efficient and Resilient Buildings and Construction Sector*; Springer: Berlin/Heidelberg, Germany, 2023. [CrossRef]
6. Xiaoqing, Z.; Fang, L.; Weiqing, C.; Shurong, Y.; Yunfei, D. Investigation and study on energy consumption of public buildings in Guangzhou. *Build. Sci.* **2007**, *23*, 76–80.
7. Zhen, L. *Chongqing High-Level Office Buildings Air Conditioning Load Characteristic and Energy Consumption Analysis Research*; College of Faculty of Urban Construction and Environment Enginerring Chongqing University: Chongqing, China, 2007.
8. Zhao, D.; Zhong, M.; Zhang, X.; Su, X. Energy consumption predicting model of VRV (Variable refrigerant volume) system in office buildings based on data mining. *Energy* **2016**, *102*, 660–668. [CrossRef]
9. Hu, Y.; Zhang, Y.; Liu, X.; Li, H. Development and demonstration of a method to detect refrigerant charge level for variable refrigerant volume systems. *Appl. Therm. Eng.* **2023**, *235*, 121354. [CrossRef]
10. Helbling, M.; Meierrieks, D. Global warming and urbanization. *J. Popul. Econ.* **2023**, *36*, 1187–1223. [CrossRef]
11. Liu, L.; Pan, X.; Jin, L.; Liu, L.; Liu, J. Association analysis on spatiotemporal characteristics of block-scale urban thermal environments based on a field mobile survey in Guangzhou, China. *Urban Clim.* **2022**, *42*, 101131. [CrossRef]
12. Kabeel, A.E.; Abdelgaied, M.; Sathyamurthy, R.; Arunkumar, T. Performance improvement of a hybrid air conditioning system using the indirect evaporative cooler with internal baffles as a pre-cooling unit. *Alexandria Eng. J.* **2017**, *56*, 395–403. [CrossRef]
13. Yau, Y.H.; Pean, H.L. The performance study of a split type air conditioning system in the tropics, as affected by weather. *Energy Build.* **2014**, *72*, 1–7. [CrossRef]
14. Chow, T.T.; Lin, Z.; Yang, X.Y. Placement of condensing units of split-type air-conditioners at low-rise residences. *Appl. Therm. Eng.* **2002**, *22*, 1431–1444. [CrossRef]
15. Chow, T.T.; Lin, Z.; Liu, J.P. Effect of condensing unit layout at building re-entrant on split-type air-conditioner performance. *Energy Build.* **2002**, *34*, 237–244. [CrossRef]
16. Nada, S.A.; Said, M.A. Solutions of thermal performance problems of installing AC outdoor units in buildings light wells using mechanical ventilations. *Appl. Therm. Eng.* **2018**, *131*, 295–310. [CrossRef]
17. Xue, H.; Chou, S.K.; Zhong, X.Q. Thermal environment in a confined space of high-rise building with split air conditioning system. *Build. Environ.* **2004**, *39*, 817–823. [CrossRef]
18. Cheng, Z.; Huang, Z.; Ma, Y. Influence of pitch angle of the louver leaves on outdoor unit's operating conditions of air conditioners. *HV&AC* **2009**, *39*, 133–135.
19. Yang, Y. The effect of shutter's parameter on performance of VRF air conditioning system. *Build. Energy Environ.* **2017**, *36*, 34–37.
20. Liu, W.; Xie, X.; Li, N. Experimental research on influence of louver angle on performance of air conditioner's outdoor unit. *Refrig. Air-Cond.* **2017**, *17*, 37–39.
21. Yang, Y.; Cheng, Y.; Liu, B.; He, K.; Hai, X. CFD analysis and optimization of thermal environment for outdoor unit of air conditioner at groove in a commercial building. *HV&AC* **2016**, *46*, 33–35.
22. Chow, T.T.; Lin, Z.; Wang, Q.W. Flow analysis of condenser cooling air delivery via building light well. *Appl. Therm. Eng.* **2001**, *21*, 831–843. [CrossRef]
23. Nada, S.A.; Said, M.A. Performance and energy consumptions of split type air conditioning units for different arrangements of outdoor units in confined building shafts. *Appl. Therm. Eng.* **2017**, *123*, 874–890. [CrossRef]
24. Yiqi, L. The Influence of the Operating Environment and Thermal Environment on the Performance of the Outdoor Air-Conditioner. Master's thesis, Harbin University of Commerce, Harbin, China, 2018.
25. Zhou, D.; Shao, X.; Zhang, X.; Li, X.; Shi, W.; Wang, B. Numercial simulation of thermal environment for outdoor units at building re-entrant. *J. Guangzhou Univ.* **2010**, *9*, 17–22.
26. Zhao, A.; Li, Q.; Meng, Q.; Meng, Q. Experimental Study on the Performance of Air-Cooled Condensing Unit in Groove. In Proceedings of the 2nd International Conference on Sustainable Energy and Environmental Engineering (SEEE 2016), Xiamen, China, 18–19 December 2016. [CrossRef]
27. Tao, Y.; Ying, Z.; Ying, S.; Peng, T. Simulation and optimization of thermal environment for outdoor units of air conditioner in Groove of a high-rise building. *J. Build. Energy Effic.* **2022**, *50*, 72–77.
28. Huang, S.; Cao, Y.; Li, X.; Pei, Z.; Ni, X.; Wang, S.; Zhou, J.; Wang, Q. CFD simulation and optimization of multi-split air conditioning outdoor unit scheme for a super high-rise building. *HV&AC* **2017**, *47*, 77–82.
29. Avara, A.; Daneshgar, E. Optimum placement of condensing units of split-type air-conditioners by numerical simulation. *Energy Build.* **2008**, *40*, 1268–1272. [CrossRef]
30. Xue, H.; Xu, B.; Wu, J.; Wei, Y. Prediction of temperature rise near condensing units in the confined space of a high-rise building. *Build. Environ.* **2007**, *42*, 2480–2487. [CrossRef]
31. Duan, R.; Wang, X.; Song, Y.; Liu, J. Influence of Air-conditioning Outdoor Unit Arrangement Strategy on Energy Consumption. *Procedia Eng.* **2016**, *146*, 350–358. [CrossRef]
32. Harithkhan, V.M.; Nissanka, I.D.; Manthilake, M.M.I.D. System performance of split type A/C units in high-rise residential buildings with different condenser arrangements. In Proceedings of the MERCon 2022—Moratuwa Engineering Research Conference, Moratuwa, Sri Lanka, 27–29 July 2022; pp. 1–6. [CrossRef]
33. Jiang, Y. Research on the thermal environment around outdoor unit of split air conditioner. Master's thesis, Tianjin University of Commerce, Tianjin, China, 2013.

34. Bruelisauer, M.; Meggers, F.; Saber, E.; Li, C.; Leibundgut, H. Stuck in a stack—Temperature measurements of the microclimate around split type condensing units in a high rise building in Singapore. *Energy Build.* **2014**, *71*, 28–37. [CrossRef]
35. Zhang, J. Simulation and Analysis to the Thermal Environment Around the Outdoor Units of VRV Air-Conditioner System. Master's thesis, Dalian University of Technology, Dalian, China, 2009.
36. Jing, Z.; Kun, Z.; Angui, L.; Xiaohong, N. Analysis on optimizing array of air-conditioning condensers. *J. Refrig.* **2008**, *29*, 12–17.
37. Wei, X.; Jiang, P. Simulation and analysis of thermal environment of VRF air-conditioning outdoor units with horizontal array arrangement. *J. Zhejiang Sci.-Technol. Univ.* **2021**, *45*, 149–156.
38. Chow, T.; Lin, Z.; Wang, Q. Effect of building re-entrant shape on performance of air-cooled condensing units. *Energy Build.* **2000**, *32*, 143–152. [CrossRef]
39. Choi, S.H.; Lee, K.S.; Kim, B.S. Effects of stacked condensers in a high-rise apartment building. *Energy* **2005**, *30*, 968–981. [CrossRef]
40. Li, H.; Liu, W. Study of position of outdoor units formulti-splits air-conditioning system. *Refrig. Air-Cond.* **2019**, *11*, 82–87.
41. Zi, X.; Wu, S. Optimization design on air-conditionging scheme for on high-rise building. *Refrig. Air-Cond.* **2017**, *17*, 59–64.
42. Fei, G.; Bo, L.; Ming, X. Numerical research on effects of outside unit placement on unit performance for variable refrigerant volume air conditioning systems in multi-storied buildings. *HV&AC* **2007**, *5*, 97–100.
43. Umar, T. Making future floating cities sustainable: A way forward. *Proc. Inst. Civ. Eng. Urban Des. Plan.* **2020**, *173*, 214–237. [CrossRef]
44. Tippu, J.; Saravanasankar, S.; Sankaranarayanan, B.; Ali, S.M.; Qarnain, S.S.; Karuppiah, K. Towards Sustainability: Analysis of Energy Efficiency Factors in Buildings of Smart Cities Using an Integrated Framework. *J. Inst. Eng. Ser. A* **2023**, *104*, 223–235. [CrossRef]
45. Chen, R.; Xu, W.; Deng, S.; Zhao, R.; Choi, S.Q.; Zhao, L. Towards the Carnot efficiency with a novel electrochemical heat engine based on the Carnot cycle: Thermodynamic considerations. *Energy* **2023**, *284*, 128577. [CrossRef]
46. R, T.; Hiremath, N.; Narayanaswamy, K.S. Numerical Studies of Air Flow Analysis in Compressed Air Filters. *J. Mines Met. Fuels* **2022**, *70*, 342. [CrossRef]
47. Zore, K.; Caridi, D.; Lockley, I. *Fast and Accurate Prediction of Vehicle Aerodynamics Using ANSYS Mosaic Mesh*; SAE Technical Paper; *Commercial Vehicle Engineering Congress*; October; SAE International: Warrendale, PA, USA, 2019. [CrossRef]
48. Canonsburg, T.D. ANSYS FLUENT Meshing Text Command List. *Knowl. Creat. Diffus. Util.* **2012**, 15317.
49. Boujelbene, M.; El-Zahar, E.R.; Seddek, L.F.; Ullah, Z.; Makinde, O.D. Viscous dissipation and variable viscosity impacts on oscillatory heat and mass transfer of gravity-driven reactive flow along heated plate. *Phys. Fluids* **2023**, *35*, 073604. [CrossRef]
50. Guo, T. Study on Layout of VRF Air-Conditioning Outdoor Units Based on CFD Simulation. Master's thesis, Xihua University, Chengdu, China, 2021.
51. Alawadhi, E.M. Effect of an incompletely closed window shutter on indoor illuminance level and heat gain. *Energy Build.* **2016**, *110*, 112–119. [CrossRef]
52. Bueno, B.; Pigeon, G.; Norford, L.K.; Zibouche, K.; Marchadier, C. Development and evaluation of a building energy model integrated in the TEB scheme. *Geosci. Model Dev.* **2012**, *5*, 433–448. [CrossRef]
53. Zheng, Z.; Xiao, J.; Yang, Y.; Xu, F.; Zhou, J.; Liu, H. Optimization of exterior wall insulation in office buildings based on wall orientation: Economic, energy and carbon saving potential in China. *Energy* **2024**, *290*, 130300. [CrossRef]
54. Liu, S.; Du, Y.; Zhang, R.; He, H.; Pan, A.; Ho, T.C.; Zhu, Y.; Li, Y.; Yip, H.L.; Jen, A.K.Y.; et al. Perovskite Smart Windows: The Light Manipulator in Energy-Efficient Buildings. *Adv. Mater.* **2024**, *36*, 2306423. [CrossRef]

buildings

MDPI

Article

Experimental Investigation on Cooling Performance of Water Spray Window

Xiu Yang [1,2], Jiang He [1,2,*], Yuanyao Zhang [2], Guanghong Huang [2], Junli Wang [2] and Linghui Liao [2]

1 School of Architecture, South China University of Technology, Guangzhou 510640, China; yangxiu@gxu.edu.cn
2 College of Civil and Architectural Engineering, Guangxi University, Nanning 530004, China
* Correspondence: 20100081@gxu.edu.cn; Tel.: +86-130-8797-5659

Abstract: The cooling performance of the energy-saving water spray window cooling system under summer conditions in Nanning city in southern China, is experimentally examined in this study. By constructing two identical test rooms for comparison, in the experiment we analyzed the effects of water temperature (22 °C, 26 °C, 30 °C), spray frequency (8, 10, 12, 15 min intervals), glass structure (single-layer, double-layer, triple-layer), air conditioning temperature (26 °C, 27 °C, 28 °C, 29 °C), and outer lamination thickness (30, 50, 100, 200 mm) on the cooling effect of windows, aiming to determine the optimal operating conditions. The experimental outcomes demonstrate that the appropriate operating mode for the water spray window system involves using the coldest water source (22 °C), with a spray interval of 8 min, a three-layer air cavity sprayed glass construction, an air cavity thickness set at 100 mm, and the best air conditioning temperature (26 °C). The study found that the water spray system can reduce the outer glass surface temperature of the window by 6 °C and the inner surface temperature by 2 °C. Moreover, the more glass layers and the thicker the lamination, the higher the energy saving rate; with a maximum energy saving rate of 35.19%. The water spray window has good adaptability and significant energy-saving effects in Southeast Asia. By scientifically selecting energy-saving glass types and fine-tuning operating modes, it is expected that efficient building energy conservation in hot climates can be achieved.

Keywords: building energy saving; cooling performance; spray evaporative cooling; surface temperature; energy saving rate

1. Introduction

With the growing severity of climate warming, the importance of effective indoor temperature control has become increasingly prominent. The windows of building facades were widely regarded as a core element affecting building energy conservation. In the southern China and ASEAN regions, where summers are long and characterized by high temperatures and humidity, the air trapped within window lamination was heated by solar radiation and the glass, becoming a potential factor for indoor temperature increase [1,2]. Therefore, effectively reducing the glass and lamination temperature has significant implications for building energy efficiency, especially in the context of deepened global carbon emission policy reforms [3,4]. Optimizing the cooling performance of the windows is a crucial step towards achieving this goal.

In recent years, with the rise of environmental awareness, green energy-saving glass cooling technology has garnered significant attention due to its substantial energy-saving benefits [5]. Especially in the field of window technology, innovative techniques such as phase-change material cooling technology [6,7], heat-reflective film technology [8,9], water spray evaporation technology [10,11], and high-pressure mist spray [12–14] have become the focus of current research. Among these techniques, heat-reflective film technology reduces heat absorption by reflecting solar radiation. However, although it effectively reduces heat conduction, it cannot lower the temperature of the glass itself, and may lead to light pollution problems. Phase-change material (PCM) cooling technology, on the other hand, relies on the heat absorption and release characteristics of PCM at specific temperature ranges, effectively regulating the surface temperature of the glass. Nevertheless, this technology still shows limitations in rapid cooling. Water spray evaporation technology utilizes the principle of water evaporation to reduce the temperature of the glass. This technology has gained attention due to its simple working principle and low maintenance cost; however, its evaporation efficiency is low and it consumes a significant amount of water, limiting its application scenarios. In contrast, high-pressure mist spray can produce fine water droplets, which significantly enhance evaporation efficiency and reduce water consumption, effectively lowering the temperature of the glass and its surrounding environment. This improvement also leads to a notable decrease in air conditioning cooling load, as shown in Table 1. Therefore, the environmental cooling characteristics of high-pressure mist spray have demonstrated superiority in practical applications [15].

Table 1. Comparing common green energy-saving window cooling methods.

Cooling Method	Working Principle and Advantages	Disadvantages
Thermal reflection film technology	Reduce heat absorption by reflecting the sunlight, improve indoor comfort, significant energy saving effect.	Leads to light pollution, is not friendly to the outdoor environment, and cannot solve the problem of high temperature.
Phase-change ma-terial cooling technology	The phase change material can absorb and release a large amount of heat to reducing the glass temperature, pollution-free, stable cooling.	High cost, technology maturity needs to be improved, and application scenarios are limited, which cannot quickly cool down.
Water spray and evaporation technology	Using the principle of water evaporation and heat absorption to reduce the glass temperature, its working principle is simple, low-maintenance cost.	Low efficiency of evaporation, high water consumption, prone to mold growth.
High-pressure spray evaporative cooling technique	Through the evaporation process of fine water mist, effectively absorb and take away the heat from the glass surface, so as to achieve rapid cooling.	High evaporation efficiency, large water consumption, and certain requirements for water quality.

Abdulrazak Qahtani et al. have investigated the cooling performance of the exterior surfaces of the glass curtain wall in Malaysia [16,17], and have found that the formed fluid water films could effectively lower the surface temperature by 2 to 4 °C, thereby decreasing the indoor temperature by 2 to 4 °C. However, the influence of external wind on the glass surface was significant, leading to unstable effects and a high consumption of water, thus limiting its application. Hence, the application of inner spray in the air lamination has become a consensus. Ye et al. attempted to introduce spray measures in the glass layer [18], testing the application effects of spraying from different directions on the window layer. The results showed that moderate spraying could reduce window temperature and reduce air conditioning usage time. Contrary to the downward ventilation in the wind tower mist cooling system [19,20], the sun evaporated the water vapor in the glass laminate, which then ventilated upwards and carried away the heat from the laminate. The size of the droplet was a crucial parameter; reducing the size of the droplet could enhance the cooling efficiency. The water droplets produced by high-pressure spray were typically smaller, ranging from 10 to 50 μm in size. The temperature of the mist was also lower than the temperature of the water [21]. Researchers have tested the thermal performance of high-pressure mist cooling glass roof lights, revealing that the temperature of the surface of the glass was 8.0 °C lower than the ambient temperature, while the sunlight transmittance decreased by 14% [22].

In the southern China and southeast Asian regions, abundant rainfall and high humidity characterizes the summer months [23], providing the local population with a wealth of low-temperature renewable water resources primarily consisting of rainwater and air conditioner condensation water [24,25]. Firstly, abundant rainfall was one of the primary sources of renewable water. The construction of sponge cities could store a large amount of rainfall in underground reservoirs, which are currently used for irrigation, road washing, and other purposes [26,27] and have not been extensively utilized. By constructing an effective collection and storage system complemented with appropriate filtration and disinfection treatment, these rainwater resources could be utilized for urban mist cooling. When sprayed into the air, water vapor absorbed heat, effectively lowering the temperature of the surrounding air. Secondly, high humidity conditions in the air environment were favorable for the production of air conditioning condensate water. In daily life and work environments, frequent use of air conditioners generated a large amount of condensate water which, after treatment, could meet the quality standards for drinking water [28] and was a green, environmentally friendly, and high-quality source of low-temperature water.

In this context, this study proposes a high-pressure mist spray cooling system specifically designed for building windows, which aims to significantly reduce the temperature of the exterior window glass during the summer. This system utilizes the evaporation process of mist droplets to absorb heat from the air, coupled with the formation of running water on the glass surface that efficiently removes accumulated heat, effectively reducing the heat gain on the exterior and air lamination of the window. This technology not only substantially enhances the thermal performance of exterior windows but also reduces the cooling energy consumption in buildings, and has a positive impact on mitigating urban heat islands [29–31]. Currently, there is a noticeable lack of experimental research that verifies the cooling performance and energy-saving benefits of the high-pressure mist spray cooling system during the summer in this region. Hence, the adaptability of this system requires thorough evaluation.

This study undertakes a detailed analysis of the thermal performance of the cooling system and probes into its cooling mechanism, with the ultimate goal of optimizing the cooling effect. The research outcomes are expected to have considerable practical application value. The study leverages an outdoor test room in Nanning, located in the Guangxi region of China, as a case study. It compares various window forms, encompassing single-layer, double-layer, and triple-layer glass; employing a quantitative analysis method. Through rigorous evaluation of variables such as water temperature, spray frequency, window structure, air conditioning temperature, and air layer thickness, this study elucidates the optimal operation mode for water spray windows.

2. Materials and Methods

2.1. Experimental Setup

To evaluate the cooling and energy-saving effects of water spray windows, an experiment was conducted at the West Campus Experimental Base of Guangxi University. This area experiences a summer season from May to October, with a duration of over half a year [32], characterized by the hot and humid climates typical of southern China and southeast Asia. The experiment was conducted in the outdoor parking lot of the experimental base, with the northern side facing the structural test hall, and the southern side located nearby the campus road. The site, where two identical test rooms (2 m long × 2 m wide × 2 m high) were constructed, was relatively open. To ensure the accuracy of the test data, the distance between the two test rooms was set at 6 m, avoiding any overlapping effects between them. The walls and roofs of both rooms were constructed with metal sandwich panels, with 60 mm thick polystyrene foam in the middle. This significantly enhanced the thermal and sound insulation performance. The flooring was made of wooden material and, in order to prevent water ingress into the room, there was a 15 mm high airspace between the floor and the ground. The western room (R1), designated as the experimental group, was equipped with a water spray cooling system. Meanwhile, the eastern room (R2) was set up as the contrast group, with standard glass windows installed. All windows were placed on the south-facing walls, and were 1.2 m high and 1.0 m wide. Furthermore, a small meteorological station was set up on the eastern side of the test room, to monitor and record real-time meteorological data in the experimental area. Both rooms were equipped with a 2600 W air conditioning system, aiming to test the energy consumption of air conditioners under different configurations; as shown in Figure 1.

Figure 1. Appearance of the test room. (R1 represents the test room, R2 represents the contrast room).

2.2. Spray System

The water spray window was composed of an ordinary glass window and spray devices. The water spray device included an atomizer, water pipe, atomization spray head, water tank, etc. It was installed outside the window or lamination to reduce the temperature of the glass surface. The water storage bucket was equipped with an automatic water replenishment system, without manual operation. The main source of water was municipal feed water, and air conditioning condensate was collected into the water storage bucket. The sprayer had the function of intelligently setting the spray time and frequency, and was connected to the timing socket to automatically start and close the system. The parameters of the pump and nozzle are listed in Table 2. After the system was started, the high-pressure sprayer delivered the water from the bucket to the atomized nozzle. The high-pressure water was atomized into fine water mist through the special nozzle, sprayed onto the glass surface, and evaporated or formed a water film downward, taking away the heat from the glass surface; as shown in Figure 2. The ejected water droplets descended from the bottom of the window. The water flowing down the glass was collected, filtered, and transported to the storage bucket, where it could be recycled. To compare the cooling efficiency of various structural glasses, single-layer, double-layer, and triple-layer glass windows were selected for experimental testing. The single-layer glass had a thickness of 6 mm. The double-layer glass window consisted of two 6 mm glass panes with a 12 mm laminate layer in between. The triple-layer glass window included two 12 mm laminate layers with three 6 mm glass panes. Moreover, to compare the cooling effect of the inner layer, the outer layer was also modified, including a sprayer and a water pipe, with a thickness of 30 mm, 50 mm, 100 mm, and 200 mm; as shown in Figure 3.

Table 2. Main components of the spray system and their parameters.

Components	Model	Specifications
High-pressure pump	FOG MACHINE1.2 L	Power: 750 W
Spray nozzles	hongbaoshi01	Working pressure: >3 MP Droplet particle size: 10–30 μm

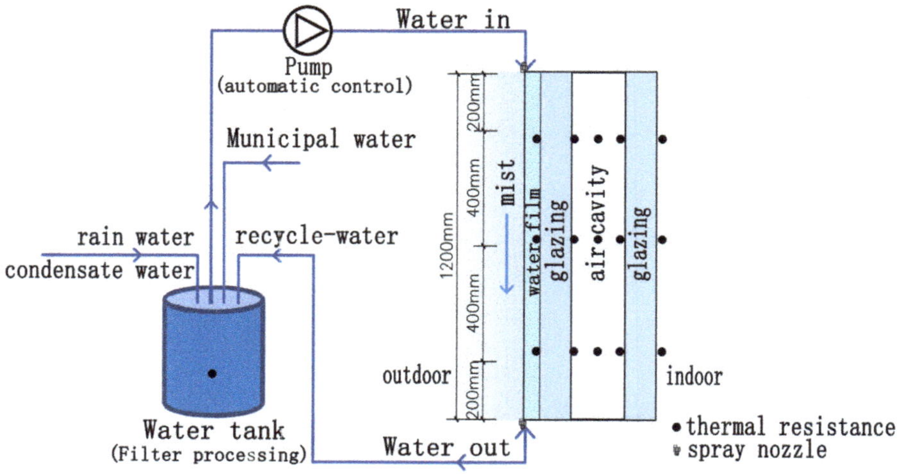

Figure 2. Flow path diagram of the water spray window.

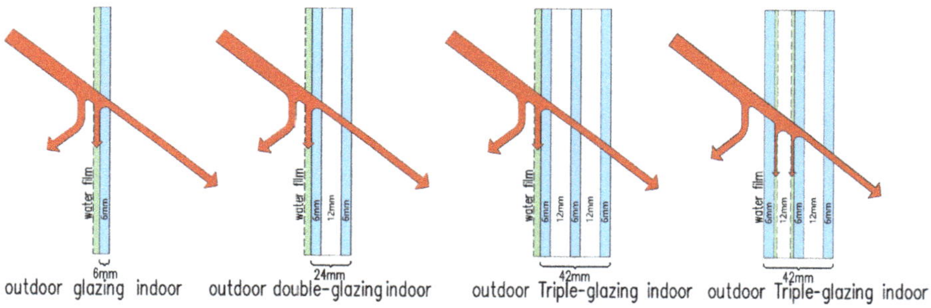

Figure 3. Schematic diagram of the glass structure.

From the upper and lower nozzles, the water vapor released had smaller particles and a lower temperature. A large amount of mist obstructed solar radiation from entering the indoor space, with droplet diameters ranging from 50 to 100 microns. The contact of water vapor with the glass surface and its evaporation absorbed a substantial amount of heat, thereby reducing the temperature of the laminate. The water droplets sprayed onto the glass formed a downward-flowing water film that absorbed heat from the glass surface. Multiple factors contributed to the reduction in window temperature, improving the indoor and outdoor environment. This reduction in temperature led to a decrease in the cooling load required by air conditioning systems, which in turn ultimately resulted in lowered energy consumption within the indoor space, as shown in Figure 4a.

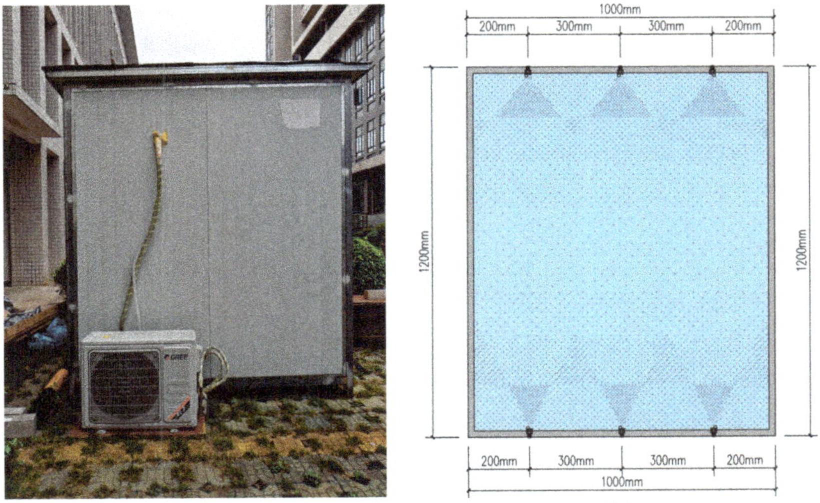

(a) (b)

Figure 4. (**a**) Layout of air conditioning; (**b**) layout of spray nozzles.

2.3. Data Measurement and Acquisition

The experiment utilized a T-type thermocouple to measure the temperature of the water body and glass surface. The measurement range of the thermocouple was between −50 °C and 200 °C, with a measurement accuracy of ±0.01 °C, as illustrated in Table 3. The layout of the nozzle is shown in Figure 4b. Since the temperature sensor had adhered to the glass surface and was directly exposed to solar radiation, its measurement accuracy may be affected. To reduce the interference of solar radiation on temperature measurement, silicon grease was applied to the sensor, which was then covered with reflective aluminum foil. All sensors were calibrated uniformly before the experiment began. Additionally, all critical parameters were recorded by data recorders, which stored data via a laptop connection or USB. The experiments were conducted between July and October of 2021 and 2023, with the climate conditions throughout the testing period remaining mostly clear. The daily solar radiation typically exceeded 20,000 W/m^2, with the climate being hot and humid. The average environmental temperature exceeded 30 °C. Under these conditions, the municipal water supply system provided water at a temperature of approximately 30 °C, with the temperature of the condensation water being maintained within the range of 16 to 20 °C. The insulation performance of the water supply pipeline was well insulated.

Table 3. Experimental instrument and specifications.

Instruments	Model	Measuring Range	Accuracy
Temperature sensors	Thermocouple	−50–200 °C	±0.01 °C
Date logger	THMA	-	±0.01 °C

2.4. Thermal Performance Analysis

Upon application of a spray on the exterior surface of the glass, a layer of flowing water film was formed. When solar radiation struck this surface, part of the radiation was reflected, while the remaining portion penetrated through the glass into the interior. The remaining radiation was absorbed by the window glass and the flowing water layer. Under these conditions, the energy–water exchange equilibrium is shown in Figure 5. Furthermore, the energy exchange equation between the water film and the surface is given as [11]:

$$\alpha_c(T_a - T_w) + \varepsilon\left(R_L - \sigma T_w^4\right) - \alpha_w(T_w - T_s) - l\frac{\alpha_c}{C_m}(X_s - X_a) = C_w W_q \frac{\partial T_w}{\partial z} \quad (1)$$

The energy balance equation for the water film and the flowing water layer is as follows:

$$aR_s + \alpha_w(T_w - T_s) = -\lambda\frac{\partial T}{\partial \chi} \quad (2)$$

where a—solar absorptance;

α_c is considered to be a function of wind speed and is given by the Jurges equation. α_w is determined as follows:

$$\alpha_w = \frac{N_u \lambda_w}{Z} \quad (3)$$

where N_u is computed from the following equation (Johnson–Rubesin equation):

$$N_u = 0.0296 R_e^{0.8} P_r^{0.6} \quad (4)$$

R_e is determined as $R_e = U_w \frac{z}{v}$.

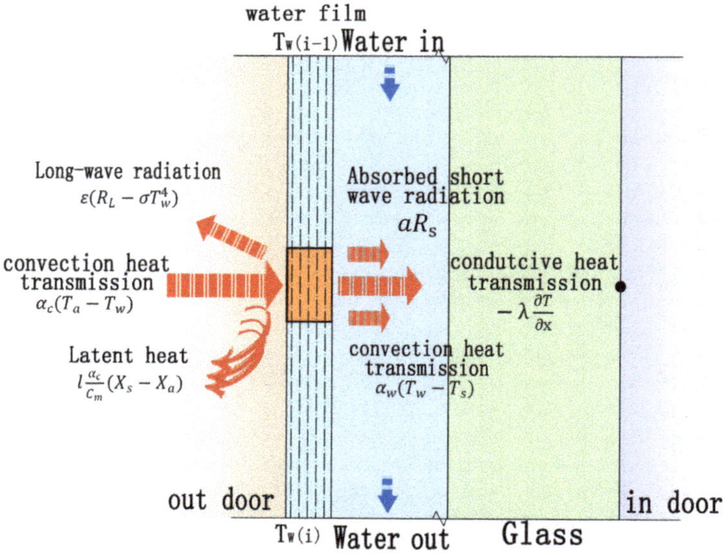

Figure 5. Energy flow paths on the surface with water film.

Upon ensuring that the specifications of the comparison room and the test room are consistent in terms of appearance, size, envelope structure, functional layout, setting parameters for air conditioning cooling equipment, meteorological parameters, etc., statistical data on the air conditioning system energy consumption of the comparison room and the test room over the same period were collected and calculated. The energy conservation rate was calculated based on the difference in air conditioning energy consumption between the comparison room and the test room divided by the air conditioning energy consumption of the comparison room. The current experiments primarily focus on assessing the energy-saving effects of spray cooling technology on glass windows, and the energy consumption of spray equipment such as water pumps has not yet been considered. Relevant research will be conducted in subsequent studies. The formula for calculating the energy conservation rate is as follows:

$$\eta = \frac{Q_c - Q_{wfw}}{Q_c} \times 100\% \tag{5}$$

In which Q_c represents the comparison of the electricity consumption of the air conditioner, Q_{wfw} represents the electricity consumption of the test air conditioner, and η represents the relative energy efficiency between the two rooms.

3. Results and Discussion

3.1. Cooling Effects of Water Temperatures

To investigate the impact of different water temperatures on the surface temperature of single-layer glass, a series of spray cooling experiments were conducted from 10–17 October 2021. During the experiments, both external windows were equipped with single-layer glass, and their water temperatures were set to 22 °C, 26 °C, and 30 °C, respectively, and the air conditioning temperature was set to 26 °C. The total duration of temperature changes was recorded for 60 min, with the spray duration lasting for 10 min. During the test period, both glass panes were exposed to direct sunlight, with solar radiation values exceeding 600 w/m^2. The wind speed was less than 0.5 m/s, and the temperature of the glass exterior surface exceeded 35 °C. The corresponding experimental results are presented in Figure 6.

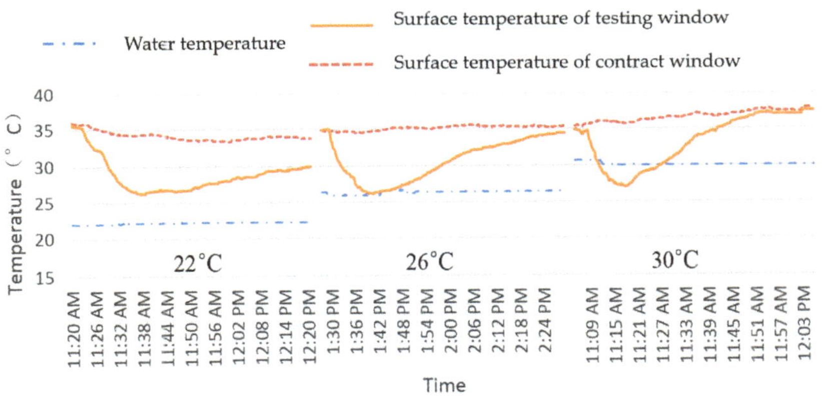

Figure 6. The temperature of the glass surface under different temperature water spray conditions.

The first set of experiments conducted on 10 October 2021, at 11:20 a.m. used 22 °C water as the spray medium to treat the surface of the test glass. Observation records indicated that the surface temperature of the test glass decreased from 35 °C to 26.2 °C after 10 min of spray treatment, with a cooling rate of 8.8 °C. After 60 min, the surface temperature of the test glass was lower by 4 °C compared to the control glass. In another experiment group, 26 °C water was used as the spray medium, and spray treatment began at 1:30 p.m. on 17 October 2021. Results showed that the surface temperature of the test glass decreased from 35 °C to 26.4 °C after 10 min of spray treatment, with a cooling rate of 8.6 °C. After 60 min, the surface temperature of the test glass was the same as that of the control glass. In the third set of experiments, 30 °C water was used as the spray medium, and spray treatment began at 11:07 a.m. on 17 October 2021. After 10 min of spray treatment, the surface temperature of the test glass decreased from 35 °C to 27 °C, with a cooling rate of 8 °C. After approximately 40 min, the surface temperature of the treated glass increased to match that of the control glass.

The experimental results indicate that the cooling effect of spray on the glass surface was practically equivalent under three different water temperatures. However, the duration of cooling was positively correlated with water temperature. Specifically, a longer cooling duration could be achieved using water at a lower temperature, such as 22 °C, exceeding 60 min. In contrast, the cooling duration would be shortened to around 40 min using water at 30 °C. Considering this, prioritizing lower-temperature water for spray evaporation enhanced cooling efficiency.

3.2. Cooling Effect at Different Periods of a Typical Day

To investigate the specific impacts of water mist spraying on the outdoor surface temperature of windows at different times of the day, the period from 6:00 a.m. on 11 August to 6:00 a.m. on 12 August (a typical day in summer) was chosen to conduct a water mist spraying cooling experiment on the outdoor surface of single-pane glass windows. During the experimental process, the air conditioning temperature was set to 26 °C. The operation of the system was divided into seven selected time periods; specifically, from 9:00 a.m. to 11:00 a.m., 12:00 p.m. to 2:00 p.m., 3:00 p.m. to 5:00 p.m., 6:00 p.m. to 8:00 p.m., 9:00 p.m. to 11:00 p.m., 12:00 a.m. to 2:00 a.m., and 3:00 a.m. to 5:00 a.m. Each time period represents a single 2 min spray and a 12 min interval between sprays. Throughout this series of experiments, detailed records were kept of the meteorological data on the day of testing, the temperature of the water supply, and the temperature changes on the outdoor surface of the windows. These data are shown in Figure 7.

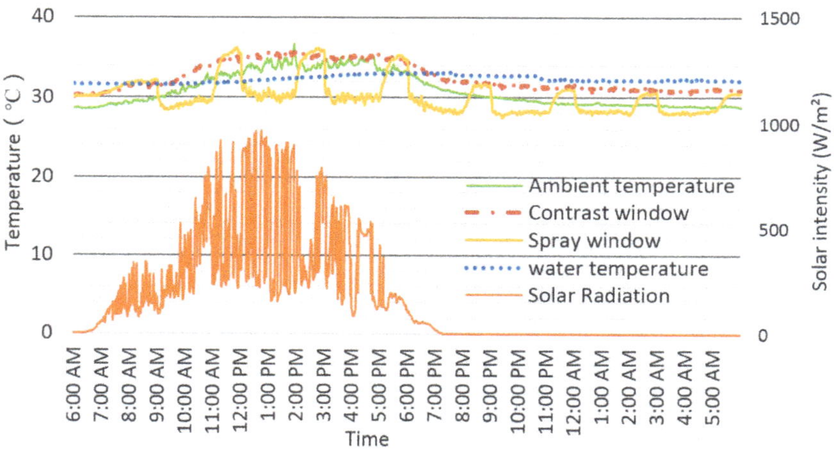

Figure 7. Glass surface temperature at different time periods on the test day.

The highest temperature of the day occurred between 12:00 p.m. and 17:00 p.m., where the cooling effect of glass surfaces was most significant, with up to a 6 °C decrease. The average temperature of the cooled glass was maintained at 30.5 °C. In the morning, from 9:00 a.m. to 11:00 a.m., and in the afternoon, from 18:00 p.m. to 20:00 p.m., the cooling effect of glass surfaces was approximately 3.5 °C, with an average temperature of 29 °C. During these two time periods, the fluctuations in solar radiation intensity were significant, resulting in a temperature change exceeding 2 °C. This suggested a positive correlation between the cooling effect of glass surfaces and the intensity of solar radiation. At night, from 21:00 p.m. to 5:00 a.m., although there was no influence from solar radiation, the temperature of the glass surface remained above 30 °C due to factors such as the temperature of the air layer and other influencing factors. Windows treated with fogging had a cooling effect of approximately 3 °C, with the temperature of the glass surface maintained at around 28 °C. This temperature—lower than the outdoor temperature—fell within the human comfort range, making it beneficial to reduce the heat transfer efficiency of windows and the cooling demand of the air conditioning system; thereby reducing energy consumption, as illustrated in Table 4.

Table 4. Glass surface temperature reduction and average temperature in each time period.

Time Period	6:00~8:00	9:00~11:00	12:00~14:00	15:00~17:00	18:00~20:00	21:00~23:00	0:00~2:00	3:00~5:00
Highest drop (°C)	2.4	2.8	3.5	3.8	3.5	3.2	3.1	3.1
Average drop (°C)	1.9	1.7	1.9	1.7	1.9	2.0	2.3	2.3
Temperature after cooling (°C)	28.5	31.3	33.2	34.9	32.8	30	29	28.4

The results indicated that during the three hottest periods of the day, the evaporation cooling effect did not persist once the spraying ended. The glass rapidly absorbed heat from the surrounding environment and solar radiation, causing its temperature to rise rapidly. During a certain period, the temperature of the glass even surpassed that of the unsprayed window surface. Therefore, to maintain the efficiency of spray cooling, a reasonable spraying interval should be set to avoid the negative effects caused by the sudden rise in temperature. Moreover, although the cooling effects of spraying at night may not be significant, it helps maintain the glass surface temperature close to 28 °C; thereby reducing the cooling demand on the air conditioning system. Particularly in areas with high nighttime temperatures, spraying at night significantly improves the indoor thermal environment of buildings.

3.3. Cooling Effect of Spray Frequency

From 23–26 July 2021, a consecutive four-day experimental study was conducted to investigate the cooling effects under various spray intervals. The focus was on a single-layer glass window, with each spray session lasting precisely 2 min and intervals strictly controlled at 8, 10, 12, and 15 min. This methodology ensured accurate and reliable data collection. The testing encompassed a full 24 h period, starting at 6:00 a.m. on the first day and ending at 6:00 a.m. on the second day. During the experimental process, the air conditioning temperature was set to 26 °C.

To gain a deeper understanding of the cooling effects throughout different times of the day, eight specific time slots were selected for detailed spray cooling tests: 6:00–8:00 a.m., 9:00–11:00 a.m., 12:00–2:00 p.m., 3:00–5:00 p.m., 6:00–8:00 p.m., 9:00–11:00 p.m., midnight-2:00 a.m., and 3:00–5:00 a.m. Additionally, meteorological data for these testing days were carefully documented, providing a comprehensive overview of the environmental conditions during the study. Refer to Figure 8 for detailed meteorological information.

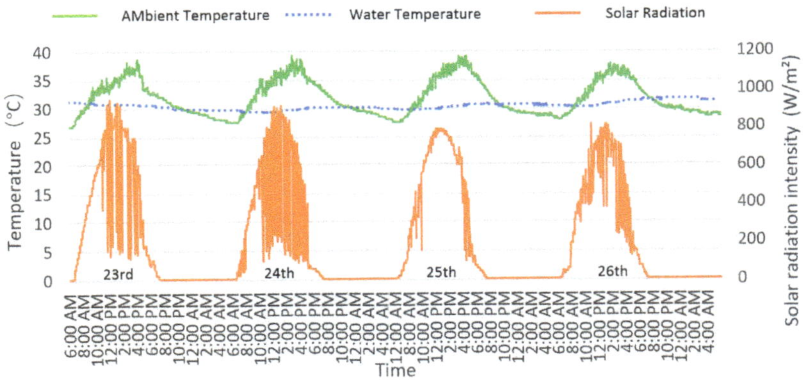

Figure 8. Meteorological data and water temperature from 23–26 July 2021.

Comparative analysis revealed a positive correlation between spray interval and cooling efficacy. Specifically, shortening the interval to 8 min yielded a maximum 4 °C cooling on the window surface, with more stable temperature fluctuation. In contrast, the cooling effect at 10- and 12-min intervals was around 3 °C, while 15-min intervals showed a limited effect, especially during the daytime, when surface temperature rose rapidly after spraying. Nighttime cooling effect was only about 1 °C, with limited efficacy. See Figure 9 for a detailed comparison.

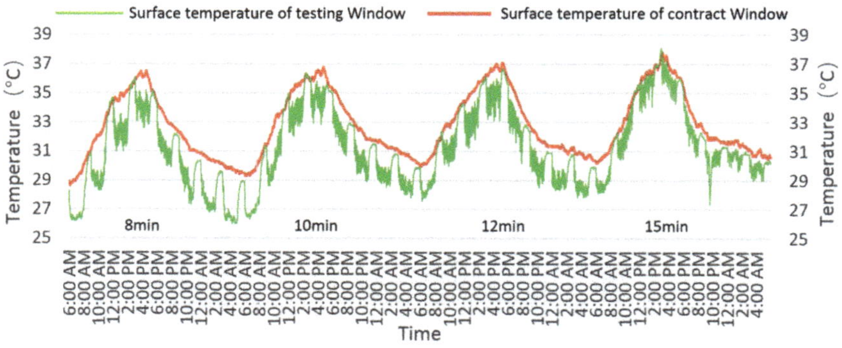

Figure 9. Comparative analysis of window surface temperatures over four days.

Based on the analysis of experimental data, it was observed that, although reducing the spray interval enhanced cooling, the efficiency increase was not significant. Specifically, at 6 min intervals, numerous water droplets frequently persisted on the glass surface. This observation was a primary factor in determining the decision to set the spray interval start point at 8 min. Taking into account both cooling effectiveness and water conservation, an interval of 12 min appeared to be more suitable. Additionally, the experiment revealed that nighttime spraying exhibited significantly better cooling compared to daytime spraying, further validating the effectiveness of night spraying for optimal cooling. Detailed results can be found in Figure 9.

3.4. Cooling Effect of Glass Structure

To quantitatively evaluate the impact of water spray on the surface temperature of glass windows in buildings, three days in September 2022 were chosen for this study: the 12th, 20th, and 29th. The experiment was conducted on single-layer, double-layer, and triple-layer glass windows, comparing the temperature changes on the glass surfaces of windows with different layers. The experimental system ran continuously from 8:00 a.m. on the first day until 6:00 a.m. the following day, with a spray applied every 12 min (each spray lasting 2 min) to simulate actual operation conditions. The relevant meteorological data and water temperature are shown in Figure 10.

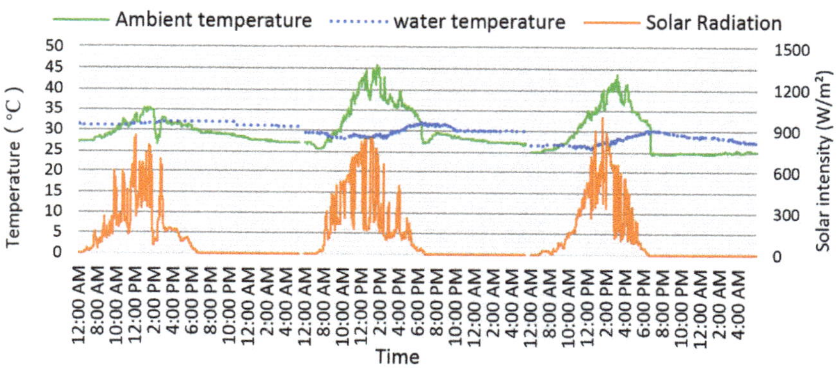

Figure 10. Meteorological data and water temperature over three days in September 2022.

During the three-day experimental cycle, the single-layer glass exhibited the most noticeable cooling amplitude, thereby making it highly susceptible to external temperature fluctuations, resulting in significant temperature variations. On the other hand, the double-layer glass demonstrated a comparatively stable cooling trend accompanied by fewer temperature fluctuations. Moreover, it exhibited significantly better cooling performance at night in comparison to during the day. However, because of its low thermal conductivity, the triple-layer glass offered excellent insulation. The introduction of two air layers served as a thermal buffer, causing a slight reduction in the cooling effects at night when compared to daytime. It is important to mention that, apart from a significant temperature decrease during the 12 to 18 h period, there were minimal temperature differences between the triple-layer glass and the interior window surface during other times; as clearly depicted in Figure 11.

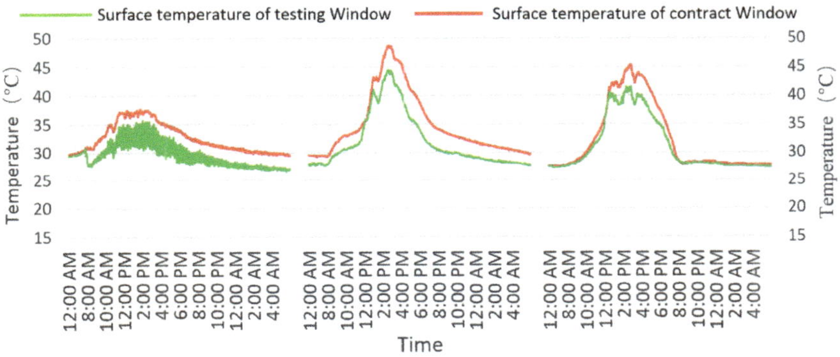

Figure 11. Comparative analysis of window surface temperatures for different window types over three days.

Experimental verification has confirmed that the temperature on a window's inner surface was significantly influenced by the thermal transmittance coefficient. When this coefficient was low, the impacts of spray cooling on the indoor environment were comparatively minor. Among different types of glass, single-layer glass exhibited the most pronounced cooling effect; with double-layer glass following; and triple-layer glass showing the least effect. This was primarily determined by their respective thermal insulation properties. It is worth noting that double-layer glass demonstrated a comparatively stable cooling effect alongside considerable energy-saving potential. Considering that most newly constructed buildings in southern China are equipped with double-layer windows, spray evaporation cooling holds promising potential for application in that region. Furthermore, triple-layer glass demonstrated remarkable cooling abilities even under high-temperature spray conditions.

3.5. *Effect of Air Conditioning Setting Temperature on Energy-Saving Rate*

To investigate the potential impact of the water spray flow on the energy consumption of air conditioning systems, comprehensive experiments were conducted in October 2022. In these experiments, three different types of windows were chosen for testing: single-layer, double-layer, and triple-layer glass windows. Additionally, two air conditioning devices were installed in two test room, with temperature settings varying between 26 °C, 27 °C, 28 °C, and 29 °C. The operational timings were maintained consistently with previous tests.

Upon analyzing the experimental results, it became evident that, as the set temperature increased, the relative energy efficiency of the test room also rose proportionately. Specifically, the energy efficiency enhancement achieved through spray cooling on single-layer glass windows increased significantly from 18.16% to 22.89%. This finding underscores the high susceptibility of single-layer glass windows to external temperature fluctuations and highlights the effectiveness of spray technology in reducing the exterior surface temperature, thereby minimizing heat infiltration into the indoor environment.

There was also a moderate improvement in energy efficiency observed in spray cooling for double-layer glass windows, which rose from 6.23% to 13.19%. Although double-layer glass windows exhibited lower energy consumption compared to single-layer ones, the enhancement in energy efficiency was relatively modest, indicating a limited role of spray technology in boosting the efficiency of these windows.

A substantial increase in energy efficiency was noticed in the test room with triple-layer glass windows when spray technology was applied, jumping from 20.03% to 26.58%. Contrary to previous experimental data that suggested a minimal effect of spray action on glass surface temperatures, the spray's exceptional thermal insulation properties significantly contributed to reducing indoor heat, thereby increasing the proportion of heat dissipated through the water spray. This remarkable impact on energy efficiency underscores the value of employing spray technology, particularly for triple-layer glass windows. A detailed analysis of these performance indicators is presented in Table 5.

Table 5. Daily energy-saving rates of three different types of windows at four air conditioning temperatures.

Air Conditioning Temperature (°C)	Single-Layer Window (%)	Double-Layer Window (%)	Triple-Layer Window (%)
26	18.16	6.23	20.03
27	19.69	8.24	24.45
28	19.96	12.13	26.51
29	22.89	13.19	26.58
Average Value	20.17	9.95	24.40

When comparing energy efficiency data between nighttime and the entire day, significant disparities were noted. Precisely, the single-layer window demonstrated a 4% boost in energy efficiency during the night relative to daytime, whereas the double-layer window revealed a 1.85% enhancement during the night in comparison to daytime. In contrast, the triple-layer window experienced a 9.58% drop in energy efficiency during the night when juxtaposed with daytime, as illustrated in Table 6.

Table 6. Night energy-saving rates of three different types of windows at four air conditioning temperatures.

Air Conditioning Temperature (°C)	Single-Layer Window (%)	Double-Layer Window (%)	Triple-Layer Window (%)
26	20.60	10.08	18.98
27	21.30	10.30	17.16
28	26.79	12.54	12.54
29	27.91	14.31	10.61
Average Value	24.15 (rise by 4%)	11.80 (rise by 1.85%)	14.82 (drop by 9.58%)

The augmented energy efficiency of single-layer and double-layer windows during nighttime hints that the cooling impact of the foggy water spray is more pronounced at night rather than during the day, thereby affirming their nighttime cooling capabilities. On the other hand, the diminished energy efficiency of the triple-layer window during the night mirrors the inherently low energy consumption of the test room during this period, thereby restricting the cooling effect of the foggy water spray. This decline in energy efficiency suggests that triple-layer windows predominantly aid in energy conservation during the sweltering daytime hours.

3.6. Effect of Spray on the Cooling and Energy Consumption of Different Lamination Thickness

To evaluate the impact of aerosol cooling on the surface temperatures of glass layers with varying thicknesses and spaces between them, a series of aerosol cooling experiments were conducted on specified dates in 29 September and 1, 4, and 6 October 2023. The samples used in these experiments were triple-glazed windows, with the outer glass layers varying in thickness: 30 mm, 50 mm, 100 mm, and 200 mm, while the inner glass layer remained a constant 12 mm thick. The air spaces between the glass layers varied accordingly. The spray cooling system was activated at the same time as in previous experiments. Relevant meteorological data and experimental results are presented in Figure 12.

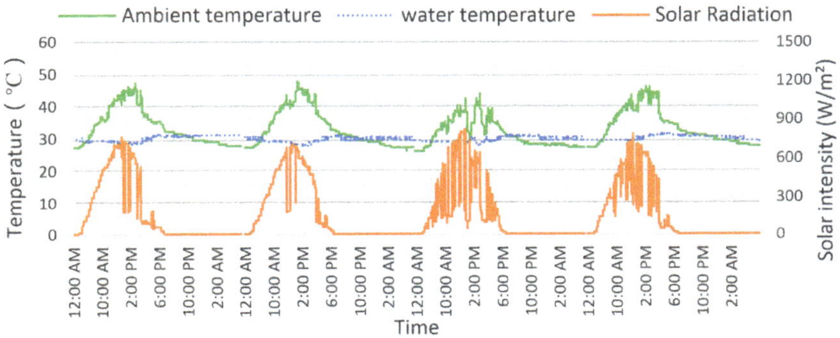

Figure 12. Meteorological data and water temperature over four days during the testing period.

The experimental findings revealed that, as the thickness of the outer glass layer increased from 30 mm to 50 mm, there was a notable decrease in the surface temperature of the sprayed window. However, no substantial difference in cooling effect was observed when the outer glass thickness was increased to 100 mm compared to the 50 mm thickness. It is worth mentioning that the lowest cooling efficiency was achieved with a 200 mm thick outer glass layer. Upon analyzing the average cooling range data, it became apparent that the sample with a 50 mm thick outer glass layer and its corresponding air space exhibited superior cooling performance compared to the thicker glass layers and their respective spaces. Surprisingly, the cooling effects were very similar for both the 100 mm and 200 mm glass thicknesses and their associated spaces. As depicted in Figure 13.

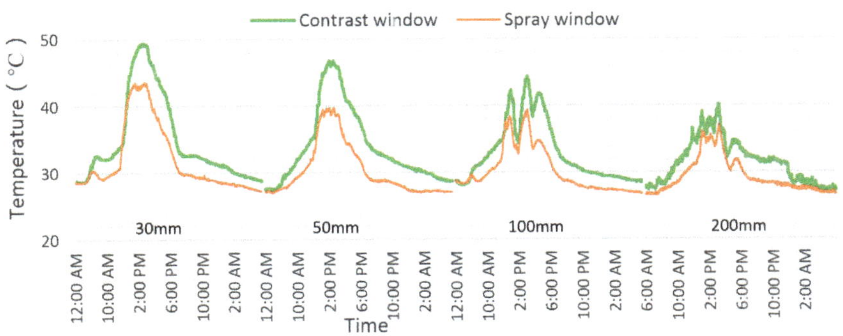

Figure 13. Comparison of inner surface temperatures between contrast window and test window.

To evaluate the influence of different thicknesses of air layers on air conditioning energy consumption under spray conditions, experiments were conducted on 30 September, 2 October, 5 October, and 7 October 2022. The air conditioner temperature in both test rooms was set at 27 °C, and the system opening time was kept consistent with previous experiments. The relevant meteorological data and water temperature are shown in Figure 14.

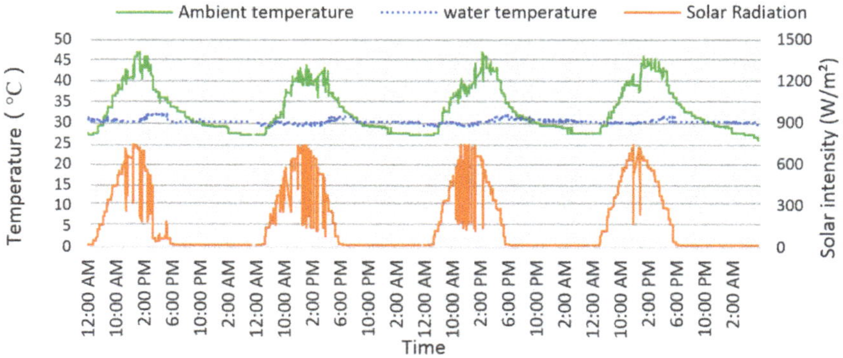

Figure 14. Meteorological data and water temperature recorded across a four-day testing phase.

Through experimental data analysis, it was observed that, as the thickness of the corrugated layer increased, the overall energy efficiency exhibited an ascending trend. When the thickness was increased from 30 mm to 50 mm, the improvement in energy efficiency was relatively small. However, when the thickness was increased from 50 mm to 100 mm, the energy efficiency rose by 3%. Furthermore, when the thickness was increased from 100 mm to 200 mm, the improvement in energy efficiency reached 5%. Overall, increasing the thickness of the air layer promoted the contact area between the spray and air, thereby enhancing cooling efficiency and boosting energy savings. These findings are shown in Table 7. Notably, the energy efficiency at a thickness of 200 mm showed the best performance; indicating that, given suitable conditions, it is advisable to prioritize increasing the thickness of the air layer.

Table 7. List of daily energy-saving rates of four lamination thickness.

Air Conditioning Temperature (°C)	30 mm (%)	50 mm (%)	100 mm (%)	200 mm (%)
27	26.38	27.10	30.25	35.19

During the analysis of nighttime energy efficiency rates, it was observed that, as the thickness of the intermediate layer increased gradually from 50 mm, the energy efficiency rose gradually. However, upon increasing the thickness from 100 mm to 200 mm, the energy efficiency rate did not continue to ascend as anticipated; but rather, it exhibited a reversal, dropping even below the daytime energy efficiency rate. After further investigating this phenomenon, it was primarily attributed to the fact that the cooling potential of the night-time itself is not high. As illustrated in Table 8, the natural decrease in outdoor temperature leads to a reduction in the indoor–outdoor temperature difference. Consequently, the thicker intermediate layer does not significantly contribute to nighttime energy efficiency; thereby resulting in a lack of notable cooling effects.

Table 8. List of night energy-saving rates of four lamination thickness.

Air Conditioning Temperature (°C)	30 mm (%)	50 mm (%)	100 mm (%)	200 mm (%)
27	31.95 (rise by 21%)	32.18 (rise by 18%)	35.19 (rise by 16%)	30.49 (drop by 13%)

3.7. Comparison of Results, Error Analysis, and Measurement Uncertainty

In the conduct of this study, the importance of result comparing, error analysis, and the uncertainty of measuring parameters such as water temperature, pressure, and humidity were recognized in ensuring the reliability, accuracy, and validity of experimental data. Therefore, measures were implemented throughout the experimental design and data processing to validate results, analyze potential error sources, and assess measurement uncertainty.

3.7.1. Comparison of Results

After rigorous analysis, the water spray window cooling system has demonstrated outstanding performance. In terms of cooling effect, it has successfully lowered the temperatures of both the interior and exterior window surfaces through the utilization of high-pressure water spray technology, resulting in an improved indoor thermal environment. This cooling effect was comparable to that achieved by spray evaporation cooling techniques reported in the literature. For instance, Document [22] reported a surface cooling value of approximately 8 °C for skylights, which is closely aligned with the 6 °C reduction achieved by our water spray window. Additionally, Document [33] cites an optimal spray interval of 10/15 min, while the current study found a preferred frequency of 2/12 min. This discrepancy can be attributed to regional climate variations; yet both studies emphasized the importance of frequency control in enhancing cooling efficiency. Furthermore, while relevant literature seldom discusses comparisons of experimental water temperature and interlayer thickness, the current findings provide a foundation for utilizing condensate water as a water source and employing triple-layer glass interlayer spray windows. In regards to energy consumption and savings, the water spray window system significantly mitigates the cooling load on air conditioning systems, thus reducing the overall energy consumption of buildings. This is achieved by effectively blocking solar radiation and lowering window temperatures.

Extensive experimental testing and comparative analyses have revealed the system's excellent stability and reliability, enabling it to consistently and effectively reduce window temperatures and enhance indoor thermal comfort. These observations align closely with the stability findings of spray evaporation cooling technologies, further validating the reliability and practicality of water spray cooling technique.

3.7.2. Error Analysis and Measurement Uncertainty

A thorough analysis was conducted to identify potential error sources. Attention was paid to the inherent error of temperature sensors, especially thermocouples used for glass surface temperature measurements. Aluminum foil was applied to reflect sunlight and minimize direct heating errors. Additionally, high-accuracy sensors from reputable manufacturers were selected. The uncertainty associated with measuring parameters such as temperature, pressure, and humidity was carefully assessed. Consideration was given to equipment limitations, environmental fluctuations, and random errors during operation. Through rigorous error propagation analysis and data interpretation, the impact of these uncertainties on conclusions was minimized. Multiple validation methods were employed to confirm the reliability and effectiveness of experimental results. Comparisons with previous studies, independent replicate experiments, and internal consistency checks were conducted.

While measures were implemented to minimize errors, validate results, and assess measurement uncertainty, certain limitations remain. These may include experimental conditions, sample size considerations, and measurement equipment precision. However, the comprehensive approach has provided reliable and valid results within the study scope.

In summary, while various measures were implemented to reduce potential error sources, some degree of uncertainty remains in the experimental results. However, the overall trends and conclusions provide valuable insights into the cooling effects of water spray window.

4. Conclusions

The present study has conducted a comprehensive investigation into the cooling performance of water spray windows under varying conditions, and their potential for energy savings. Through a rigorous experimental comparison analysis, the impacts of water temperature, spraying frequency, and glass structure on the cooling effectiveness of the system were examined. Additionally, the influence of air conditioning set temperatures and layer thickness on building energy efficiency was explored. Utilizing the experimental data, a multi-faceted analysis was conducted, encompassing thermal environment regulation and energy consumption reduction, leading to the following conclusions:

(1) Water spray windows, in contrast to conventional glass windows, exhibit notable cooling effects and enhanced temperature stability. This technology is capable of achieving a cooling effect of up to 6 °C during the day and lowering the temperature by approximately 3 °C at night. This significantly mitigates the rise in window surface temperature and substantially reduces temperature fluctuations within the building, thereby elevating indoor comfort. The use of lower temperature water extends the cooling effects, while the utilization of condensate and rainwater further enhances renewable energy efficiency, resulting in reduced energy consumption costs.

(2) The substantial benefits of energy savings have been affirmed by experimental data. By adjusting to optimal spraying frequencies and air conditioning set temperatures, the mist-spraying window system can effectively decrease energy consumption, leading to energy efficiency improvements ranging from 6% to 27%. Additionally, there exists a positive correlation between energy efficiency and air conditioning set temperature. It is worth noting that double-layer glass windows, when compared to single-layer ones, offer more significant energy savings during nighttime use, which especially highlights the tremendous energy-saving potential of mist technology.

(3) The advantages of insulating spray in triple-layer windows: Research has revealed that the insulating spray within triple-layer windows possesses superior performance both in terms of energy saving and temperature reduction when compared to outdoor spray. The findings indicate that, up to a certain thickness, increasing the thickness of the insulating layer enhances energy efficiency; particularly within the 30–50 mm range. Significant energy savings are observable at night. However, a decrease in efficiency is noticed when the thickness reaches 200 mm, suggesting the need to consider an optimal thickness range during the design phase.

(4) Applicability: Given the widespread use of large-scale exterior windows in commercial, office buildings, and modern industrial plants; coupled with prolonged air conditioning usage, the introduction of spray-on window technology presents an innovative solution for temperature regulation and energy savings. Spray-on window technology offers significant advantages in temperature control, humidity adjustment, and dust suppression, especially in environments such as modern factories, where precise environmental conditions are crucial for production processes. This is especially the case in the southern China and ASEAN regions, where the unique climatic conditions create a highly favorable environment for applying spray-evaporation cooling technology. By leveraging these conditions, more lightweight, energy-efficient window systems with improved fogging efficiency can be developed; to elevate living and working comfort, subsequently enhancing the quality of life for residents and the productivity of industrial operations.

Author Contributions: Conceptualization, X.Y.; methodology, J.H.; software, J.W.; validation, L.L., Y.Z. and G.H.; formal analysis, Y.Z.; investigation, G.H.; resources, X.Y.; data curation, X.Y.; writing—original draft preparation, X.Y.; writing—review and editing, X.Y.; visualization, J.W.; supervision, J.W.; project administration, X.Y.; funding acquisition, J.H. All authors have read and agreed to the published version of the manuscript.

Funding: This research was funded by National Natural Science Foundation of China (NSFC) (51968003) and The College Students' Innovation and Entrepreneurship Training Program project of Guangxi University (No. S202210593323 and No. S202310593002).

Data Availability Statement: The original contributions presented in the study are included in the article, further inquiries can be directed to the corresponding author.

Conflicts of Interest: The authors declare no conflicts of interest.

Nomenclature

T	inside temperature [K]
T_a	outdoor air temperature [K]
T_w	temperature of water film [K]
T_s	surface temperature [K]
R_L	long-wave radiation from surrounding [W/m^2]
C_m	humid air specific heat [J/ (kg K)]
C_w	water specific heat [J/ (kg K)]
X_s	absolute humidity mixing ratio at temperature $\boldsymbol{T_s}$ [kg/kg(DA)]
X_a	absolute humidity mixing ratio at temperature T_a [kg/kg(DA)]
W_q	water supply [kg/m^2 h]
z	distance from water outlet [m]
R_s	short-wave (solar) radiation [W/m^2]
λ_w	water thermal conductivity [W/ (m K)]
Greek symbols	
α_c	surface convection coefficient with air [W/ (m^2 K)], the subscripts "c" means "convection"
α_w	surface convection coefficient with water [W/ (m^2 K)], the subscripts "w" means "water"
ε	emissivity
σ	Stefan-Boltzmann constant [W/ (m^2 K4)]
l	latent heat [W/ (m K)]
λ	thermal conductivity [W/ (m K)]
ν	kinematic viscosity $\left[m^2/s\right]$

References

1. Li, C.; Lyu, Y.; Li, C.; Qiu, Z. Energy performance of water flow window as solar collector and cooling terminal under adaptive control. *Sustain. Cities Soc.* **2020**, *59*, 102152. [CrossRef]
2. Jankovic, A.; Goia, F. Control of heat transfer in single-story mechanically ventilated double skin facades. *Energy Build.* **2022**, *271*, 112304. [CrossRef]
3. Onat, N.C.; Kucukvar, M. Carbon footprint of construction industry: A global review and supply chain analysis. *Renew. Sustain. Energy Rev.* **2020**, *124*, 109783. [CrossRef]
4. Nong, D.; Simshauser, P.; Nguyen, D.B. Greenhouse gas emissions vs CO_2 emissions: Comparative analysis of a global carbon tax. *Appl. Energy* **2021**, *298*, 117223. [CrossRef]
5. Ulpiani, G. Water mist spray for outdoor cooling: A systematic review of technologies, methods and impacts. *Appl. Energy* **2019**, *254*, 113647. [CrossRef]
6. Ismail, K.A.; Lago, T.G.; Lino, F.A.; Mondlane, M.V.; Teles, M.P. Experimental investigation on ventilated window with reflective film and development of correlations. *Sol. Energy* **2021**, *230*, 421–434. [CrossRef]
7. Huang, H.-Y.; Hu, W.-C.; Chen, C.-K.; Lin, T.-H.; Lin, F.-Y.; Cheng, C.-C.; Su, T.-C.; Yu, P.-Y. Evaluation of the Effects of Window Films on the Indoor Environment and Air-Conditioning Electricity Consumption of Buildings. *Energies* **2024**, *17*, 1388. [CrossRef]
8. Al-Yasiri, Q.; Alktranee, M.; Szabó, M.; Arıcı, M. Building envelope-enhanced phase change material and night ventilation: Effect of window orientation and window-to-wall ratio on indoor temperature. *Renew. Energy* **2023**, *218*, 119263. [CrossRef]
9. Khetib, Y.; Alotaibi, A.A.; Alshahri, A.H.; Rawa, M.; Cheraghian, G.; Sharifpur, M. Impact of phase change material on the amount of emission in the double-glazed window frame for different window angles. *J. Energy Storage* **2021**, *44*, 103320. [CrossRef]
10. Qahtan, A.; Keumala, N.; Rao, S.; Abdul-Samad, Z. Experimental determination of thermal performance of glazed façades with water film, under direct solar radiation in the tropics. *J. Affect. Disord.* **2011**, *46*, 2238–2246. [CrossRef]
11. He, J.; Hoyano, A. A numerical simulation method for analyzing the thermal improvement effect of super-hydrophilic photocatalyst-coated building surfaces with water film on the urban/built environment. *Energy Build.* **2008**, *40*, 968–978. [CrossRef]
12. Ye, T.Q. Experimental Study on Thermal Insulating Performance of Passive Evaporative Cooling Room with Double-Skin Facade. *Build. Energy Effic.* **2013**, *41*, 30–34. [CrossRef]

13. Sotelo-Salas, C.; del Pozo, C.E.; Esparza-López, C.J. Thermal assessment of spray evaporative cooling in opaque double skin facade for cooling load reduction in hot arid climate. *J. Build. Eng.* **2021**, *38*, 102156. [CrossRef]
14. Al Touma, A.; Ouahrani, D. Evaporatively-cooled façade integrated with photovoltaic thermal panel applied in hot and humid climates. *Energy* **2019**, *172*, 409–422. [CrossRef]
15. Kumar, V.V.; Raut, N.; Akeel, N. An experimental investigation on passive and hybrid roof cooling systems with a double skin envelope. *Energy Sources Part A Recover. Util. Environ. Eff.* **2021**, 1–13. [CrossRef]
16. Qahtan, A.M. An empirical study on the use of water film in glazed buildings to reduce solar radiation transmittance in the tropics. *J. Green Build.* **2015**, *10*, 141–160. [CrossRef]
17. Qahtan, A.; Rao, S.; Keumala, N. The effectiveness of the sustainable flowing water film in improving the solar-optical properties of glazing in the tropics. *Energy Build.* **2014**, *77*, 247–255. [CrossRef]
18. Ye, T.; Ke, X. Study on Thermal Performance of Passive Evaporative Cooling Double-Skin Facade. In Proceedings of the 2011 International Conference on Measuring Technology and Mechatronics Automation (ICMTMA), Shanghai, China, 6–7 January 2011; pp. 593–596.
19. Kang, D.; Strand, R.K. Analysis of the system response of a spray passive downdraft evaporative cooling system. *J. Affect. Disord.* **2019**, *157*, 101–111. [CrossRef]
20. Liu, M.; Jimenez-Bescos, C.; Calautit, J.K. Passive heat recovery wind tower: Assessing the overheating risk in summertime and ventilation heat loss reduction in wintertime. *Sustain. Energy Technol. Assess.* **2023**, *58*, 103310. [CrossRef]
21. Otović, M.; Milovančević, U.; Genić, S.; Mitrović, N.; Černicin, V.; Otović, S.; Antić, M. Testing of the Influence of the Spray Cooling System on the Air Cooler Efficiency. In Proceedings of the 51st International HVAC&R Congress and Exhibition, Online, 2–4 December 2020; pp. 11–16.
22. Wang, J.; Meng, Q.; Yang, C.; Ren, P.; Santamouris, M. Spray optimization to enhance the cooling performance of transparent roofs in hot-humid areas. *Energy Build.* **2023**, *286*, 112929. [CrossRef]
23. Ismail, Z.; Go, Y.I. Fog-to-Water for Water Scarcity in Climate-Change Hazards Hotspots: Pilot Study in Southeast Asia. *Glob. Chall.* **2021**, *5*, 2000036. [CrossRef] [PubMed]
24. Dhamodharan, P.; Ayalur, B.K.; Prabakaran, R.; Kumar, A.S.; Choi, G.S.; Kim, S.C. Current state of research in air-conditioning condensate utilization and challenges: A review approach. *Appl. Therm. Eng.* **2023**, *233*, 121171. [CrossRef]
25. Chan, L.S. Numerical study on the thermal performance of water flow window fed with air-conditioning condensate. *Energy* **2023**, *263*, 125633. [CrossRef]
26. Hawken, S.; Sepasgozar, S.; Prodanovic, V.; Jing, J.; Bakelmun, A.; Avazpour, B.; Che, S.; Zhang, K. What makes a successful Sponge City project? Expert perceptions of critical factors in integrated urban water management in the Asia-Pacific. *Sustain. Cities Soc.* **2021**, *75*, 103317. [CrossRef]
27. Yin, D.; Chen, Y.; Jia, H.; Wang, Q.; Chen, Z.; Xu, C.; Li, Q.; Wang, W.; Yang, Y.; Fu, G.; et al. Sponge city practice in China: A review of construction, assessment, operational and maintenance. *J. Clean. Prod.* **2020**, *280*, 124963. [CrossRef]
28. Dhamodharan, P.; Ayalur, B.K.; Judefelix, J.; Prabakaran, R.; Kim, S.C. Energy saving potential in radiant cooling system by utilizing air-conditioning condensate: A strategy for green building rating. *Appl. Therm. Eng.* **2023**, *236*, 121492. [CrossRef]
29. Cai, Z.; La Sorte, F.A.; Chen, Y.; Wu, J. The surface urban heat island effect decreases bird diversity in Chinese cities. *Sci. Total Environ.* **2023**, *902*, 166200. [CrossRef]
30. Younes, J.; Ghali, K.; Ghaddar, N. Diurnal Selective Radiative Cooling Impact in Mitigating Urban Heat Island Effect. *Sustain. Cities Soc.* **2022**, *83*, 103932. [CrossRef]
31. Lin, Y.; Wang, Z.; Jim, C.Y.; Li, J.; Deng, J.; Liu, J. Water as an urban heat sink: Blue infrastructure alleviates urban heat island effect in mega-city agglomeration. *J. Clean. Prod.* **2020**, *262*, 121411. [CrossRef]
32. Li, Y.; He, J. Evaluating the improvement effect of low-energy strategies on the summer indoor thermal environment and cooling energy consumption in a library building: A case study in a hot-humid and less-windy city of China. *Build. Simul.* **2021**, *14*, 1423–1437. [CrossRef]
33. Wang, Y.; He, F.; Lv, Z.; Yang, W.; Wang, Q. Experimental investigations on thermal performance of spray cooling double skin façade in hot humid climate region. *Energy Build.* **2022**, *277*, 112605. [CrossRef]

buildings

MDPI

Article

Strategies for Mitigating Urban Residential Carbon Emissions: A System Dynamics Analysis of Kunming, China

Jian Xu, Yujia Qian, Bingyue He, Huixuan Xiang, Ran Ling and Genyu Xu *

Urban Construction and Digital City Teaching Experiment Center, School of Architecture and Planning, Yunnan University, Kunming 650550, China; xujian@ynu.edu.cn (J.X.); qianyujia@itc.ynu.edu.cn (Y.Q.); hebing@stu.ynu.edu.cn (B.H.); xianghuixuan@itc.ynu.edu.cn (H.X.); lingran@itc.ynu.edu.cn (R.L.)
* Correspondence: xugenyu@ynu.edu.cn

Abstract: To effectively combat environmental challenges, it is necessary to evaluate urban residential building carbon emissions and implement energy-efficient, emission-reducing strategies. The lack of a specialized carbon emission monitoring system complicates merging macro- and micro-level analyses to forecast urban residential emissions accurately. This study employs a system dynamics (SD) model to examine the influence of social, economic, energy, and environmental factors on carbon emissions in urban residences in Kunming, China. The SD model forecasts household carbon emissions from 2022 to 2030 and establishes three scenarios: a low-carbon scenario (LCS), a medium low-carbon scenario (MLCS), and a high low-carbon scenario (HLCS) to assess emission reduction potentials. It predicts emissions will climb to 4.108 million tons by 2030, significantly surpassing the 2014 baseline, with economic growth, urbanization, residential energy consumption, and housing investment as key drivers. To curb emissions, the study suggests enhancing low-carbon awareness, altering energy sources, promoting research and development investment, and expanding green areas. The scenarios indicate a 5.1% to 16.1% emission reduction by 2030 compared to the baseline. The study recommends an 8.3% to 11.4% reduction in MLCS as a practical short-term target for managing urban residential emissions, offering a valuable SD approach for optimizing carbon strategies and aiding low-carbon development.

Keywords: carbon emissions; urban residential buildings; system dynamics modeling; Kunming; low-carbon scenarios

1. Introduction

The sixth assessment report of the Intergovernmental Panel on Climate Change (IPCC) highlights the significant contribution of urban areas to global carbon emissions [1]. Currently, over half of the global population resides in urban regions, which are responsible for approximately 71–76% of global end-use energy consumption and 67–76% of CO_2 emissions attributable to energy use in cities [2]. This trend suggests an ongoing and likely increase in energy consumption and carbon emissions from residential buildings, particularly considering the growing urban population.

Residential carbon emissions can be categorized into five key phases: manufacturing of building materials, construction, usage, maintenance, and finally, demolition and disposal [3]. For instance, in 2020, the residential building sector of China accounted for 45.5% of the total national energy consumption and 50.9% of its carbon emissions. The production of building materials and building operation phases are the most significant contributors, accounting for 28.2% and 21.7% of emissions, respectively [4]. Notably, the operational phase tends to increase its emissions proportionally over time, eventually becoming the dominant source over the lifecycle of the building [5]. Therefore, it is crucial to study the carbon emissions of urban residential buildings during the use phase and, on this basis, formulate energy-saving and emission reduction strategies for urban residential buildings during the operation phase to reduce carbon emissions.

A deeper look shows that various macro factors, including urbanization, demographics, climate parameters, economic development, and energy consumption, have been identified as significant drivers of emissions from urban residential buildings. Studies, such as those by Burleyson et al. [6], Wang et al. [7], and Huo et al. [8], highlight how urbanization affects household energy patterns and thereby impacts carbon emissions, with noticeable regional variations. Climate conditions also play a crucial role in influencing construction intensity and energy consumption [9,10].

On the micro-level, research presents a more nuanced picture. There is evidence suggesting that higher household incomes may lead to increased energy consumption [11,12], while other studies propose that wealthier households tend to adopt energy-saving practices [13,14]. Notably, household energy use patterns in China differ significantly between urban and rural areas and across different regions [15], depending on things like building types, heating systems, population density, and educational levels [16,17]. However, much of the existing research tends to isolate these factors, overlooking their interconnected and cumulative effects.

In framing emission reduction strategies, it becomes clear that variations in lifestyle and usage patterns can lead to significant disparities in household energy consumption, with differences of up to 300% [18]. This highlights the importance of considering both micro- and macro-level factors, like income, consumption habits, urban expansion, and population dynamics [19,20], in designing low-carbon pathways. Additionally, building design and usage patterns have important influences on carbon abatement outcomes. For instance, regarding heating mode selection, centralized heating entails higher energy use [21]. However, after accounting for per capita metrics, centralized systems demonstrate superior energy savings over distributed heating alternatives. To address these complexities, some scholars have incorporated policy effects like carbon trading into their analyses [22,23].

In this context, Kunming, as Yunnan Province's political, economic, and cultural center, emerges as a focal point in China's low-carbon initiative. Despite setting explicit targets for carbon emission reduction, Kunming faces ongoing challenges with high residential carbon emissions, exacerbated by its growing population and economic growth. Current research on the residential carbon footprint of Kunming predominantly adopts a static approach, neglecting the dynamic evolution of emissions and the underlying mechanisms that drive these changes. This study proposes the development of a system dynamics (SD) model to forecast future emission patterns dynamically, delving into the intricate interplay between various factors—economy, society, energy, and environment—at both macro- and micro-levels. This approach marks a departure from traditional static analysis, highlighting the dynamic interactions within complex systems and offering fresh insights into the mechanisms underlying residential carbon emission trends. Such insights are crucial for enhancing the accuracy of emission predictions and the efficacy of emission reduction strategies. Moreover, this research is poised to inform policy formulation, providing a solid foundation for crafting low-carbon development strategies in Kunming and beyond, thereby underscoring its significance for informed planning and policymaking.

2. Methodology

2.1. Research Framework

To construct an SD model to simulate the residential carbon emissions system in Kunming, China, this study adhered to the methodology illustrated in Figure 1. Initially, it delineated the boundaries of the system to match the research scope, pinpointed crucial factors within the residential carbon emissions framework, and explained the interrelationships among these factors. Subsequently, feedback loops among subsystem variables were established as the foundation for crafting both the causal loop diagram and the system flow diagram. To complete the construction of the model, this stage involved creating system equations using database data and then conducting rigorous testing and calibration. The next phase entailed devising four experimental scenarios: the baseline, low-carbon scenario (LCS), medium low-carbon scenario (MLCS), and high low-carbon scenario (HLCS), each reflecting the status of Kunming and the long-term strategic vision of the government. Finally, the study executed these scenarios, analyzed the simulation outcomes, and derived relevant conclusions.

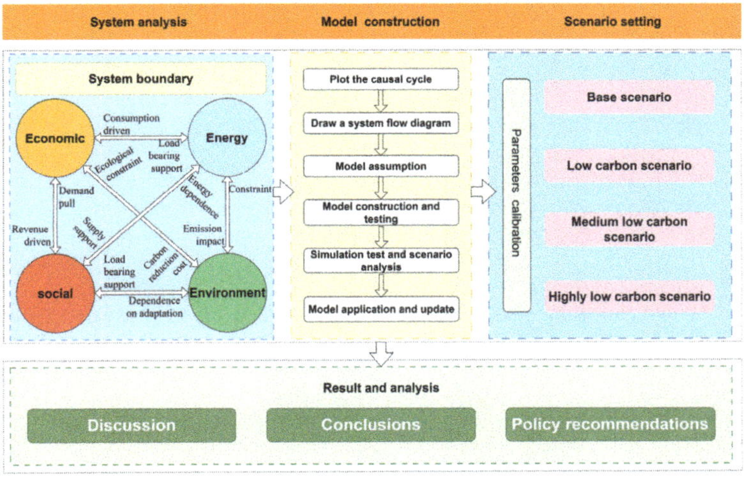

Figure 1. Research framework.

2.2. Study Area

Kunming, the capital of southwest China (Figure 2), is experiencing rapid urbanization. In 2021, its population reached 8,873,500, with an urbanization rate of 79.6%, and this urban expansion has led to a residential area growth of 105 million square meters [24]. Kunming has a plateau monsoon climate at an elevation of around 2000 m. It experiences ample sunshine year-round, substantial diurnal temperature variations yet minor seasonal differences, four distinct seasons without frigid winters or sweltering summers [25], an annual average temperature of 16.5 °C, as well as adequate and suitable sunlight exposure, which has earned it the nickname "Spring City" [26]. The energy consumption patterns of the city are heavily influenced by its climate, which is characterized by mild winters and cool summers. This climate allows for natural ventilation in summer and minimal heating in winter, with electricity, rather than air conditioning, being the primary energy consumer.

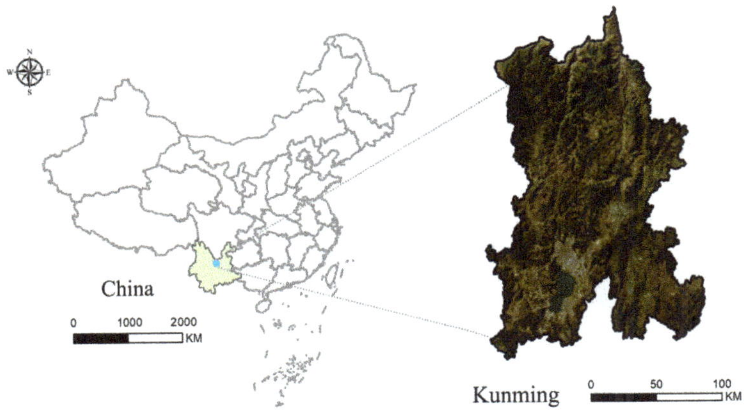

Figure 2. Location of the study area.

The geographical location and climatic conditions of Kunming have shaped a distinctive energy consumption pattern and residential building energy use. As a national low-carbon pilot city of China, Kunming has made notable early achievements in reducing carbon emissions and has established ambitious objectives for the future. This achievement underlines the necessity for well-thought-out strategies for continued carbon reduction efforts. Moreover, the endeavors in carbon reduction in Kunming offer invaluable insights for other Chinese cities with analogous climates, such as Guiyang and Zunyi, that are on a similar low-carbon journey. However, despite the benefits derived from its geographic, climatic, and policy framework, Kunming faces challenges due to an increasing demand for energy-intensive home appliances and rapid urbanization, which have escalated domestic energy usage. This situation highlights the critical need for a detailed investigation into the urban residential carbon emissions in Kunming. Such an analysis is essential to crafting effective local low-carbon strategies, aiming to achieve broader zero-carbon targets.

2.3. Carbon Emission Accounting Method

In the building industry of China, the lack of a dedicated carbon emission monitoring system necessitates the use of various computational techniques for carbon emission estimation. Among these, micro-methods, notably the computer simulation technique and actual measurement technique, are commonly employed. The actual measurement method is apt for industries with significant emissions and intricate emission characteristics [27], and computer simulation predominantly focuses on micro-level studies of specific systems and processes, often struggling with the complexity of macro-level objects [28,29]. Therefore, life cycle assessment (LCA) methods can be used to provide in-depth analysis at the individual level and scale up to assess larger systems. However, despite its comprehensive approach, LCA is constrained by the challenges of data acquisition and workload [30,31].

Alternatively, the IPCC carbon emission factor method is widely embraced in the construction sector for its simplicity, accessibility of data, and broad applicability [32,33]. It adeptly addresses both macro- and micro-level analysis requirements. This study adopts the IPCC method for macro-level carbon emission assessment, utilizing the statistical data available in the regional yearbooks of Kunming. The IPCC coefficients are calculated using Equation (1):

$$C = \sum E_i \times \mu_i \quad (1)$$

where C denotes total carbon emissions; E_i denotes the consumption of the ith energy source; and μ_i denotes the carbon emission factor of the ith energy source.

As shown in Table 1, data from the Kunming Statistical Yearbook [34] and Energy Statistical Yearbook (2014–2021) [35] form the basis of this study analysis, with conversion coefficients for various energy sources normalized to standard coal equivalents as per the 2020 China Energy Statistical Yearbook. The electricity emission factor is sourced from [36,37], and this study applied an emission factor of 0.5271 kg CO_2/kWh for Kunming.

Table 1. Standard coal coefficient of various energy sources (reference value). [34,35].

Energy Source	Average Status Heat Generation	Conversion Factor of Standard Coal
Raw coal	20,934 kJ/kg (5000 kcal/kg)	0.7143 kgce/kg
Petroleum	38,979 kJ/kg (9310 kcal/kg)	1.3300 kgce/m^3
Liquefied petroleum gas	50,242 kJ/kg (12,000 kcal/kg)	1.7143 kgce/kg
Coke oven gas	18,003 kJ/kg (4300 kcal/kg)	0.6143 kgce/m^3

Figure 3 illustrates the trend in urban residential carbon emissions in Kunming from 2014 to 2021, highlighting a continual upward trajectory from 1.839 million tons to 3.102 million tons. This represents an average annual growth rate of 18.6%. The escalation in emissions is closely linked to the enhancement of living standards and the rapid pace of urbanization. Notably, a decline in emissions was observed in 2017 and 2018, aligning with significant changes in the city's energy infrastructure. This period marked the extensive substitution of piped artificial gas with natural gas, signaling a transition to more environmentally friendly energy sources. Despite the adoption of natural gas, a cleaner energy alternative, aggregate carbon emissions continued to escalate, driven by factors such as population growth, heightened living standards, and an increasing demand for energy services.

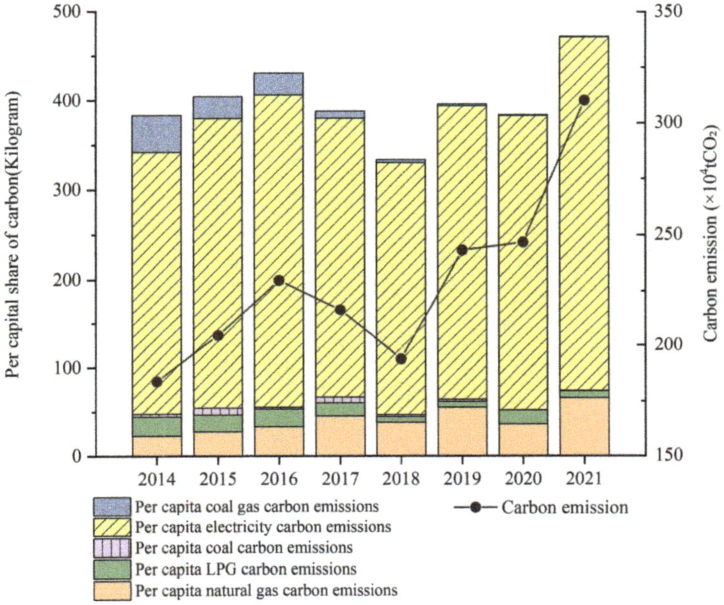

Figure 3. Urban residential carbon emissions and per capita carbon emissions from different energy sources in Kunming, China, spanning the years 2014 to 2021.

The findings also indicate a shift in the composition of carbon emission sources. Prior to 2017, gas and electricity constituted about 90% of the total residential carbon emissions in Kunming. After 2018, the share of natural gas and electricity exceeded 95%, highlighting a continuous decline in emissions from coal, liquefied petroleum gas (LPG), and gas. This trend underscores the importance of energy structure in emission reduction strategies, although the ongoing rise in emissions suggests that improvements in energy structure alone may not fully offset the environmental impact of other factors driving increased energy demand.

2.4. Influencing Factors

In this study, the SD model was used to conduct an in-depth analysis of carbon emissions from residential buildings. Central to this analysis is the identification of key influencing factors. By sorting and summarizing the literature [8–10], a preliminary set of influencing factors was determined. Grey correlation analysis [38] is then employed to further refine and validate these factors. The grey correlation analysis method is mainly used to judge the degree of correlation through the similarity of the trend of the data series, identify the advantages and disadvantages of the judged object, and does not reflect the absolute level [39]. This technique directly calculates using original data without normalization, thus offering strong reliability [40]. By calculating the grey correlation coefficients between diverse factors and carbon emissions, the correlation between each pair can be determined. The results reveal that only residential floor area completed, urban road area per capita, and annual technology contract transaction amount have low correlations, while the grey correlations of other factors exceed 0.7, indicating strong correlations [40,41]. Therefore, this meticulous approach enables the identification of 22 distinct influencing factors, categorized across four subsystems, which are detailed in Appendix A. Prominent among these are the energy consumption expenditures of residents, energy prices, and heating degree days (Figure 4). Each of these factors plays a significant role in impacting carbon emissions. A significant positive correlation exists between Figure 4a,c, revealing the energy consumption of residents and heating degree days as key drivers of carbon emissions. Figure 4b shows energy prices and per capita emissions negatively correlating short-term; however, long-term trends demonstrate decoupling between the two. This decoupling suggests energy price reforms alone have limited efficacy to improve efficiency and mitigate emissions. Therefore, a combination of other regulatory, technical, and economic policy tools is necessary to effectively achieve the set carbon reduction targets.

2.5. Model Building

Forrester invented SD modeling in 1956, which uses structural function analysis based on computer simulation technology to address complex dynamic feedback systems [42]. Combining cybernetics, information theory, and system theory, SD modeling is a way to look at problems in social, economic, ecological, and other complex systems that are dynamic, nonlinear, and have feedback at more than one level. Recent applications [43–45] in urban energy consumption and building carbon dioxide emissions have been noted. Notably, Murray et al. [46] and Liu et al. [47] utilized this methodology for analyzing energy usage and carbon emissions in Changsha City, underscoring its utility in supporting policy decision-making by unraveling the complex mechanisms driving building carbon emissions. Future expansions of this model could encompass a broader range of influencing factors, improved spatial and temporal resolution, and integration with statistical and econometric models for enhanced accuracy. However, research on building carbon emissions using this approach remains relatively limited.

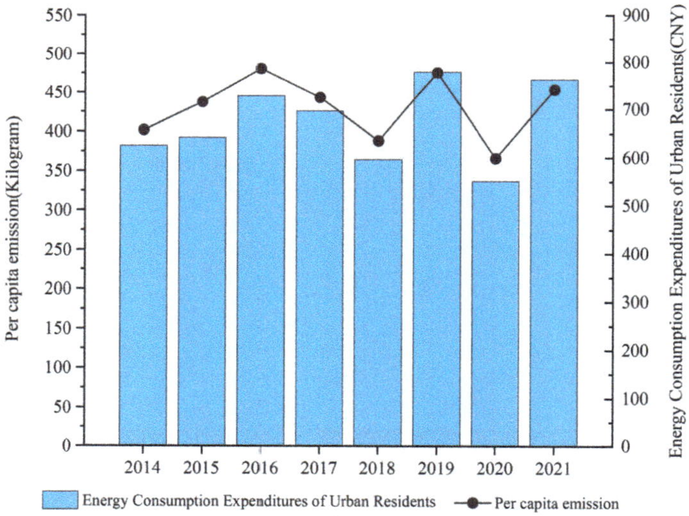

(a)

Figure 4. *Cont.*

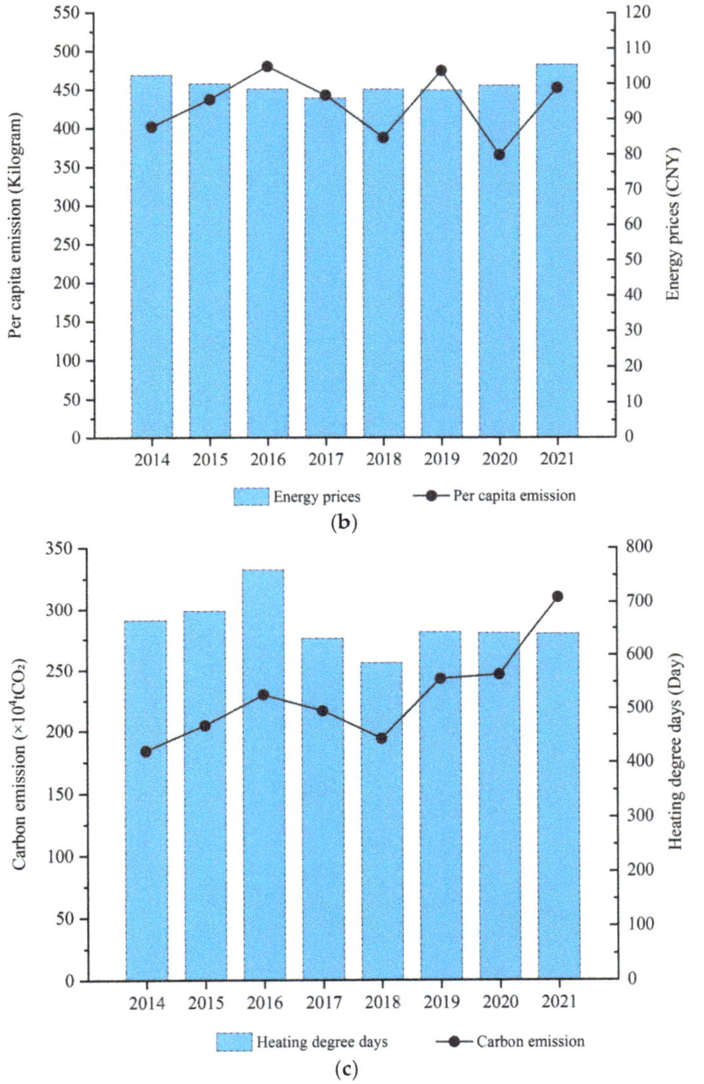

Figure 4. Key influencing factors of carbon emissions from residential buildings in Kunming, China. (**a**) The energy consumption expenditure of residents; (**b**) energy prices; (**c**) heating degree days.

This research conducts a thorough analysis of the intricate framework underlying residential carbon emissions through the lens of SD. It aims to unravel the multitude of factors that drive residential carbon emissions and investigates their interconnections. Identifying key variables, the study leverages feedback mechanisms inherent in the subsystems to construct causal loop diagrams and stock flow diagrams. These diagrams elucidate the mechanisms through which each factor impacts residential carbon emissions.

Regression fitting enabled the calibration of the model parameters, allowing for the simulation of the urban residential carbon emissions trajectory of Kunming for the years 2022–2030. This simulation process is not static; it involves meticulous adjustments to the parameters and influencing factors to accurately reflect the evolving trends of carbon emissions under a variety of potential scenarios. Subsequently, the article provides a detailed analysis of the most effective mitigation strategies. By changing the parameters and contributing factors of the model, the paper puts forward feasible suggestions for reducing emissions. These recommendations are further substantiated through simulations that project the trends of household carbon emissions under diverse scenarios. The structural framework of the SD model in this study is shown in Figure 5, which provides a visual representation of the model's complexity and its operational dynamics.

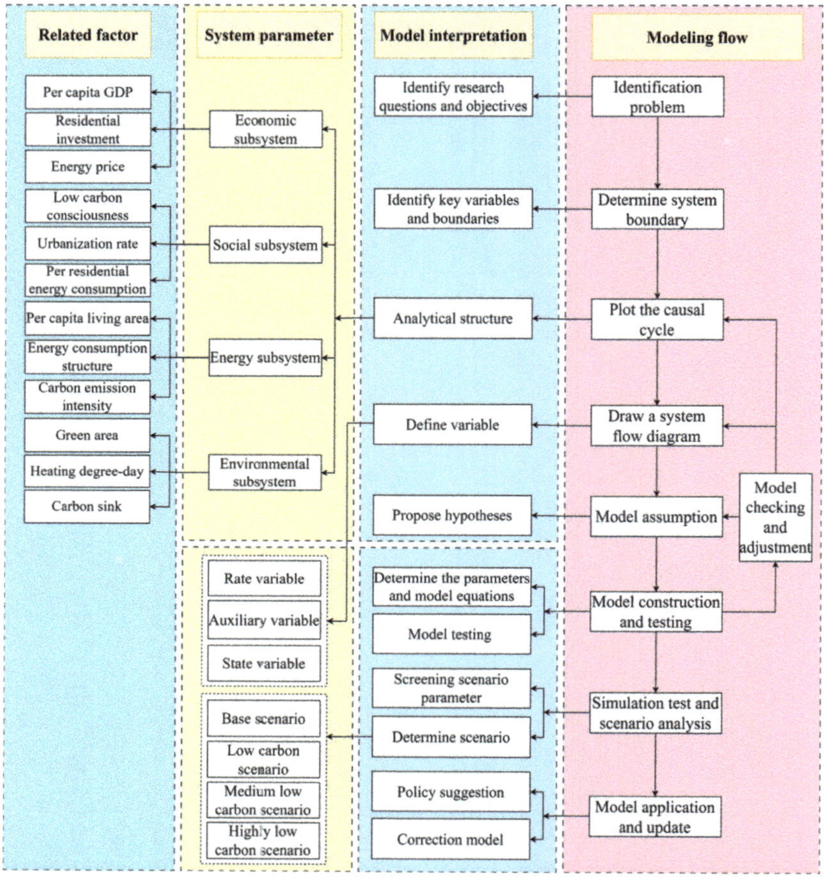

Figure 5. The framework of the system dynamics (SD) model.

For this study, the temporal boundary spans from 2014 to 2030. Among them, 2014–2021 is the historical benchmark period of the model. Kunming carried out the natural gas replacement project from 2014 to 2017, and the energy structure changed, thus affecting residential carbon emissions. The data in this stage are representative and typical and can provide a basis for the estimation and calibration of model parameters; 2022–2030 is the simulation prediction stage. The year 2030 is a key node in the realization of the carbon peak goal of China, and the selection of this period is helpful to evaluate the low-carbon transformation process of Kunming's residential sector and provide decision-making reference for formulating relevant policies. From 2014 to 2030, the time span is relatively long, considering the short- and medium-term development trend, which is conducive to a comprehensive analysis of the dynamic effects of various influencing factors. The model operates with an annual time step, offering a detailed yet manageable temporal resolution. Spatially, the model focuses on urban residences within Kunming City, specifically targeting the carbon emissions generated during the use phase of these residences. This spatial boundary ensures that the model remains focused on urban residential emissions while capturing the relevant dynamics within the specified geographical context.

The causal loop diagram (Figure 6) is a key component in this study because it demonstrates how various elements of Kunming's residential carbon emission system interact and affect one another. This study has categorized this system into four distinct subsystems: economic, social, energy, and environmental. Each subsystem encompasses a set of variables and relationships, which are meticulously analyzed and synthesized to delineate the causal dynamics of residential carbon emissions.

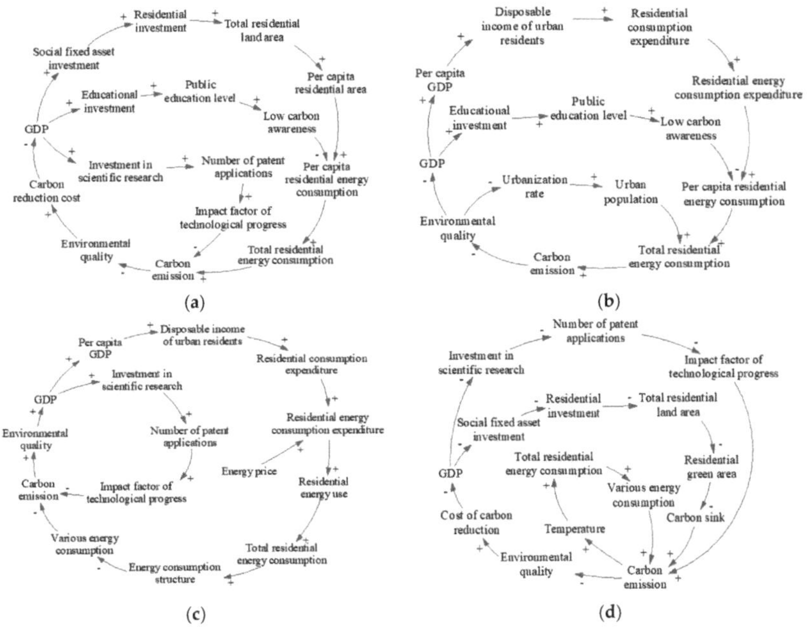

Figure 6. Causal loop diagrams of (**a**) economic subsystems, (**b**) social subsystems, (**c**) energy subsystems, and (**d**) environmental subsystems. +: incremental effect, −: reducing effect.

In terms of economic subsystems, the study explores how factors like industry investment and economic growth influence residential carbon emissions, along with the associated feedback loops. For instance, Figure 6a traces the path from gross domestic product (GDP) growth to increased social fixed assets and residential investments, leading to an expansion in the total residential land area and per capita residential area. This expansion correlates with an increase in per capita residential energy consumption and, consequently, total residential energy consumption, which in turn impacts carbon emissions and environmental quality. The feedback loop continues with the costs of carbon reduction influencing GDP. Similarly, GDP boosts educational investment, enhancing public education levels and awareness of low-carbon practices, thereby influencing per capita residential energy consumption and total residential energy consumption, which further affects carbon emissions and environmental quality. Another critical aspect involves GDP driving investment in scientific research, leading to an increase in patent applications and the impact factor of technological progress, which also feeds back into the GDP through its influence on carbon emissions and environmental quality.

The primary causative relationship observed is that economic growth not only stimulates societal development and the expansion of residential areas, contributing to increased carbon emissions from residential houses, but also promotes investments in science and education. These investments, in turn, are instrumental in devising strategies to reduce carbon emissions through various avenues, where modifications in any of the variables can exert a direct influence on others, potentially reducing carbon emissions favorably. This reasoning is also used in the feedback loops shown in Figure 6b–d, which show the social, energy, and environmental subsystems, respectively. This shows a complete way to understand and deal with the changes in Kunming, China, residential carbon emissions.

Creating a stock–flow diagram is an important part of this study because it shows exactly how the different parts of the system are connected logically and how dynamic feedback works. This diagrammatic representation is crucial for facilitating a comprehensive analysis and enabling an effective simulation of the system's behavior. In the context of the SD model in this study, the flows within the system act as catalysts, triggering changes in the stocks. Figure 7 presents the stock flow diagram specifically for residential carbon emissions. This diagram was created using the knowledge obtained from the causal loop diagram presented in Figure 5. It meticulously maps out the flow of variables and their cumulative impact on the stock of carbon emissions, thereby offering a visual and analytical tool to understand and predict the patterns and trends of carbon emissions in residential settings.

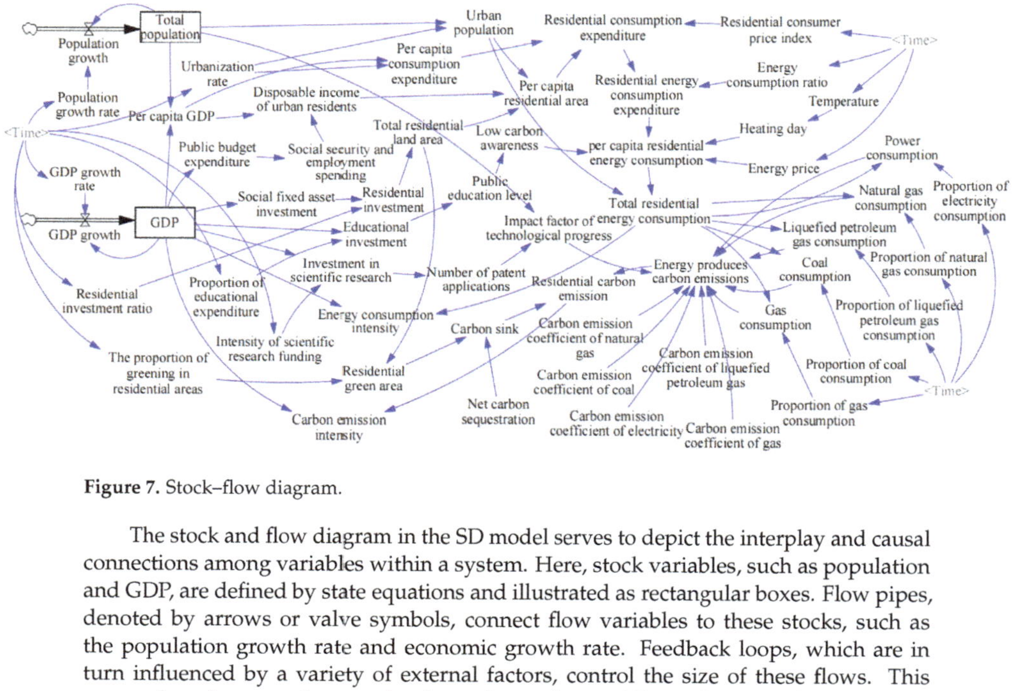

Figure 7. Stock–flow diagram.

The stock and flow diagram in the SD model serves to depict the interplay and causal connections among variables within a system. Here, stock variables, such as population and GDP, are defined by state equations and illustrated as rectangular boxes. Flow pipes, denoted by arrows or valve symbols, connect flow variables to these stocks, such as the population growth rate and economic growth rate. Feedback loops, which are in turn influenced by a variety of external factors, control the size of these flows. This approach underscores the complex dynamics at play, enabling a deeper understanding of the system's behavior.

This research adopts three foundational methodologies to craft the equations that determine the initial values and variable parameters, forming the crux of the model under investigation, with comprehensive details provided in Appendix B.

The first method, constant determination, leverages an extensive array of sources, including existing research, the Kunming City Development Strategy Plan [48], and the Kunming City Statistical Yearbook [34], among others. This technique anchors the model in the empirical reality of Kunming, China, ensuring it accurately mirrors the city's current development trajectory.

Secondly, the study employs table function entries to systematically gather variable data over time, introducing a vital temporal element to the analysis. This approach is instrumental in capturing the dynamic evolution of variables. During the mathematical formulation phase, the study judiciously selects the most fitting function type—whether logarithmic, exponential, or linear—based on data trends and patterns. This careful selection is crucial for precisely depicting the interrelationships among variables.

For addressing complex interactions between variables, especially where multiple independent variables impact a dependent variable amidst a mix of linear and nonlinear relationships, the study turns to regression analysis. This statistical technique allows for an in-depth examination of these intricate dynamics. The Statistical Package for the Social Sciences (SPSS) Statistics 27 software, known for its statistical analysis capabilities, was utilized for this purpose, providing a reliable tool for dissecting and interpreting the multifaceted relationships within the model. To ascertain the reliability of the regression analysis outcomes, a battery of statistical tests was executed using SPSS, with the findings detailed in Appendix B Table A1. The reliability was first assessed by evaluating the adjusted R-squared, Durbin–Watson (DW) statistics, and the correlation coefficient between independent and dependent variables. This assessment confirmed the strong fit and explanatory power of the regression model, as evidenced by adjusted R-squared values exceeding 0.8, DW values approximating the optimal 2, and the correlation coefficient close to 1 [49,50]. Following this, the analysis delved into investigating potential multicollinearity among the independent variables. The results, indicated by variance inflation factor (VIF) values below 2, demonstrated the absence of significant multicollinearity [51], thereby affirming the accuracy of the regression analysis.

2.6. Model Validation

To ensure the validity and reliability of the SD model, this study has implemented a series of rigorous testing techniques, including boundary testing, structural testing, and historical testing. These methods are essential for a comprehensive reevaluation and integration of the boundaries and subsystem feedback loops of the model. They help make sure that the equations of the model are correct in showing how different variables are related to each other, that parameter value estimates are based on good evidence, and that the quantitative framework is consistent.

The functionality and structural feasibility of the model have been further evaluated using the Vensim 10.1.2 software, a specialized tool for SD modeling [52]. The capability of this software to operate the model effectively attests to the practicality of the boundary and structure of the model. A pivotal aspect of the model validation process in research is historical testing, which involves comparing the simulation results of the model with historical data. The credibility of the model is reinforced when the discrepancy between these two datasets falls within the acceptable error margin of 15 percent as per SD standards [53].

Table 2 demonstrates this aspect by presenting a comparison of four key variables: total residential energy consumption, total residential carbon emissions, per capita disposable income, and energy consumption expenditures. The simulation results for these variables exhibit errors significantly lower than 15 percent, all falling under the 10 percent threshold, thus underscoring the accuracy of the model. Consequently, the model not only passes the historical test but also proves its utility in simulating the regulation of residential carbon emissions in Kunming under various policy scenarios. To evaluate how variations in pivotal parameters influence outcomes, sensitivity analysis was employed, with the findings presented in Appendix B Table A2. The simulation outcomes reveal a low sensitivity to parameter changes, signifying that the model exhibits robust stability. This characteristic affirms the model's reliability for conducting scenario simulations, indicating its capacity to provide consistent predictions under varying conditions. This stability is crucial for the model's application in strategic planning and policy development, allowing for confident use in exploring potential future scenarios and their implications.

Table 2. Comparison of historical test values and modeling simulation values.

Year	Total Carbon Emissions (×10⁴ tons)			Total Energy Consumption of Urban Dwellings (×10⁴ kg of Standard Coal)		
	Actual Value	Analog Value	Error (%)	Actual Value	Analog Value	Error (%)
2014	183.90	175.79	−4.41	63,860.22	61,367.40	−3.90
2015	204.53	200.36	−2.04	66,060.51	64,953.10	−1.68
2016	229.72	235.12	2.35	73,715.49	75,579.30	2.53
2017	216.53	228.81	5.67	68,773.18	72,669.80	5.67
2018	193.77	200.75	3.60	59,051.58	61,263.10	3.75
2019	243.11	252.94	4.04	75,167.47	78,455.60	4.37
2020	246.66	263.90	6.99	73,286.46	78,509.00	7.13
2021	310.24	298.87	−3.67	94,492.23	91,599.80	−3.06
Year	**Per capita Disposable Income of Urban Residents (CNY)**			**Energy Consumption Expenditure of Urban Residents (CNY)**		
	Actual Value	**Analog Value**	**Error (%)**	**Actual Value**	**Analog Value**	**Error (%)**
2014	31,295	32,603.8	4.18	625.27	657.28	5.12
2015	33,955	34,176.5	0.65	642.27	646.88	0.72
2016	36,739	36,458.7	−0.76	728.59	705.00	−3.24
2017	39,788	40,350.6	1.41	696.61	737.61	5.89
2018	42,988	42,506.7	−1.12	596.19	585.44	−1.80
2019	46,289	50,356.6	8.79	779.43	764.77	−1.88
2020	48,018	46,268.5	−3.64	551.03	553.71	0.49
2021	52,523	48,916.6	−6.87	763.00	718.87	−5.78

3. Results

3.1. *Simulation Results of the CO_2 Emission Trend*

This study uses a carefully constructed simulation model to predict future patterns of carbon emissions and emissions intensity in residential areas of Kunming from 2022 to 2030 (Figure 8). The simulation results indicate a sustained increase in household carbon emissions, mostly due to ongoing urbanization and the improving living standards of its residents. The simulation predicts a significant surge in the household carbon emissions of Kunming, projecting an increase to almost 4.108 million tons by 2030. This value is nearly 2.5 times higher than the baseline level observed in 2014.

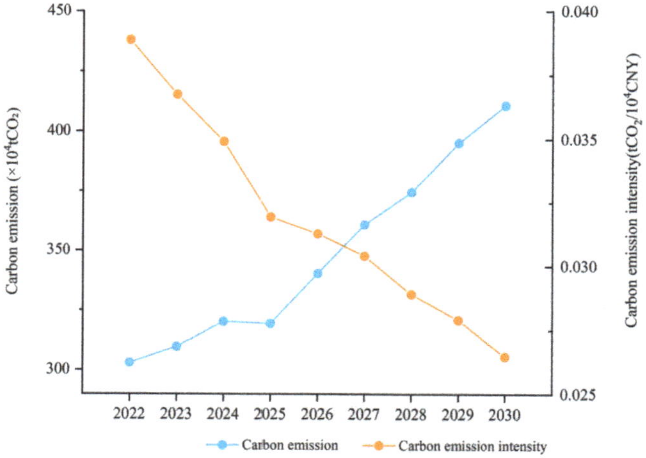

Figure 8. Carbon emission and carbon emission intensity of urban housing in Kunming, China, 2014–2021.

Moreover, the model forecasts a substantial decrease in the intensity of carbon emissions from residential sources. It is estimated that by 2025, the emission intensity will reduce to 0.0319 tons/million CNY, which represents an 18.4% decrease from the 2020 level of 0.0392 tons/million CNY. The anticipated reduction is in line with the goal of an 18% reduction, demonstrating Kunming's commitment to putting effective carbon control measures in place. Nevertheless, it is imperative to recognize the ongoing difficulties in controlling home carbon emissions, even with the introduction of many low-carbon projects. The continuous reduction in carbon emission intensity indicates advancement, but it also emphasizes the requirement for consistent and intensified endeavors in carbon management.

3.2. Simulation of Single-Factor Carbon Emissions

The optimal path to reduce emissions is determined by simulating the four main subsystems—economy, society, energy, and environment—using the scenario analysis method. This analytical approach allows for a detailed examination of how different factors influence various potential trajectories. The adjustments made in these simulations are meticulously outlined in Table 3, providing insight into the specific strategies implemented.

Table 3. Scenario modeling program settings.

Programmatic	Adjustment Range
Basic program	Keeping the system's factors constant and progressing smoothly at the pace prior to the x-factor
Option 1	Constant factors in the system, reduce factor x by 1 percentage point
Option 2	Constant factors in the system, increase factor x by 1 percentage point
Option 3	Constant factors in the system, increase factor x by 2 percentage points

Simulations in this study indicate that different factors within these subsystems have varying impacts on household carbon emissions, as detailed in Appendix C. The main factors contributing to an increase in carbon emissions include the urbanization rate, the economic growth rate, the residential investment ratio, and the level of residential energy consumption. For instance, a 1% increase in each of these factors leads to annual emission increases of approximately 82,600 tons, 69,800 tons, 3000 tons, and 12,000 tons, respectively, compared to the base scenario. Conversely, factors such as heightened low-carbon awareness, increased scientific research investment, expanded green spaces, optimized energy structures, and rising temperatures serve to inhibit the growth of carbon emissions. Compared to the base scenario, annual emissions will decrease by approximately 17,000 tons, 15,000 tons, 4000 tons, 16,700 tons, and 28,500 tons, respectively, for every 1% increase in these factors.

However, it is imperative to note that examining the contribution of a single factor in isolation is not sufficient, given the complexity and interactivity of the household carbon emission system. A comprehensive scenario incorporating various combinations of factors is required to fully understand and determine the most effective path for reducing carbon emissions.

Table 4 presents the results of analyzing the increase in carbon emissions under various scenarios, selecting those scenarios that maximize carbon emission factors in comparison to the base scenario. The sequence of influences on the increase in residential emissions is as follows: accelerating economic development, urbanization, residential energy consumption, and residential investment. Notably, accelerating the GDP by 1% above the initial growth rate would lead to a rise in household carbon emissions of around 10,000 tons in 2022 and 152,000 tons by 2030. This indicates that there is a positive and growing correlation between the GDP growth rate and carbon emissions over time. The annual increase in carbon emissions will amount to 80,000 tons, corresponding to a 1% rise in the urbanization rate. Conversely, the increase in residential investment has a minimal impact, resulting in an average annual emission rise of only 0.3 million tons. These findings indicate that the primary factor driving the increase in carbon emissions from residential sources is the significant growth of the economy, which concurrently leads to improved living standards and higher income for residents. The rapid increase in urbanization leads to a surge in the need for energy and housing in urban regions, resulting in a corresponding increase in residential energy usage and carbon emissions [53]. The impact of residential investment is minimal, primarily due to the shift towards environmentally friendly and low-carbon residential projects as well as the revitalization of existing neighborhoods in Kunming, China.

Table 4. Carbon emissions under different emission increase scenarios (10,000 tons).

Year of Modeling	Basic Program	Accelerating the Urbanization Process	Accelerating Economic Development	Increased Residential Investment	Increasing the Level of Residential Energy Consumption
2022	303.12	311.00	304.10	303.43	304.30
2023	309.88	317.82	311.97	310.19	311.08
2024	320.42	328.47	323.78	320.74	321.62
2025	319.62	327.49	324.25	319.92	320.82
2026	340.54	348.62	346.71	340.84	341.74
2027	361.06	369.40	369.06	361.36	362.27
2028	374.69	383.17	384.72	374.99	375.91
2029	395.35	404.10	407.78	395.65	396.59
2030	410.82	419.76	426.02	411.12	412.08

Considering the uncontrollable nature of temperature, the four main strategies for reducing residential carbon emissions are increased research investment, energy consumption structure optimization, raising low-carbon consciousness, and expanding green areas. Table 5 illustrates the relative influence of various strategies on the reduction of residential carbon emissions. The strategies are categorized in order of their effectiveness: enhancing awareness of low-carbon practices, optimizing the energy infrastructure, increasing investment in scientific research, and fostering the development of green environments. The three primary scenarios—increasing scientific research investment, improving the energy structure, and elevating the low-carbon consciousness of residents—demonstrate similar degrees of impact on reducing carbon emissions. Specifically, these strategies lead to annual emission reductions of 17,000 tons, 16,700 tons, and 15,000 tons, respectively, illustrating their substantial potential for diminishing carbon emissions comparably.

Table 5. Carbon emissions under different emission reduction scenarios (10,000 tons).

Year of Modeling	Basic Program	Raising Low-Carbon Awareness	Increased Investment in Research	Optimizing the Energy Mix	Expansion of Green Areas
2022	303.12	302.13	302.20	301.72	302.72
2023	309.88	308.78	308.86	308.45	309.48
2024	320.42	319.18	319.28	318.93	320.01
2025	319.62	318.26	318.39	318.10	319.20
2026	340.54	338.97	339.13	338.90	340.12
2027	361.06	359.24	359.44	359.31	360.64
2028	374.69	372.63	372.87	372.85	374.26
2029	395.35	392.97	393.28	393.41	394.92
2030	410.82	408.10	408.49	408.79	410.38

The analysis indicates that raising public awareness about low-carbon practices plays a crucial role in influencing both governmental and industrial policies towards energy-saving and emission-reducing building designs and regulations [54]. Such increased awareness also encourages residents to adopt more sustainable consumption habits and lifestyles, which are key to reducing carbon emissions in residential buildings. The optimization of the energy structure, particularly the shift towards natural gas, emerges as a significant factor due to its composition mainly of methane, which boasts a higher hydrocarbon ratio than coal and results in lower carbon dioxide emissions during combustion. Additionally, the thermal efficiency of natural gas power generation typically exceeds 60%, contributing to lower emissions and improved conversion efficiency [55,56]. The bolstering of financial resources for scientific research promotes the development and application of low-carbon technologies, which, along with effective management and increased public awareness, collectively contribute to the mitigation of the rise in residential carbon emissions.

However, the analysis also reveals that expanding green areas in residential spaces has a relatively minor impact on reducing carbon emissions in comparison to the other strategies. This limited effect is partly due to the current selection of green species and their carbon sequestration rate. Therefore, simply increasing the proportion of greening is insufficient. To maximize the impact on carbon emission reduction, it is imperative to enhance carbon sequestration efficiency, select tree species with high carbon-sequestering capabilities, and implement greening measures intensively.

3.3. Simulation of Multi-Factor Carbon Emissions

The residential carbon emission system is characterized by its complexity and dynamic nature, fluctuating with economic and societal growth. As such, regulating a single factor becomes increasingly challenging, necessitating a synergistic approach of multiple policies to support the low-carbon transformation of cities. According to the analysis in Section 3.2 of the study, factors like economic development, urbanization processes, energy consumption levels and structures, low-carbon awareness, and scientific research investments have a significant impact on changes in household carbon emissions in Kunming. After an in-depth examination of these critical factors, the study formulates three scenarios, drawing on references [57–60] to precisely depict the prevailing conditions in Kunming. As shown in Table 6, these scenarios rigorously simulate the influence of several policy interventions on the trend of residential carbon emissions in Kunming, with a specific focus on energy conservation and emission reduction.

Table 6. Multi-factor carbon emission scenario modeling options for Kunming, China.

Norm	Base Case	Low-Carbon Scenario (LCS)	Medium Low-Carbon Scenario (MLCS)	High Low-Carbon Scenario (HLCS)
GDP growth rate	Leave the original data unchanged	Reduction of 0.5 percent per year	1% reduction per year	2% reduction per year
Urbanization rate	Leave the original data unchanged	Reduction of 0.5 percent per year	1% reduction per year	2% reduction per year
Energy price	Leave the original data unchanged	1% increase per year	2% increase per year	3% increase per year
Energy consumption structure	Leave the original data unchanged	1.5% increase for natural gas, no change for gas, 1% decrease for LPG, 0.5% decrease for electricity	2% increase for natural gas, no change for gas, 1% decrease for LPG, 1% decrease for electricity	3% increase for natural gas, no change for gas, 1% decrease for LPG, 2% decrease for electricity
Investment in education	Leave the original data unchanged	1% increase per year	2% increase per year	3% increase per year
Investment in scientific research	Leave the original data unchanged	1% increase per year	2% increase per year	3% increase per year

The LCS involves a moderate reduction in the rate of growth and urbanization while increasing investment in education and scientific research. The focus is on promoting the development of low-carbon consciousness and technological progress, implementing gradual energy pricing modifications to manage residential energy usage, and increasing the proportion of renewable energy sources. This scenario proposes achieving efficient regulation of carbon emissions from residential areas by implementing a well-coordinated strategy that includes policies, technological advancements, and structural modifications.

The MLCS is an extension of the LCS that implements more rigorous methods to reduce emissions. Additionally, it further slows down urbanization and prioritizes the equitable progress of both economic expansion and carbon emission reduction. This scenario entails increased investment in education and scientific research, modifications to energy prices, and further optimization of the energy consumption structure.

The HLCS is characterized by a modest development rate for Kunming's economy and urbanization. This entails maintaining lower values for the GDP growth rate and urbanization rate. Building upon the MLCS, the government will enhance funding in the fields of education and scientific innovation, implement higher energy tariffs, and streamline the energy infrastructure. The HLCS embodies a bold strategy for achieving low-carbon development, with a focus on substantial changes in economic and urban growth patterns.

Every scenario is specifically created to examine a plan that combines economic growth with the reduction of residential carbon emissions in a sustainable manner. This provides valuable information on feasible approaches for achieving low-carbon development in Kunming, China.

Figure 9 presents the outcomes of a systematic simulation of various scenarios, projecting residential carbon emissions and carbon emission intensity in Kunming from 2022 to 2030. The base scenario, assuming the continuation of current policies and development trends, forecasts an average annual growth rate of 3.9% in the total residential carbon emissions of Kunming. This translates to an increase of 35.53% from 3,031,200 tons in 2022 to 4,108,200 tons by 2030. In contrast, under the LCS, emissions are projected to be controlled at 3.8951 million tons in 2030, marking a 32.34% increase over 2022 with an average annual growth rate of 3.6%. The LCS also anticipates a cumulative emission reduction of 1.2888 million tons compared to the base scenario.

Figure 9. Trend chart of (**a**) carbon emission and (**b**) carbon emission intensity changes in different scenarios for Kunming, China.

Similarly, the MLCS is expected to limit emissions to 3.7153 million tons in 2030, a 26.6% increase from 2022, with a modest average annual growth rate of 2.9%. The HLCS shows a more pronounced effect on reducing carbon emissions, with emission reductions ranging from 307,000 tons in 2022 to 659,000 tons by 2030 compared to the base scenario.

The analysis indicates that all scenarios demonstrate a growth in residential carbon emissions, albeit at varying rates. The low-carbon scenarios, particularly the MLCS and HLCS, offer more significant emission reductions compared to the LCS, suggesting the effectiveness of these strategies in curbing the rise of residential carbon emissions in Kunming. The LCS reduces emissions by 143,200 tons annually, while the MLCS and HLCS reduce emissions by 267,900 and 307,000 tons annually, respectively.

4. Discussion

The comprehensive energy conservation and emission reduction program outlined in the Kunming City Plan aims to achieve a GDP exceeding CNY 1 trillion by 2025. Concurrently, it targets a reduction in carbon dioxide emissions and energy consumption per unit of GDP of 18% and 14%, respectively, from the 2020 levels [48]. The simulation results of the three low-carbon development scenarios formulated in this study indicate that each scenario will lead to a decrease in carbon emissions and energy demand, aligning effectively with these planning objectives. Among the scenarios, the MLCS and HLCS exhibit the most substantial impact on emissions reduction. However, the HLCS falls short of meeting the GDP target due to its lower rate of economic growth. Consequently, the MLCS emerges as the most viable option for achieving the dual objectives of the city plan of Kunming—enhancing energy efficiency and reducing carbon emissions while sustaining GDP growth.

The approach to low-carbon urban economic development in China prioritizes maintaining its current level of economic growth and urbanization, which are fundamental and cannot be compromised for carbon emission reduction [61,62]. This necessitates a balance between economic development and environmental conservation, requiring multifaceted strategies, including institutional innovation, technological advancement, and industrial restructuring. A long-term strategy integrating environmental preservation with economic growth is essential. This involves setting phased emission reduction targets across various time frames, creating a systematic roadmap for residential low-carbon development, and implementing these plans at different administrative levels.

Moreover, urbanization should be viewed as an opportunity for transitioning to a low-carbon economy. Balancing urbanization rates with carbon emission increases is crucial. Aggressive energy and environmental policies should be adopted to enhance energy efficiency and structural optimization. Additionally, government incentives such as financial subsidies or tax breaks can motivate businesses to invest in low-carbon technology research and development [63]. This will not only improve the energy efficiency of household appliances but also accelerate the low-carbon development of residential buildings throughout their lifecycle. The government should also lead in enhancing incentive and constraint mechanisms, fostering collaboration among businesses, the public, and other societal actors to promote energy conservation and emission reduction in the residential sector.

Overall, this research offers a valuable exploration of the SD approach for simulating household carbon emissions, considering the geographical characteristics of Kunming. The modeling framework and findings provide a foundational basis for further studies and can be adapted to other cities with similar highland climates, serving as an important reference for policy development and research expansion. However, the models in this study are based on methodological assumptions made to simplify the dynamic complexity associated with characterizing real-world building stocks and energy performance. These assumptions inherently limit the model, warranting caution in its application. Future work could focus on model expansion and refinement to enhance its flexibility and applicability. This might include a thorough evaluation of the effects of potential socio-economic factors on carbon emissions, such as the impact of household income levels on energy usage and transportation choices, the lifestyle and consumption patterns differences between urban and rural dwellers, the level of public awareness and acceptance of environmental conservation and low-carbon lifestyles, as well as the influence of regional industrial structures and energy supply configurations on macroeconomic scales. By quantifying these key factors' modes of action, a more precise evaluation of the economic and social viability of various emission reduction strategies can be achieved.

5. Conclusions

This study utilized an SD model to simulate urban residential carbon emissions in Kunming, China, for the period 2022–2030 under various scenarios. The key findings from the simulations are summarized as follows:

Growth in Urban Residential Carbon Emissions: The urban residential carbon emissions of Kunming have exhibited a significant increase, with an average annual growth rate of nearly 10% between 2014 and 2021, culminating in an overall rise of 68.7%. This trend is predominantly attributed to the expansion of residential areas and energy consumption driven by urbanization. Despite a temporary decrease in emissions during 2017 and 2018 due to changes in energy use structure, the overall pattern of growth remains consistent.

Projected Increase in Carbon Emissions: Predictions in the study indicate a steep rise in carbon emissions from 2022 to 2030, with an expected peak of 4.108 million tons—almost 2.5 times the 2014 baseline. This underscores the ongoing and future increase in domestic energy consumption and associated carbon emissions.

Influencing Factors and Emission Reduction Strategies: The most significant factors influencing urban residential carbon emissions include economic growth, urbanization, residential energy consumption, and residential investment. Conversely, the key drivers for emission reduction are, in order of effectiveness: enhancing low-carbon awareness, modifying the energy structure, increasing investment in scientific and technological research and development, and expanding green space areas.

Efficacy of Comprehensive vs. Single-Factor Regulation: The study reveals that comprehensive regulation strategies are more effective in reducing emissions compared to single-factor approaches. The MLCS and HLCS demonstrate a higher potential for emission reduction than the standard LCS, particularly the HLCS, which can potentially reduce emissions by up to 20%. However, the HLCS does not align with the 2025 GDP target of CNY 1 trillion. Therefore, it is recommended that, in the short term, the MLCS scenario be targeted to achieve a reduction of about 10% by 2030, balancing economic and environmental pressures. In the medium and long term, investment in science and technology should increase, energy use structure should be optimized, the low-carbon concept of the public should be cultivated, and there should be preparation for achieving a higher level of emission reduction.

This study introduces a pioneering approach to scrutinizing the residential carbon emissions in Kunming, China, a topic that has been relatively underexplored, particularly due to the city's distinctive highland geography and climate. Through the development of an SD model, this study illuminates the intricate interactions within the complex system of residential carbon emissions in Kunming, thereby offering a more nuanced and accurate depiction of their formation process. To find the best way to reduce emissions while still promoting economic growth and meeting emission reduction goals, the study also started to construct different scenarios that showed how residential carbon emissions might change in the future. These scenarios were used to compare the effects of different emission reduction strategies using numbers. This aids not only in crafting long-term visions but also in devising phased action plans. The methodology and model presented here hold the potential for adaptation across other cities, enabling the development of tailored emission reduction strategies that resonate with local specificities. Moving forward, there is potential to broaden the model to include more emissions sectors, such as industry and transportation, and allow for a comprehensive evaluation of emission reduction initiatives across the entire city.

Author Contributions: Conceptualization, J.X. and G.X.; methodology, Y.Q.; software, Y.Q.; validation, J.X., G.X. and Y.Q.; investigation, R.L., Y.Q. and B.H.; resources, J.X. and G.X.; data curation, Y.Q.; writing—original draft preparation, Y.Q.; writing—review and editing, J.X. and G.X.; visualization, Y.Q., B.H. and H.X.; supervision, J.X. and G.X.; project administration, J.X. and G.X. All authors have read and agreed to the published version of the manuscript.

Funding: This research was funded by the National Natural Science Foundation of China, grant number 51878591.

Data Availability Statement: The original contributions presented in the study are included in the article, further inquiries can be directed to the corresponding author.

Conflicts of Interest: The authors declare no conflicts of interest. The funders had no role in the design of the study; in the collection, analyses, or interpretation of data; in the writing of the manuscript; or in the decision to publish the results.

Appendix A. Grey Relational Degree Analysis of Influencing Factors

Subsystem	Influencing Factor	Grey Correlation Degree (γ_n)
Economic subsystem	GDP	0.761
	Investment in fixed assets	0.830
	Per capita GDP	0.819
	Public budget expenditure	0.844
	Education expenditure	0.821
	Expenditure on science and technology	0.824
	Energy price	0.722
	Residential investment	0.756
Social subsystem	Urbanization rate	0.799
	Urban population	0.881
	Per capita disposable income of urban residents	0.861
	Per capita living area of urban residents	0.752
	Per capita consumption expenditure of urban residents	0.736
	Residential energy consumption expenditure of urban residents	0.815
	Residential consumption expenditure of urban residents	0.868
Energy subsystem	Energy structure	0.799
	Carbon emission intensity	0.730
	Per capita residential energy consumption	0.730
	Public education level (per 100,000 people enrolled in institutions of higher learning)	0.843
	Applied technology scientific and technological achievements	0.767
Environmental subsystem	Heating degree day	0.741
	Residential green space	0.886

Appendix B. The Main Equations of the SD Systems

(1) Population growth rate = WITH LOOKUP (Time, ([(2014, 0)-(2030, 0.3)], (2014, 0.018515), (2015, 0.00830099), (2016, 0.00649399), (2017, 0.0113687), (2018, 0.01823), (2019, 0.1862), (2020, 0.003592), (2021, 0.0055), (2022, 0.0065), (2023, 0.0075), (2024, 0.0085), (2025, 0.0095), (2026, 0.0105), (2027, 0.0115), (2028, 0.0125), (2029, 0.0135), (2030, 0.012))), Units: %

(2) GDP growth rate = WITH LOOKUP (Time, ([(2014, 0)-(2030, 0.3)], (2014, 0.0692191), (2015, 0.0832317), (2016, 0.129571), (2017, 0.0718991), (2018, 0.243711), (2019, 0.0398262), (2020, 0.0725758), (2021, 0.0795), (2022, 0.0819), (2023, 0.0878), (2024, 0.0899), (2025, 0.088), (2026, 0.0911), (2027, 0.0915), (2028, 0.0934), (2029, 0.0953), (2030, 0.0945))), Units: %

(3) Urbanization rate = WITH LOOKUP (Time, ([(2014, 0.6)-(2030, 1)], (2014, 0.698198), (2015, 0.714703), (2016, 0.731405), (2017, 0.748306), (2018, 0.763604), (2019, 0.78), (2020, 0.7967), (2021, 0.805), (2022, 0.8133), (2023, 0.8216), (2024, 0.8299), (2025, 0.8382), (2026, 0.8465), (2027, 0.8548), (2028, 0.8631), (2029, 0.8714), (2030, 0.8797))), Units: %

(4) Proportion of natural gas consumption = WITH LOOKUP (Time, ([(2014, 0)-(2030, 0.5)], (2014, 0.1077), (2015, 0.1335), (2016, 0.1526), (2017, 0.2376), (2018, 0.2429), (2019, 0.2912),

(2020, 0.204), (2021, 0.2937), (2022, 0.2889), (2023, 0.292), (2024, 0.31), (2025, 0.33), (2026, 0.356), (2027, 0.369), (2028, 0.382), (2029, 0.386), (2030, 0.395))), Units: %

(5) Proportion of electricity consumption = WITH LOOKUP (Time, ([(2014, 0)-(2030, 1)], (2014, 0.540148), (2015, 0.606754), (2016, 0.617335), (2017, 0.62122), (2018, 0.686736), (2019, 0.657694), (2020, 0.710263), (2021, 0.670625), (2022, 0.678), (2023, 0.67715), (2024, 0.66159), (2025, 0.64584), (2026, 0.62217), (2027, 0.61234), (2028, 0.60076), (2029, 0.59599), (2030, 0.59142))), Units: %

(6) Proportion of liquefied petroleum gas consumption = WITH LOOKUP (Time, ([(2014, 0)-(2030, 0.2)], (2014, 0.0962861), (2015, 0.0854082), (2016, 0.0857058), (2017, 0.0702644), (2018, 0.0372152), (2019, 0.0309252), (2020, 0.080279), (2021, 0.0311655), (2022, 0.029), (2023, 0.027), (2024, 0.025), (2025, 0.021), (2026, 0.019), (2027, 0.016), (2028, 0.015), (2029, 0.016), (2030, 0.012))), Units: %

(7) The proportion of coal consumption = WITH LOOKUP (Time, ([(2014, 0)-(2030, 0.04)], (2014, 0.00997455), (2015, 0.0243855), (2016, 0.00541693), (2017, 0.0211978), (2018, 0.00927345), (2019, 0.00838114), (2020, 0.000788273), (2021, 0.00129286), (2022, 0.001), (2023, 0.00095), (2024, 0.00091), (2025, 0.00086), (2026, 0.00083), (2027, 0.00076), (2028, 0.00074), (2029, 0.00071), (2030, 0.00068))), Units: %

(8) Proportion of gas consumption = WITH LOOKUP (Time, ([(2014, 0)-(2030, 0.3)], (2014, 0.245922), (2015, 0.149968), (2016, 0.138979), (2017, 0.0497589), (2018, 0.023837), (2019, 0.0118216), (2020, 0.0046909), (2021, 0.00320352), (2022, 0.0031), (2023, 0.0029), (2024, 0.0025), (2025, 0.0023), (2026, 0.002), (2027, 0.0019), (2028, 0.0015), (2029, 0.0013), (2030, 0.0009))), Units: %

(9) Residential consumer price index = WITH LOOKUP (Time, ([(2014, 0)-(2030, 110)], (2014, 104.6), (2015, 100.3), (2016, 102.9), (2017, 99.2), (2018, 99.8), (2019, 99.8), (2020, 99.3), (2021, 99.6), (2022, 99.5), (2023, 99.1), (2024, 98.6), (2025, 98.1), (2026, 98.3), (2027, 98.6), (2028, 98.2), (2029, 98.3), (2030, 98.4))), Units: %

(10) Temperature = WITH LOOKUP (Time, ([(2014, 9)-(2030, 20)], (2014, 10.6167), (2015, 10.4167), (2016, 9.56667), (2017, 11), (2018, 11.5), (2019, 10.8667), (2020, 10.8833), (2021, 10.9), (2022, 10.9), (2023, 11.1), (2024, 11.3), (2025, 11.4), (2026, 11.1), (2027, 11), (2028, 11.1), (2029, 11.1), (2030, 11.3))), Units: °C

(11) Residential investment ratio = WITH LOOKUP (Time, ([(2014, 0.2)-(2030, 0.5)], (2014, 0.266056), (2015, 0.2504), (2016, 0.237705), (2017, 0.251047), (2018, 0.26082), (2019, 0.327293), (2020, 0.333349), (2021, 0.338772), (2022, 0.3339), (2023, 0.3406), (2024, 0.3472), (2025, 0.3536), (2026, 0.3596), (2027, 0.3652), (2028, 0.3707), (2029, 0.3757), (2030, 0.3806))), Units: %

(12) Proportion of educational expenditure = WITH LOOKUP (Time, ([(2014, 0)-(2030, 1)], (2014, 0.0232728), (2015, 0.0233531), (2016, 0.0253728), (2017, 0.0243264), (2018, 0.0249763), (2019, 0.0209537), (2020, 0.0212946), (2021, 0.0200042), (2022, 0.0215), (2023, 0.0225), (2024, 0.0235), (2025, 0.0245), (2026, 0.0255), (2027, 0.0265), (2028, 0.0275), (2029, 0.0285), (2030, 0.0295))), Units: %

(13) Intensity of scientific research funding = WITH LOOKUP (Time, ([(2014, 0)-(2030, 0.2)], (2014, 0.0154377), (2015, 0.0185718), (2016, 0.0191562), (2017, 0.0188281), (2018, 0.0188097), (2019, 0.0173027), (2020, 0.0179483), (2021, 0.01875), (2022, 0.0205), (2023, 0.0226), (2024, 0.0243), (2025, 0.0255), (2026, 0.0276), (2027, 0.0296), (2028, 0.031), (2029, 0.0328), (2030, 0.0339))), Units: %

(14) GDP = INTEG (GDP growth, 3712.99), Units: one hundred million CNY

(15) Total population = INTEG (population increase, 655.258), Units: 10,000 people

(16) GDP growth = GDP × GDP growth rate, Units: 100 million CNY/year

(17) Population growth = Population growth rate × Total population, Units: 10,000 people/year

(18) Per capita GDP = GDP × 10,000/total population, Units: CNY

(19) Residential investment = Social fixed asset investment × residential investment ratio, Units: 100 million CNY

(20) Research fund input = GDP × Research fund input intensity, Units: 100 million CNY

(21) Education input = GDP × Proportion of education expenditure input, Units: 100 million CNY

(22) Public budget expenditure = 492.307 × ln (GDP) − 3458, Units: 100 million CNY

(23) Social fixed asset investment = 2202.73 × LN (GDP) − 14,668.9, Units: one hundred million CNY

(24) Consumption Expenditure per Urban Resident = 0.228 × GDP per capita + 117,664 × Share of Urban Population − 76,233, Units: CNY

(25) Residential Energy Consumption = Residential Consumption Expenditure × Energy Consumption Ratio, Units: CNY

(26) Number of patent applications = 311.768 × Investment in research funds − 11,215.2, Units: pieces

(27) Public education level = 0.004 × Education input + 0.282, Units: Dmnl

(28) Urban disposable income = 0.384 × GDP per capita + 106.265 × Social Security and Employment Expenditures + 3665.93, Units: CNY

(29) Electricity Consumption = (Total Energy Consumption of Urban Residence × Percentage of Electricity Consumption)/0.123, Units: 10^4 KWh

(30) Natural Gas Consumption = (Total Energy Consumption of Urban Residence × Percentage of Natural Gas Consumption)/1.33, Units: 10^4 m^3

(31) LPG consumption = (Total energy consumption of urban housing × Proportion of LPG consumption)/1.714, Units: Ten thousand kilograms

(32) Carbon sink = Green area of residential area × Net carbon sequestration × 0.001, Units: 10^4 tons

(33) Total energy consumption of urban dwellings = 0.981 × Residential domestic energy use + 28.98 × Heating days − 18,064, Units: 10^4 kg of standard coal

(34) Residential consumption expenditure = (0.147 × Consumption expenditure per capita of urban residents + 492.114 × Living space per capita − 19,410.2) × Consumer price index of residential category/100, Units: CNY

(35) Technological Progress Impact Factor = 1 − Number of Patent Applications/Total Population/10^4, Units: Dmnl

(36) Low Carbon Awareness = 1 − Number of Patent Applications/Total Population/104, Units: Dmnl

(37) Carbon emissions from energy generation = (Natural gas consumption × Natural gas carbon emission factor + Liquefied petroleum gas consumption × Liquefied petroleum gas carbon emission factor + Coal consumption × Coal carbon emission factor + Electricity consumption × Electricity carbon emission factor + Coal gas consumption × Coal gas carbon emission factor) × Technological Progress Impact Factor × 0.001, Units: 10^4 tons

(38) Urban Residential Carbon Emissions = Carbon Emissions from Energy Generation − Carbon Sinks, Units: 10^4 tons

(39) Carbon Emission Intensity = Urban Residential Carbon Emissions/GDP, Units: tons/10^4 CNY.

(40) Residential energy consumption per capita = (0.129 × Residential energy consumption + 0.122 × Heating days − 0.513 × (Energy prices−100) − 29.694) × Low-carbon awareness, Units: kg standard coal.

(41) Social security and employment expenditure = 0.14 × Public budget expenditure − 14.727, Units: 100 million CNY

(42) Residential land area = 2.657 × Residential investment + 14,977.395, Units: 10 thousand square meters

(43) Per capita living area = 0.154 × (residential area/urban population) + 0.370 × per capita disposable income of urban residents × 0.0001 + 36.629, Units: Square meters per person

(44) Residential consumption expenditure of urban residents = (0.147 × per capita consumption expenditure of urban residents + 492.114 × per capita living area − 19,410.154) × Residential Consumer Price Index/100, Units: CNY

Table A1. Equation test data of the SD model.

Equation	The Adjusted R-Squared	DW	Correlation Coefficient	VIF
Social fixed-asset investment	0.821			
Public budget expenditure	0.925			
Social security and employment expenditure	0.915	2.401	0.963	1
Number of patent applications	0.967	1.432	0.986	1
Public education level	0.869	2.249	0.942	1
Residential land area	0.921	1.609	0.966	1.377
Per capita living area	0.912	1.22	0.961	1.377
Consumption Expenditure per urban resident	0.966	1.1	0.861	1.487
Residential consumption expenditure of urban residents	0.915	2.223	0.966	1.137
Residential energy consumption per capita	0.818	1.313	0.844	1.084
Technological Progress Impact Factor	0.899	2.1	0.977	1

Table A2. Sensitivity test results of the SD model (%).

Variable	Increase (10%)	Decrease (−10%)
Urbanization rate	2.5	2.5
Population growth rate	5	0.16
Energy prices	3.8	3.8
Proportion of scientific research expenditure	2.8	2.8
Proportion of educational expenditure	3	3
Greening area ratio	1.3	1.3
Residential investment ratio	1.1	1.1
GDP growth rate	3.3	2.4
Heating days	8.3	8.3
Natural gas consumption	8.55	8.55
Liquefied petroleum gas consumption	1.8	1.8
Coal consumption	0.9	0.9
Electricity consumption	2.26	2.26
Gas consumption	0.44	0.44

Appendix C. Impacts of Different Factors on Carbon Emissions

Table A3. The impact of urbanization rate on carbon emissions.

Urbanization Rate	Carbon Emission (×104 t CO_2)			
Year	Basic program	Option 1	Option 2	Option 3
2022	303.12	295.35	311.00	319.00
2023	309.88	302.05	317.82	325.87
2024	320.42	312.49	328.47	336.62
2025	319.62	311.86	327.49	335.46
2026	340.54	332.57	348.62	356.81
2027	361.06	352.84	369.40	377.84
2028	374.69	366.32	383.17	391.76
2029	395.35	387.72	404.10	412.95
2030	410.82	401.99	419.76	428.80

Table A4. The impact of education input on carbon emissions.

Education Input	**Carbon Emission (×10^4 t CO_2)**			
Year	Basic program	Option 1	Option 2	Option 3
2022	303.12	304.12	302.13	301.13
2023	309.88	310.98	308.78	307.67
2024	320.42	321.67	319.18	317.94
2025	319.62	320.98	318.26	316.91
2026	340.54	342.12	338.97	337.40
2027	361.06	362.88	359.24	357.43
2028	374.69	376.76	372.63	370.56
2029	395.35	397.74	392.97	390.58
2030	410.82	413.54	408.10	405.38

Table A5. The impact of energy prices on carbon emissions.

Energy Prices	**Carbon Emission (×10^4 t CO_2)**			
Year	Basic program	Option 1	Option 2	Option 3
2022	303.12	304.30	301.94	300.76
2023	309.88	311.08	308.69	307.49
2024	320.42	321.62	319.22	318.02
2025	319.62	320.82	318.42	317.21
2026	340.54	341.74	339.34	338.14
2027	361.06	362.27	359.85	358.64
2028	374.69	375.91	373.47	372.25
2029	395.35	396.59	394.12	392.88
2030	410.82	412.08	409.56	408.31

Table A6. The impact of GDP growth rate on carbon emissions.

GDP Growth Rate	**Carbon Emission (×10^4 t CO_2)**			
Year	Basic program	Option 1	Option 2	Option 3
2022	303.12	302.14	304.10	305.08
2023	309.88	307.80	311.97	314.08
2024	320.42	317.11	323.78	327.20
2025	319.62	315.10	324.25	328.99
2026	340.54	334.57	346.71	353.07
2027	361.06	353.38	369.06	377.38
2028	374.69	365.14	384.72	395.23
2029	395.35	383.60	407.78	420.92
2030	410.82	396.56	426.02	442.20

Table A7. The impact of residential investment ratio on carbon emissions.

Residential Investment Ratio	**Carbon Emission (×10^4 t CO_2)**			
Year	Basic program	Option 1	Option 2	Option 3
2022	303.12	302.81	303.43	303.75
2023	309.88	309.57	310.19	310.51
2024	320.42	320.11	320.74	321.05
2025	319.62	319.32	319.92	320.22
2026	340.54	340.25	340.84	341.14
2027	361.06	360.76	361.36	361.66
2028	374.69	374.40	374.99	375.29
2029	395.35	395.06	395.65	395.95
2030	410.82	410.52	411.12	411.43

Table A8. The impact of energy structures on carbon emissions.

Energy Structures	**Carbon Emission (×10^4 t CO_2)**			
Year	Basic program	Option 1	Option 2	Option 3
2022	303.12	301.72	300.50	298.05
2023	309.88	308.45	307.20	304.69
2024	320.42	318.93	317.61	314.99
2025	319.62	318.10	316.78	314.12
2026	340.54	338.90	337.46	334.57
2027	361.06	359.31	357.77	354.69
2028	374.69	372.85	371.24	368.01
2029	395.35	393.41	391.70	388.28
2030	410.82	408.79	407.01	403.45

Table A9. The impact of scientific research input on carbon emissions.

Scientific Research Input	**Carbon Emission (×10^4 t CO_2)**			
Year	Basic program	Option 1	Option 2	Option 3
2022	303.12	304.05	302.20	301.27
2023	309.88	310.90	308.86	307.84
2024	320.42	321.56	319.28	318.14
2025	319.62	320.85	318.39	317.16
2026	340.54	341.96	339.13	337.71
2027	361.06	362.69	359.44	357.82
2028	374.69	376.51	372.87	371.06
2029	395.35	397.43	393.28	391.20
2030	410.82	413.15	408.49	406.15

Table A10. The impact of heating days on carbon emissions.

Heating Days	**Carbon Emission (×10^4 t CO_2)**			
Year	Basic program	Option 1	Option 2	Option 3
2022	303.12	305.90	300.34	297.57
2023	309.88	312.70	307.07	304.25
2024	320.42	323.25	317.60	314.52
2025	319.62	322.45	316.79	313.69
2026	340.54	343.37	337.72	334.89
2027	361.06	363.91	358.21	355.36
2028	374.69	377.57	371.82	368.94
2029	395.35	398.27	392.44	389.52
2030	410.82	413.78	407.86	404.63

Table A11. The impact of greening area ratio on carbon emissions.

Greening Area Ratio	**Carbon Emission (×10^4 t CO_2)**			
Year	Basic program	Option 1	Option 2	Option 3
2022	303.12	303.52	302.72	302.32
2023	309.88	310.29	309.48	309.07
2024	320.42	320.83	320.01	319.60
2025	319.62	320.04	319.20	318.79
2026	340.54	340.97	340.12	339.70
2027	361.06	361.49	360.64	360.21
2028	374.69	375.13	374.26	373.83
2029	395.35	395.79	394.92	394.48
2030	410.82	411.27	410.38	409.93

References

1. IPCC. *Climate Change 2023: Synthesis Report. Contribution of Working Groups I. II and III to the Sixth Assessment Report of the Intergovernmental Panel on Climate Change*; IPCC: Geneva, Switzerland, 2023; pp. 35–115. [CrossRef]
2. Rong, P.; Zhang, Y.; Qin, Y.; Liu, G.; Liu, R. Spatial differentiation of carbon emissions from residential energy consumption: A case study in Kaifeng, China. *J. Environ. Manag.* **2020**, *271*, 110895. [CrossRef]
3. Chau, C.K.L.; Leung, T.M.; Ng, W.Y. A review on life cycle assessment, life cycle energy assessment and life cycle carbon emissions assessment on buildings. *Appl. Energy* **2015**, *143*, 395–413. [CrossRef]
4. Geng, J.J.; Wang, J.J.; Huang, J.G.; Zhou, D.; Bai, J.; Wang, J.Y.; Zhang, H.; Duan, H.B.; Zhang, W.B. Quantification of the carbon emission of urban residential buildings: The case of the Greater Bay Area cities in China. *Environ. Impact Assess. Rev.* **2022**, *95*, 106775. [CrossRef]
5. You, F.H.; Hu, D.; Zhang, H.; Guo, Z.; Zhao, Y.; Wang, B.; Yuan, Y. Carbon emissions in the life cycle of urban building system in China—A case study of residential buildings. *Ecol. Complex.* **2011**, *8*, 201–212. [CrossRef]
6. Burleyson, C.D.; Iyer, G.; Hejazi, M.; Kim, S.; Kyle, P.; Rice, J.S.; Smith, A.D.; Taylor, Z.T.; Voisin, N.; Xie, Y.L. Future western US building electricity consumption in response to climate and population drivers: A comparative study of the impact of model structure. *Energy* **2020**, *208*, 118312. [CrossRef]
7. Wang, Q.; Lin, J.; Zhou, K.; Fan, J.; Kwan, M.P. Does urbanization lead to less residential energy consumption? A comparative study of 136 countries. *Energy* **2020**, *202*, 117765. [CrossRef]
8. Huo, T.F.; Li, X.H.; Cai, W.G.; Zuo, J.; Jia, F.Y.; Wei, H.F. Exploring the impact of urbanization on urban building carbon emissions in China: Evidence from a provincial panel data model. *Sustain. Cities Soc.* **2020**, *56*, 102068. [CrossRef]
9. Gan, L.; Liu, Y.; Shi, Q.; Cai, W.; Ren, H. Regional inequality in the carbon emission intensity of public buildings in China. *Build. Environ.* **2022**, *225*, 109657. [CrossRef]
10. Wu, L.F.; Huang, G.M.; Fan, J.L.; Zhang, F.C.; Wang, X.K.; Zeng, W.Z. Potential of kernel-based nonlinear extension of Arps decline model and gradient boosting with categorical features support for predicting daily global solar radiation in humid regions. *Energy Convers. Manag.* **2019**, *183*, 280–295. [CrossRef]
11. Trotta, G. Factors affecting energy-saving behaviours and energy efficiency investments in British households. *Energy Policy* **2018**, *114*, 529–539. [CrossRef]
12. Wan, C.; Shen, G.Q.; Choi, S. The moderating effect of subjective norm in predicting intention to use urban green spaces: A study of Hong Kong. *Sustain. Cities Soc.* **2018**, *37*, 288–297. [CrossRef]
13. Shi, H.X.; Wang, S.Y.; Zhao, D.T. Exploring urban resident's vehicular PM2.5 reduction behavior intention: An application of the extended theory of planned behavior. *J. Clean. Prod.* **2017**, *147*, 603–613. [CrossRef]
14. Liu, X.; Wang, Q.C.; Wei, H.H.; Chi, H.L.; Ma, Y.T.; Jian, I.Y. Psychological and Demographic Factors Affecting Household Energy-Saving Intentions: A TPB-Based Study in Northwest China. *Sustainability* **2020**, *12*, 836. [CrossRef]
15. Fan, J.B.; Ran, A.B.; Li, X.M. A Study on the Factors Affecting China's Direct Household Carbon Emission and Comparison of Regional Differences. *Sustainability* **2019**, *11*, 4919. [CrossRef]
16. Li, J.B.; Huang, X.J.; Yang, H.; Chuai, X.W.; Li, Y.; Qu, J.S.; Zhang, Z.Q. Situation and determinants of household carbon emissions in Northwest China. *Habitat Int.* **2016**, *51*, 178–187. [CrossRef]
17. Fremstad, A.; Underwood, A.; Zahran, S. The Environmental Impact of Sharing: Household and Urban Economies in CO_2 Emissions. *Ecol. Econ.* **2018**, *145*, 137–147. [CrossRef]
18. Paone, A.; Bacher, J.P. The Impact of Building Occupant Behavior on Energy Efficiency and Methods to Influence It: A Review of the State of the Art. *Energies* **2018**, *11*, 953. [CrossRef]
19. Hu, Z.; He, J.; Wang, Y. Factor analysis and enlightenment of household carbon emissions in Japan based on IPAT-LMDI extension model. *Resour. Sci.* **2018**, *40*, 1831–1842.
20. Li, Z.; Liu, Y.Z.; Zhang, Z. Carbon Emission Characteristics and Reduction Pathways of Urban Household in China. *Front. Environ. Sci.* **2022**, *10*, 896765. [CrossRef]
21. Jin, J.; Wang, Y.C.; Zheng, X.Y. District heating versus self-heating: Estimation of energy efficiency gap using regression discontinuity design. *China Econ. Q. Int.* **2021**, *1*, 208–220. [CrossRef]
22. Payne, S.; Barker, A. Carbon regulation and pathways for institutional transition in market-led housing systems: A case study of English housebuilders and zero carbon housing policy. *Environ. Plan. E-Nat. Space* **2018**, *1*, 470–493. [CrossRef]
23. Zhang, X.F.; Fan, D.C. The Spatial-Temporal Evolution of China's Carbon Emission Intensity and the Analysis of Regional Emission Reduction Potential under the Carbon Emissions Trading Mechanism. *Sustainability* **2022**, *14*, 7442. [CrossRef]
24. Li, H.; Cao, L.; Chen, Y.B.; Qiu, H.Y.; Huang, X.; Guo, W.Q.; Wang, S.T.; Sheng, W.G.; Zhou, Y.L.; Xu, Q. Emergy analysis-based study of the sustainable development of Kunming's urban eco-economic system. *Int. J. Urban Sci.* **2023**, *27*, 322–343. [CrossRef]
25. Shao, P.; Xin, J.Y.; Zhang, X.L.; Gong, C.S.; Ma, Y.J.; Wang, Y.S.; Wang, S.G.; Hu, B.; Ren, X.B.; Wang, B.Y. Aerosol optical properties and their impacts on the co-occurrence of surface ozone and particulate matter in Kunming City, on the Yunnan-Guizhou Plateau of China. *Atmos. Res.* **2022**, *266*, 105963. [CrossRef]
26. Gao, Y.; Pitts, A.; Amsterdam Univ, P. The Ancient Southern Silk Road and the Contemporary One Belt and One Road in Southwest China Case study of Kunming City in Yunnan Province. In Proceedings of the 12th International Convention of Asia Scholars (ICAS), Electr Network, 24–28 August 2021; pp. 177–182.

27. Liu, Z.; Sun, T.; Yu, Y.; Ke, P.; Deng, Z.; Lu, C.; Huo, D.; Ding, X. Near-Real-Time Carbon Emission Accounting Technology Toward Carbon Neutrality. *Engineering* **2022**, *14*, 44–51. [CrossRef]
28. Wen, Z.; Yang, Z.; Xing, Q. A comprehensive evaluation of regional carbon emission based on the composite model in China: A case study of Huaibei city (China). *Sci. Rep.* **2023**, *13*, 15387. [CrossRef] [PubMed]
29. Xiong, P.; Wu, X.; Ye, J. Building a novel multivariate nonlinear MGM(1,m,N | γ) model to forecast carbon emissions. *Environ. Dev. Sustain.* **2023**, *25*, 9647–9671. [CrossRef]
30. Oh, J.; Hong, T.; Kim, H.; An, J.; Jeong, K.; Koo, C. Advanced Strategies for Net-Zero Energy Building: Focused on the Early Phase and Usage Phase of a Building's Life Cycle. *Sustainability* **2017**, *9*, 2272. [CrossRef]
31. Schmidt, M.; Crawford, R.H. A framework for the integrated optimisation of the life cycle greenhouse gas emissions and cost of buildings. *Energy Build.* **2018**, *171*, 155–167. [CrossRef]
32. Zhong, X.; Li, X.; Li, Y. Urban Expansion and Carbon Emission Logistic Curve Hypothesis and Its Verification: A Case Study of Jiangsu Province. *Land* **2022**, *11*, 1066. [CrossRef]
33. Cheng, S.; Zhou, X.; Zhou, H. Study on Carbon Emission Measurement in Building Materialization Stage. *Sustainability* **2023**, *15*, 5717. [CrossRef]
34. People's Government of Kunming. Kunming Statistical Yearbook. Available online: https://www.km.gov.cn/zfxxgk/fdzdgknr/tjxx/jjyxqk/tjnj/ (accessed on 18 January 2024).
35. National Bureau of Statistics of China. Kunming Statistical Yearbook (2014–2021). Available online: https://www.stats.gov.cn/sj/ndsj/2022/indexeh.htm (accessed on 20 January 2024).
36. National Development and Reform Commission of China. Technology Policy for Upgrading Coal Power Units for Energy Conservation and Emission Reduction. Available online: https://www.cmenhu.cn/13448.html (accessed on 20 January 2024).
37. Zhou, G.H.; Chung, W.; Zhang, Y.X. Carbon dioxide emissions and energy efficiency analysis of China's regional thermal electricity generation. *J. Clean. Prod.* **2014**, *83*, 173–184. [CrossRef]
38. Huang, Y.S.; Shen, L.; Liu, H. Grey relational analysis, principal component analysis and forecasting of carbon emissions based on long short-term memory in China. *J. Clean. Prod.* **2019**, *209*, 415–423. [CrossRef]
39. Kim, G.; Jong, Y.; Liu, S.F. Generalized Hybrid Grey Relation Method for Multiple Attribute Mixed Type Decision Making. *J. Grey Syst.* **2014**, *26*, 142–153.
40. Sun, Y.M.; Liu, S.X.; Li, L. Grey Correlation Analysis of Transportation Carbon Emissions under the Background of Carbon Peak and Carbon Neutrality. *Energies* **2022**, *15*, 3064. [CrossRef]
41. Lv, T.G.; Hu, H.; Zhang, X.M.; Sun, L.; Li, Z.L.; Chen, Y.J.; Fu, S.F. Driving mechanism of the allometric relationship between economic development and carbon emissions in the Yangtze River Delta urban agglomeration, China. *Environ. Dev. Sustain.* **2023**. [CrossRef]
42. Forrester, J.W. System dynamics—The next fifty years. *Syst. Dyn. Rev.* **2007**, *23*, 359–370. [CrossRef]
43. Zhan, J.Y.; Wang, C.; Wang, H.H.; Zhang, F.; Li, Z.H. Pathways to achieve carbon emission peak and carbon neutrality by 2060: A case study in the Beijing-Tianjin-Hebei region, China. *Renew. Sustain. Energy Rev.* **2024**, *189*, 113955. [CrossRef]
44. Huo, T.F.; Ma, Y.L.; Xu, L.B.; Feng, W.; Cai, W.G. Carbon emissions in China's urban residential building sector through 2060: A dynamic scenario simulation. *Energy* **2022**, *254*, 124395. [CrossRef]
45. Huo, T.F.; Xu, L.B.; Feng, W.; Cai, W.G.; Liu, B.S. Dynamic scenario simulations of carbon emission peak in China's city-scale urban residential building sector through 2050. *Energy Policy* **2021**, *159*, 112612. [CrossRef]
46. Murray, P.; Marquant, J.; Niffeler, M.; Mavromatidis, G.; Orehounig, K. Optimal transformation strategies for buildings, neighbourhoods and districts to reach CO_2 emission reduction targets. *Energy Build.* **2020**, *207*, 109569. [CrossRef]
47. Liu, L.Y.; Tang, Y.L.; Chen, Y.Y.; Zhou, X.; Bedra, K.B. Urban Sprawl and Carbon Emissions Effects in City Areas Based on System Dynamics: A Case Study of Changsha City. *Appl. Sci.* **2022**, *12*, 3244. [CrossRef]
48. Kunming People's Government. The 14th Five-Year Plan for the National economic and Social Development of Kunming City and the Outline of the Long-Term Goals for 2035. Available online: https://www.km.gov.cn/c/2021-04-06/3890927.shtml (accessed on 20 January 2024).
49. Chicco, D.; Warrens, M.J.; Jurman, G. The coefficient of determination R-squared is more informative than SMAPE, MAE, MAPE, MSE and RMSE in regression analysis evaluation. *Peerj Comput. Sci.* **2021**, *7*, e623. [CrossRef] [PubMed]
50. Uyanto, S.S. Power comparisons of five most commonly used autocorrelation tests. *Pak. J. Stat. Oper. Res.* **2020**, *16*, 119–130. [CrossRef]
51. Shrestha, N. Detecting Multicollinearity in Regression Analysis. *Am. J. Appl. Math. Stat.* **2020**, *8*, 39–42. [CrossRef]
52. Liu, J.K.; Liu, Y.D.; Wang, X.T. An environmental assessment model of construction and demolition waste based on system dynamics: A case study in Guangzhou. *Environ. Sci. Pollut. Res.* **2020**, *27*, 37237–37259. [CrossRef] [PubMed]
53. Ning, X.; He, Y.; Zhang, J.; Wu, C.; Zhang, Y. Analysis of Carbon Emission Projections and Reduction Potential of Resource-Dependent Urban Agglomerations from the Perspective of Multiple Scenarios-A Case Study of Hu-Bao-O-Yu Urban Agglomeration. *Int. J. Environ. Res. Public Health* **2023**, *20*, 4250. [CrossRef] [PubMed]
54. Tian, X.; Geng, Y.; Dong, H.; Dong, L.; Fujita, T.; Wang, Y.; Zhao, H.; Wu, R.; Liu, Z.; Sun, L. Regional household carbon footprint in China: A case of Liaoning province. *J. Clean. Prod.* **2016**, *114*, 401–411. [CrossRef]
55. Oda, T.; Feng, L.; Palmer, P.; Baker, D.F.; Ott, L.E. Assumptions about prior fossil fuel inventories impact our ability to estimate posterior net CO_2 fluxes that are needed for verifying national inventories. *Environ. Res. Lett.* **2023**, *18*, 124030. [CrossRef]

56. Schernikau, L.; Smith, W.H. Climate impacts of fossil fuels in today's electricity systems. *J. S. Afr. Inst. Min. Metall.* **2022**, *122*, 133–145. [CrossRef]
57. Zheng, X.Q.; Wang, J.Y.; Chen, Y.; Tian, C.; Li, X.M. Potential pathways to reach energy-related CO_2 emission peak in China: Analysis of different scenarios. *Environ. Sci. Pollut. Res.* **2023**, *30*, 66328–66345. [CrossRef] [PubMed]
58. Sun, X.M.; Zhang, H.T.; Wang, X.Y.; Qiao, Z.K.; Li, J.S. Towards Sustainable Development: A Study of Cross-Regional Collaborative Carbon Emission Reduction in China. *Sustainability* **2022**, *14*, 9624. [CrossRef]
59. Ge, X.X.; Qi, W. Analysis of CO_2 Emission Drives Based on Energy Consumption and Prediction of Low Carbon Scenarios: A Case Study of Hebei Province. *Pol. J. Environ. Stud.* **2020**, *29*, 2185–2197. [CrossRef]
60. Ayuketah, Y.; Gyamfi, S.; Diawuo, F.A.; Dagoumas, A.S. Assessment of low-carbon energy transitions policies for the energy demand sector of Cameroon. *Energy Sustain. Dev.* **2023**, *72*, 252–264. [CrossRef]
61. Wang, J.Y.; He, D.Q. Sustainable urban development in China: Challenges and achievements. *Mitig. Adapt. Strateg. Glob. Chang.* **2015**, *20*, 665–682. [CrossRef]
62. Wu, H.L.; Wang, L.; Peng, D.X.; Liu, B.J. Input-output efficiency model of urban green-energy development from the perspective of a low-carbon economy. *Clean Energy* **2022**, *6*, 905–916. [CrossRef]
63. Liu, Y.; Cui, J.F.; Jiang, H.; Yan, H. Do county financial marketization reforms promote food total factor productivity growth?: A mechanistic analysis of the factors quality of land, labor, and capital. *Front. Sustain. Food Syst.* **2023**, *7*, 1263328. [CrossRef]

buildings MDPI

Article

Mist Spraying as an Outdoor Cooling Spot in Hot-Humid Areas: Effect of Ambient Environment and Impact on Short-Term Thermal Perception

Pin Wang *, Sumei Lu, Xiaowei Wu, Jun Tian and Ning Li

School of Environment and Civil Engineering, Dongguan University of Technology, Dongguan 523808, China; lusumei@dgut.edu.cn (S.L.); wxw@dgut.edu.cn (X.W.); tianjun@dgut.edu.cn (J.T.); wylining@163.com (N.L.)
* Correspondence: wangpin@dgut.edu.cn; Tel.: +86-152-1876-1737

Abstract: Mist spraying is an active cooling technology used to alleviate heat stress during hot summers. However, there is limited experimental research on the relationship between ambient thermal parameters and spray cooling efficiency, as well as the transient and short-term thermal perceptions of local residents. In this study, an intermittent mist spraying system was set up, and environmental measurements, coupled with questionnaire surveys, were conducted under typical high temperature and still air conditions. The aim was to investigate the relationship among environmental factors, spray cooling effects, and dynamic improvements in human thermal perception. The results showed that higher ambient temperatures resulted in a more significant cooling effect, with a maximum value of 5.68 °C. Upon entering the spraying area, people experienced a large perceptual change, with the mean thermal sensation and thermal comfort change covering 73% and 62% of the total change ranges, respectively. This study indicated that the mist spray system can be activated if the ambient temperature exceeds 32.5 °C, helping local residents maintain a physiological state close to slightly hot and neutral comfort. These findings suggest that mist spraying can be applied in environmental design as an outdoor cooling spot to mitigate urban overheating, providing valuable insights for the application of mist spray systems in actual outdoor settings in hot-humid areas.

Keywords: periodic mist spraying; urban cooling spot; outdoor microclimate; environmental impact factor; short-term thermal perception

1. Introduction

Urbanization has rapidly transformed urban structures and landscapes, resulting in a significant increase in carbon emissions [1]. Urban heat island and extreme heat have been occurring more frequently in cities across the world [2], increasing the discomfort of people and leading to a rise in the rate of heat-related mortality [3]. In hot and humid regions, high summer temperatures often fail to ensure pedestrian safety and comfort, impacting residents' daily mobility [4]. The inadequate urban pedestrian environment hinders city sustainability [5]. Therefore, integrating heat stress mitigation measures into urban outdoor design is crucial [6]. Establishing small and dispersed cool spots may mitigate urban overheating, reduce heatstroke risk, and enhance outdoor livability for people's travel and activities [7].

Most passive measures, like retro-reflective materials and vegetation, have limited controllability [8]. In contrast, mist spray systems, an active cooling technology based on evaporative cooling principles [9], offer flexibility and dynamic controls, effectively reducing heat stress [10]. Santamouris compared thermal environment improvement methods in hundreds of projects worldwide and found that spray-based evaporative cooling showed excellent cooling effects [11]. Mist spray systems have been installed in bus waiting areas, semi-outdoor public spaces in shopping malls, and significant event venues (such as the World Expo) in some cities. The installation and concealment of hydropower equipment in these locations, including the host control equipment, water filters, high-pressure water pipes, and fixed nozzle installations, provide important conditions for the application of mist spray systems and their cooling effects on the environment. During hot summers, creating scattered cool spots in open or semi-open spaces, such as station waiting areas, squares, and green belts where pergolas or pavilions can attract people to gather or briefly stop, can alleviate pedestrians' thermal discomfort through environmental design combined with spraying [12]. This can be effective even after just a few minutes, providing relief from the heat and enhancing the overall comfort of public spaces.

In recent years, the spray cooling system has gained attention in the built environment [12]. Ulpiani [13,14] and Giuseppe [15] conducted experiments in an Italian park and found that the spraying reduced air temperature by 8 °C and increased relative humidity by 7%. Sureshkumar [16] carried out experiments in India, where the temperature drop was up to 5.2 °C. Yamada [17] and Oh [18] executed experiments at the Aichi Expo and at a station in Japan. The results showed that air temperature could be reduced by 3 °C while relative humidity increased by over 23%. Huang [19] conducted mist spraying experiments at the Shanghai Expo, reducing air temperature by 5–7 °C with an ambient temperature and humidity of around 35 °C and 45%. Zhang [20] found that the mist spraying resulted in an average and maximum temperature reduction of 2.7 °C and 4.7 °C, respectively, as well as an average and maximum humidity increase of 9.4% and 15.5% at a bus stop waiting area built in the experiment in China. In these studies, the temperature reduction range was mostly within 1.0 °C to 9.0 °C, with a maximum of 15 °C [21], and the overall increase in relative humidity was within 5.3% to 29.3%, mostly related to the local climate [12].

Mist spray also affects the thermal perception of the human body. Farnham [22] conducted experiments and found that subjects' thermal sensation decreased from "Hot" (+2.7) to "Slightly cool" (−1.41) due to the outdoor mist. The research of Ulpiani [13] showed that 83% of the subjects felt uncomfortably hot outside the spraying area between 12:00 and 14:00, and the percentage dropped to 4% as they entered the mist. Zhang [20] suggested that participants' thermal sensation decreased from "between Hot and Slightly hot" (+1.44) to "near to Neutral" (+0.33), and thermal comfort increased from "almost Slightly uncomfortable" (−1.13) to "near to Slightly comfortable" (+0.76) as a result of spraying experiments. Desert [21] demonstrated that subjects experienced a reduction of 2~3 in thermal sensation in hot and humid extreme regions. Wong [23] found that the spraying could effectively achieve thermal neutrality by reducing air temperatures in Singapore. The effect of increased humidity caused by mist spraying on thermal perception has not been consistently observed in related studies. Wong [23] and Desert [21] suggested that severe discomfort was increased due to higher moisture levels of the skin, while Ulpiani [13], Farnham [22], and Vanos [24] believed that the moisture created by the spray was pleasing to the subjects. Indeed, the benefits of mist spray for enhancing physiological comfort could vary for people in different regions and climates [25]. Residents may have become accustomed to the prevailing weather conditions in their locality [26], including those living in hot-humid areas [27], especially when they are outdoors and semi-outdoors [28]. This may affect their assessment of the effect of mist spraying. Therefore, when evaluating the thermal perception and acceptability of the local population, it is essential to conduct targeted spraying experimental research in specific regions.

Most existing studies have focused on the thermal adaptation of participants to the spraying environment. Desert [21] compared subjects' thermal perception before entering the spray area and after spending 2–10 min in the mist. Participants in Ulpiani's experiments [13] stayed in the spraying area for 5–7 min before conducting the assessment. Zhang [20] had subjects remain in the mist for 10 min and complete a questionnaire evaluation every 3 min. In Wong's experiment [23], subjects completed a questionnaire at half-hour intervals after being exposed to the mist for at least 30 min. However, few studies have paid attention to the change in thermal perception at the moment of entering the mist spray. When people move into different environments, the human body is sensitive to sudden changes in local temperature, and thermal perception may undergo significant instantaneous changes [29], which may form a person's initial impression of the new environment and likely last for some time. It is necessary to study the changes in transient and short-term thermal perception when people enter the spraying area in the summer because people are more sensitive to cold stimulation [30].

The cooling efficiency of the spraying is partly determined by the setup of the mist spray equipment, including water pressure [12,13], droplet size [31,32], and the height of nozzles [13,15,33], among others. Sureshkumar [16] and Wang [10] revealed that higher pressure with smaller nozzles could produce more cooling. Yamada [34] proved that larger droplets remained in a lower position when spraying. When coming into contact with pedestrians, water spray may cause a strong sensation of wetness. Additionally, water spraying requires a larger amount of water usage. In contrast, the dispersed and smaller droplets of mist spray almost evaporate immediately after being injected, which is not only effective for cooling but also avoids unpleasant dampness for pedestrians [9]. Related studies suggested that the optimal water pressure range of high-pressure spraying was around 2 MPa–7 MPa [14,15,19], and the design value of spray volume was around 20 mL/min–50 mL/min [35,36]. Although most of these results were obtained under ideal conditions [12], they have a certain value for improving the operation performance of the spraying and can provide a useful reference for the construction of a typical mist spray system.

In addition to the setup of the mist spray equipment, outdoor thermal parameters also influence the cooling efficiency of the spraying [12]. Rising temperatures caused by solar radiation can accelerate mist evaporation [18]. Ulpiani [37] found that the intensity of solar radiation was negatively correlated with the drop in temperature during the spray experiment. He also reported that 80% of subjects felt pleasant warmth in the mist and concluded that mist spray was well suited to highly irradiated regions [13]. Meng [12] suggested that appropriate shading treatments combined with spraying could have a better cooling effect when the solar radiation intensity was too high. Wind can speed up water evaporation and lower air temperature more effectively [22,23], making mist-spray well-suited to suitable airflow conditions [13]. At the same time, wind will also make mist droplets run away from the target location, and non-directional wind could reduce the cooling efficiency of the spray system [31,38]. Giuseppe's numerical simulation analysis showed that the cooling capacity increased in calm air [15]. Ulpiani also concluded that the spray cooling effect was more apparent and easily measurable in still air conditions [9].

Several studies have shown that temperature and humidity are the primary outdoor thermal parameters affecting the effectiveness of spray cooling, with researchers demonstrating this through mathematical models [31] and numerical simulations [32]. For instance, Huang [31] used a developed mathematical model to calculate two extreme environmental conditions: one with a temperature of 38 °C and a relative humidity of 40%, and the other with a temperature of 30 °C and a relative humidity of 70%. The results indicated that higher temperatures and lower humidity resulted in higher cooling efficiency compared to lower temperatures and higher humidity. However, the mathematical model and numerical simulation are based on assumed ideal droplet evaporation conditions and a steady environment, which differ from actual outdoor spraying situations [12]. On the other hand, experimental testing is relatively commonly used and generally acknowledged as a more accurate method. Xie [39] constructed a small-scale experimental platform in a confined space and found that the higher the ambient temperature, the more significant the cooling efficiency. In an area with extremely high radiation and humidity, Desert established a three-dimensional spray experimental platform and found that mist cooling was more effective at higher ambient temperatures [21]. Huang's mist-spray testing experiments at the Shanghai World Expo also confirmed that the hotter and drier the weather, the higher the cooling reduction [19]. Nonetheless, these experimental studies have only yielded qualitative conclusions based on the comparison of a limited number of operating conditions. Due to uncontrolled environmental changes, there is a significant lack of experimental research that considers environmental factors as variables to quantitatively study the relationship between ambient thermal parameters and spray cooling efficiency.

In most experimental studies, the spray system operates continuously [20,23], which may potentially damage the equipment's lifespan, increase water consumption, and lead to excessive accumulation of water vapor in the air. A few studies have utilized spraying on/off modes, with timing variables ranging from several seconds to several minutes [9]. For instance, Ishii's experiment employed a pattern of the spray running for 2 min followed by a 3 min pause [38]. However, a short spray duration may not provide sufficient cooling of the micro-environment, and when combined with a survey, it may offer too few opportunities for individuals to enter the misted area for evaluation.

Therefore, in order to investigate the relationship between environmental factors, spray cooling effects, and improvements in human thermal perception, this study adopted intermittent spraying. The test experiment and questionnaire survey were conducted simultaneously. The dynamic effects of periodic spraying on the local thermal environment under typical high temperatures and still air conditions were analyzed. Each spraying operation period was considered a working condition for studying how the ambient thermal environment affects the efficacy of misting. Furthermore, this study compared the instantaneous thermal perception changes when people entered the mist under different ambient thermal environmental conditions and explored the dynamic changes in thermal perception of local people in hot-humid areas during their short-term stay in the spray. The findings are applicable to similar urban environments where conditions permit the installation of mist spray equipment and are expected to provide control strategies and application guidance for the efficient implementation of mist spraying as an outdoor cooling spot.

2. Materials and Methods

2.1. The Experimental Site

The experiment took place in the University Innovation Zone in Dongguan City, Guangdong Province, China. The area is surrounded by a mix of residential buildings, research centers, and office buildings. The test site was situated on the periphery of a grassy area adjacent to a pedestrian walkway, with no buildings within 25 m of the site (see Figure 1).

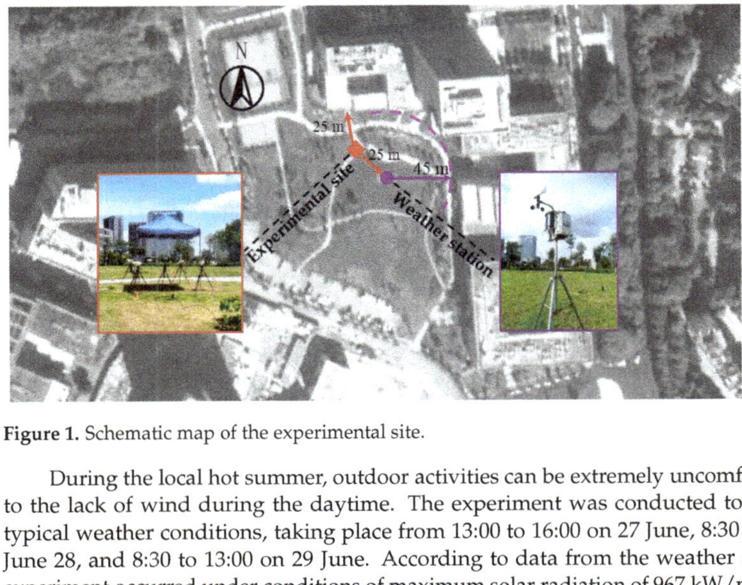

Figure 1. Schematic map of the experimental site.

During the local hot summer, outdoor activities can be extremely uncomfortable due to the lack of wind during the daytime. The experiment was conducted to align with typical weather conditions, taking place from 13:00 to 16:00 on 27 June, 8:30 to 16:00 on June 28, and 8:30 to 13:00 on 29 June. According to data from the weather station, the experiment occurred under conditions of maximum solar radiation of 967 kW/m^2, with the highest ambient temperature reaching 36.9 °C and relative humidity ranging from 48.4% to 74.5%. The average wind speed was 0.18 m/s, reflecting the typical summer heat climate in Dongguan City.

2.2. The Mist Spray System

A small square shed measuring approximately 2.2 m on each side was constructed at the experimental site with an internal height of about 2.3 m, similar in size to typical pergolas or pavilions, and was designed to replicate a semi-open, shaded outdoor space. The internal support structure of the shed top was equipped with the nozzles of the mist spray system. Experiments, including environmental measurements and questionnaire surveys, were conducted in and around this shed.

A total of 9 nozzles were arranged in a matrix in the central area of the shed top, approximately 0.35 m apart and positioned 2.35 m–2.45 m above the ground, directing the mist downwards. Each nozzle had an effective humidification diameter of about 1 m, ensuring that the spraying wet area was centered at the experimental site. The chosen nozzle height was consistent with previous research conducted by Desert [21] and Zheng [40].

Figure 2 illustrates the schematic diagram of the mist spray system. Filtered water entered the system through the inlet pipe and was pressurized to 6 Mpa. After the start-up and debugging of the host machine, the pressurized water flowed through the high-pressure outlet pipe. Subsequently, mist was ejected from high-pressure nozzles with a discharge diameter of approximately 0.15 mm. The experimental period was set to 30 min in on-off mode, with the mist system operating continuously for the first 25 min of each period and then stopping for the remaining 5 min. The three-day experiment comprised a total of 30 experimental periods.

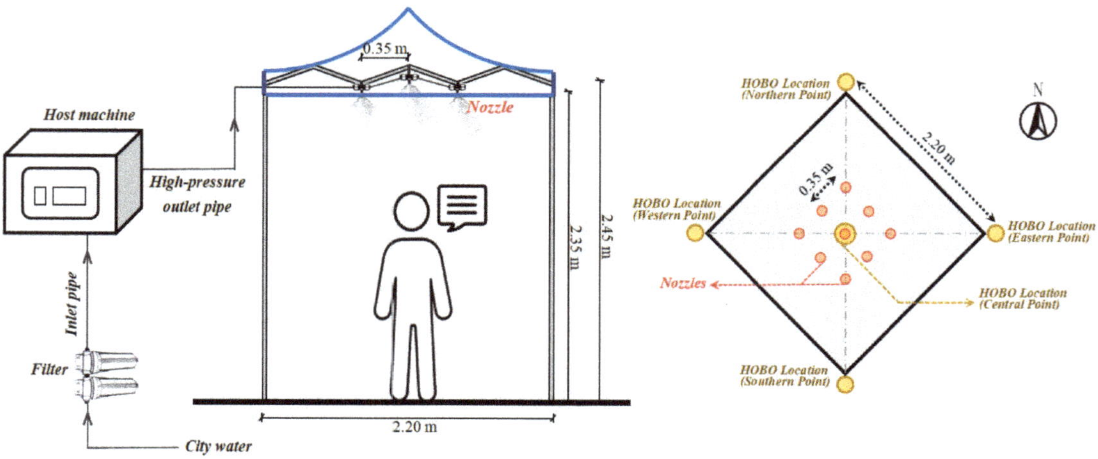

Figure 2. Schematic illustration of the mist spray system.

2.3. Environmental Measurement

The main instruments used for environmental measurements are listed in Table 1. The environmental experiment utilized the HOBO logger to measure air temperature and relative humidity, recording data at 30 s intervals. The instrument's probe was mounted on a tripod at a height of 1.1 m above the ground. Additionally, a radiation box with three blades providing shelter was placed outside the HOBO instrument (see Figure 3).

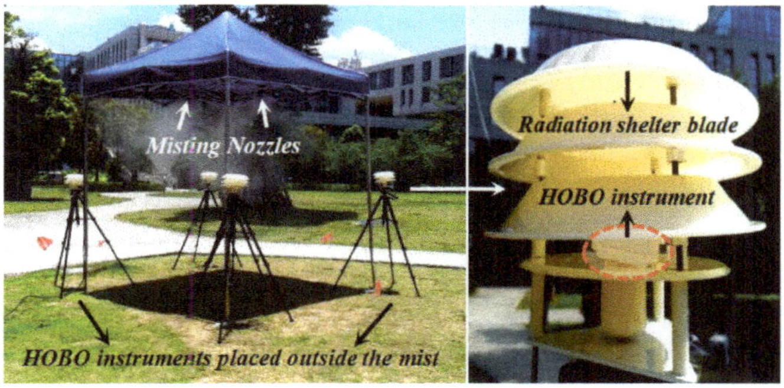

Figure 3. Hobo instruments and the radiation shelter blades.

Two groups of HOBO instruments were used in the experiment. The first group was placed in the center of the misting area to measure changes in temperature and humidity within the mist. The second group was positioned outside the misting area, beyond the four pillars of the shed, to measure ambient temperature and humidity changes surrounding the mist. Due to the low wind conditions during the experiment, the mist droplets did not drift sideways, ensuring that the second set of HOBO instruments would not be affected by the spray. This allowed for the simultaneous collection of environmental data inside and outside the misting area and the dynamic monitoring of the cooling and humidifying effects of the mist.

A weather station (Watch Dog, WD for short) was also included in this study, recording solar radiation, air temperature, relative humidity, and wind speed data at 10 min intervals. It was located in an open area 25 m away from the misting area, with no obstruction from buildings within a 45 m radius. The longer minimum time interval and lower accuracy of the WD data make it unsuitable for comparison with the HOBO data collected in the spray area. The WD data were used as a reference to verify whether the data collected by HOBO outside the mist accurately reflected the outdoor thermal environment without misting in the experiment.

Table 1. Measuring range and accuracy of Instruments used in the experiment.

Instrument	Environment Factor	Range	Accuracy
Temperature and humidity data logger (HOBO MX2301) Manufacturer: Onset, USA	Air temperature	−40~70 °C	±0.2 °C
	Relative humidity	0~100%	±2.5%
Weather station (WatchDog 2000) Manufacturer: SPECTRUM, USA	Solar radiation	0~1500 W/m^2	±5%
	Air temperature	−32~100 °C	±0.5 °C
	Relative humidity	10~100%	±3%
	Air speed	0.1~322 km/h	±5%

2.4. *Questionnaire Survey*

Dozens of participants were invited to the experimental site, and selected based on the following criteria: healthy adults who could understand the test rules, had proportional body shapes, had resided in the local city for more than 1 year, and had adapted to the local climate. An effort was made to ensure a similar number of men and women were selected as subjects.

During the hot summer daytime, people generally avoid staying outdoors for extended periods unless necessary, with the most common exception being waiting for transportation or taking a brief rest in pergolas and pavilions when feeling fatigued from walking. Zhang [20] noted that people typically wait at transportation sites for about 5 min or more. As a result, this study standardized the time for participants to stay in the mist at

5 min. The simulation replicated pedestrians entering the cool spot from an open space and staying briefly in the misting area.

Each participant was given a questionnaire form with three parts: (1) Personal information, including gender, age, and height; (2) Thermal perception survey before participants entered the mist area (see Figure 4); and (3) Thermal perception survey after participants entered the mist area (see Figure 5).

2. Please fill in the following questions before you step in the mist area.

Current date and time: ____________________

- ***How do you feel at this moment?***

Cold (−3)	Cool (−2)	Slightly cool (−1)	Neutral (0)	Slightly hot (1)	Hot (2)	Very hot (3)
☐	☐	☐	☐	☐	☐	☐

- ***Do you feel comfortable or not at this moment?***

Very uncomfortable (−3)	Uncomfortable (−2)	Slightly uncomfortable (−1)	Neutral (0)	Slightly comfortable (1)	Comfortable (2)	Very comfortable (3)
☐	☐	☐	☐	☐	☐	☐

- ***Do you think the current environment indicators need to be changed?***

	Substantially decrease (−2)	Slightly decrease (−1)	Unchanged (0)	Slightly increase (1)	Substantially increase (2)
Temperature:	☐	☐	☐	☐	☐
Humidity:	☐	☐	☐	☐	☐

Figure 4. Thermal perception survey before participants entered the mist area.

3. Please fill in the following questions when you step in the mist area, and fill in the same questions every one minute.

Current time: ____________________

- ***How do you feel at this moment?***

	Cold (−3)	Cool (−2)	Slightly cool (−1)	Neutral (0)	Slightly hot (1)	Hot (2)	Very hot (3)
Just step in:	☐	☐	☐	☐	☐	☐	☐
After 1 min:	☐	☐	☐	☐	☐	☐	☐
After 2 mins:	☐	☐	☐	☐	☐	☐	☐
After 3 mins:	☐	☐	☐	☐	☐	☐	☐
After 4 mins:	☐	☐	☐	☐	☐	☐	☐
After 5 mins:	☐	☐	☐	☐	☐	☐	☐

- ***Do you feel comfortable or not at this moment?***

	Very uncomfortable (−3)	Uncomfortable (−2)	Slightly uncomfortable (−1)	Neutral (0)	Slightly comfortable (1)	Comfortable (2)	Very comfortable (3)
Just step in:	☐	☐	☐	☐	☐	☐	☐
After 1 min:	☐	☐	☐	☐	☐	☐	☐
After 2 mins:	☐	☐	☐	☐	☐	☐	☐
After 3 mins:	☐	☐	☐	☐	☐	☐	☐
After 4 mins:	☐	☐	☐	☐	☐	☐	☐
After 5 mins:	☐	☐	☐	☐	☐	☐	☐

- ***Do you think the current environment indicators need to be changed?***

		Substantially decrease (−2)	Slightly decrease (−1)	Unchanged (0)	Slightly increase (1)	Substantially increase (2)
Just step in:	*Temperature:*	☐	☐	☐	☐	☐
	Humidity:	☐	☐	☐	☐	☐
After 1 min:	*Temperature:*	☐	☐	☐	☐	☐
	Humidity:	☐	☐	☐	☐	☐
After 2 mins:	*Temperature:*	☐	☐	☐	☐	☐
	Humidity:	☐	☐	☐	☐	☐
After 3 mins:	*Temperature:*	☐	☐	☐	☐	☐
	Humidity:	☐	☐	☐	☐	☐
After 4 mins:	*Temperature:*	☐	☐	☐	☐	☐
	Humidity:	☐	☐	☐	☐	☐
After 5 mins:	*Temperature:*	☐	☐	☐	☐	☐
	Humidity:	☐	☐	☐	☐	☐

Figure 5. Thermal perception survey after participants entered the mist area.

Thermal perception aspects included thermal sensation, thermal comfort, and preferences for temperature and humidity. Thermal sensation evaluation adopted 7-point scales from Very hot (3) to cold (−3), based on the 7-point standardized scale by ASHRAE and used in existing studies by Zhang [20]. Thermal comfort also employed 7-point scales from Very comfortable (3) to Very uncomfortable (−3). In addition, the 5-point scale from Substantially increase (2) to Substantially decrease (−2) was used to rate the preferences for temperature and humidity.

After providing their personal information, participants completed their initial assessment outside the spraying area. They then entered the mist, stood under the nozzles, and immediately performed a second assessment to obtain instantaneous thermal perception data. Subsequently, they conducted an assessment every minute until they left. Assessments were completed within the first 25 min of each experimental period; that is, the mist spraying kept running continuously for each participant during the questionnaire assessment. Each participant completed seven assessments within a 6 min timeframe, both outside and inside the mist, to monitor dynamic changes in their thermal perception.

3. Results

3.1. Experimental Results and Analysis

The measurement results from the weather station were compared with the mean results from the second group of HOBO instruments. As shown in Figure 6, the maximum differences in temperature and humidity were found to be 0.53 °C and 6.46%, respectively, while the average differences were 0.53% and 3.21%, respectively. The measurement results from the HOBO instruments showed consistency with the weather stations, indicating minimal impact from the mist. Therefore, they can be considered representative of the ambient temperature and humidity outside the mist spray area for comparison purposes.

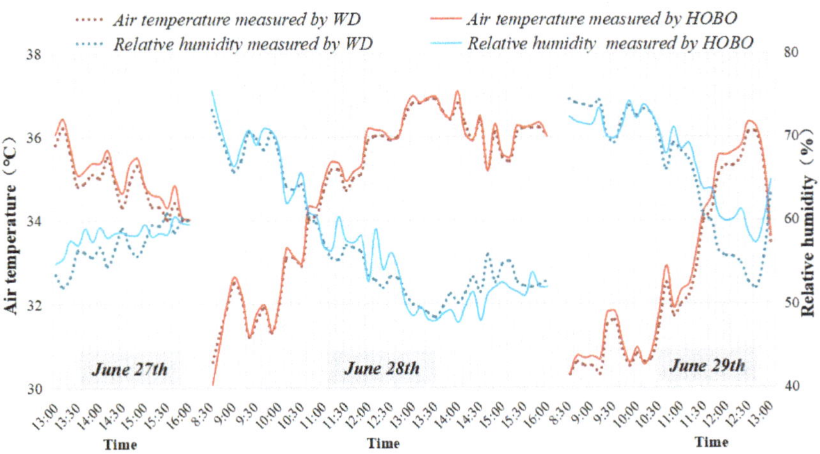

Figure 6. Comparison of the temperature and humidity between the WD and the second group of HOBO outside the mist area.

Figure 7 compares the temperature and humidity inside and outside the spraying area. The variations are essentially the same, with a decrease in air temperature and an increase in relative humidity. Additionally, the periodic changes in temperature and humidity align with the time periods of the mist spray system. At the start of each period, the air temperature drops and the relative humidity rises. When the system stops running after 25 min of each period, the air temperature increases and the relative humidity decreases, approaching the values observed outside the spray area at the end of the period. Throughout the experiment, the spraying reduces the air temperature by approximately 0.12 °C to 5.68 °C, with an average decrease of 2.42 °C. Moreover, the relative humidity in the mist area increases by around 0.04% to 22.21%, with an average humidification of 9.02%.

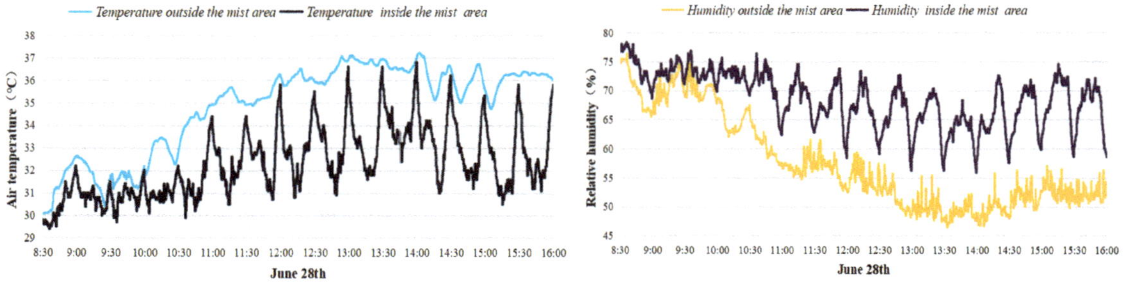

(a) (b)

Figure 7. Comparison of the temperature and humidity inside and outside the mist area during the experiment: (**a**) Air temperature on 28th June; (**b**) Relative humidity on 28th June.

Figures 8 and 9 illustrate the average cooling and humidifying effects during each spraying period. The data indicates that the average temperature difference gradually increases throughout the day. Prior to 10:00, the average temperature difference was 0.83 °C. From 10:00 to 11:00, it rises to 1.83 °C, and after 11:00, it further increases to 3.02 °C. Similarly, the average humidity difference follows a comparable pattern. Before 10:00, the average humidity difference is 3.44%, increasing to 7.84% from 10:00 to 12:00, and eventually reaching 11.71% after 12:00. The decision to activate the mist spray system early during the cooler morning hours, despite achieving some degree of cooling, raises questions about its necessity. The cost-effectiveness and energy efficiency of running the spraying should be considered. A more strategic approach may involve activating the system later in the morning when the ambient temperature is higher. Based on these results, it can be inferred that the mist spraying system demonstrates optimal cooling performance after 10:00 during the daytime in summer.

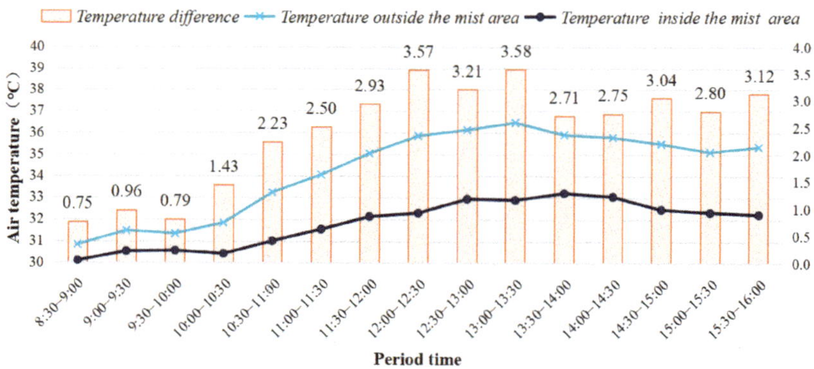

Figure 8. Comparison of the average air temperature for each period of time.

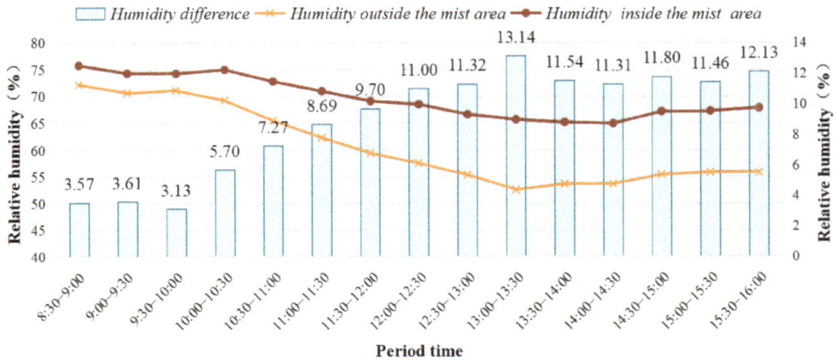

Figure 9. Comparison of the average relative humidity for each period of time.

To better illustrate the relationship between ambient thermal parameters and spray cooling efficiency, the average cooling and humidification levels under each working condition were used for comparison, considering the fluctuations in temperature and humidity during the spraying process. Each spraying operation period is considered a working condition for conducting a quantitative analysis of the relationship between ambient temperature and humidity and the magnitude of cooling and humidification, as shown in Figure 10. The x-axis represents the average ambient air temperature and the average relative humidity for each spraying period, while the y-axis represents the average decrease in air temperature and the average increase in relative humidity in the mist area within the corresponding period of spraying. The significance (p-value) is reported using a t-test, as are the graphs below in this article. The model is statistically significant, with a correlation r^2 greater than 0.9. The analysis demonstrates that as the ambient temperature increases, the cooling and humidifying effects of the mist spray become more pronounced. For every 1 °C increase in ambient air temperature beyond 31 °C, the average air temperature reduction caused by spraying increases by approximately 0.5 °C. At an ambient temperature of around 31.28 °C, the mist spray can achieve an air temperature reduction of about 1.0 °C. Meanwhile, at an ambient temperature of about 35.48 °C, the mist spray can lower the average air temperature by approximately 3.0 °C.

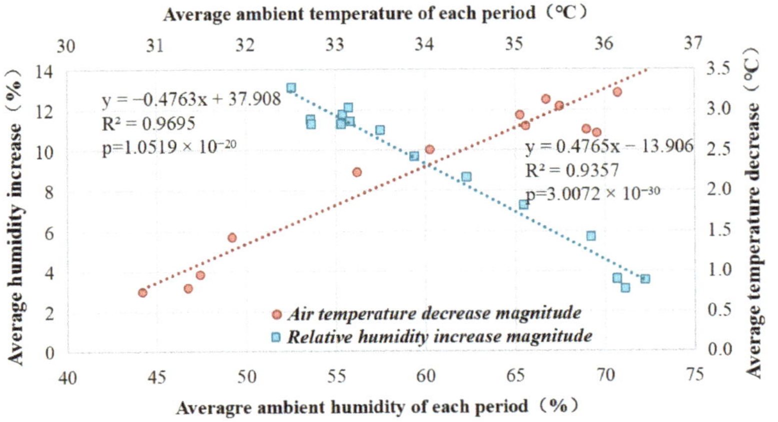

Figure 10. Relationship between ambient temperature and humidity and magnitude of cooling and humidification during spraying periods.

The spraying periods are sequentially numbered from 1 to 30 to facilitate the comparison and analysis of the dynamic changes in cooling and humidification. Periods 1–6, 7–21, and 22–30 correspond to 6 experimental periods from 13:00 to 16:00 on 27 June, 15 experimental periods from 8:30 to 16:00 on 28 June, and 9 experimental periods from 8:30 to 13:00 on 29 June, respectively. Figure 11 illustrates the variation in the temperature difference magnitude for different periods during the operation of the spray system. The graph depicts a continuous increase in the temperature difference within the first 2 min of each period. Subsequently, some curves start to fluctuate between 2 and 4 min due to the influence of the external environment and the mist spraying. The curves are categorized based on the initial ambient temperature of each period. The most rapid cooling rate is observed in the first 4 min of Period 18, which has the highest initial ambient temperature of 37.09 °C. Periods 7 and 22, with the lowest initial ambient temperatures of 30.07 °C and 30.32 °C, respectively, exhibit the most stable curves. It can be inferred that the magnitude of the temperature difference is correlated with the ambient temperature, influencing the rate at which the temperature difference increases. When the ambient temperature is below 32 °C, there is a minimal temperature difference between inside and outside the mist area, resulting in relatively flat curves. Conversely, when the ambient temperature exceeds 32 °C, the temperature difference becomes obvious, with the temperature magnitude increasing by approximately 0.5 °C to 1.0 °C per minute in the first 2 min.

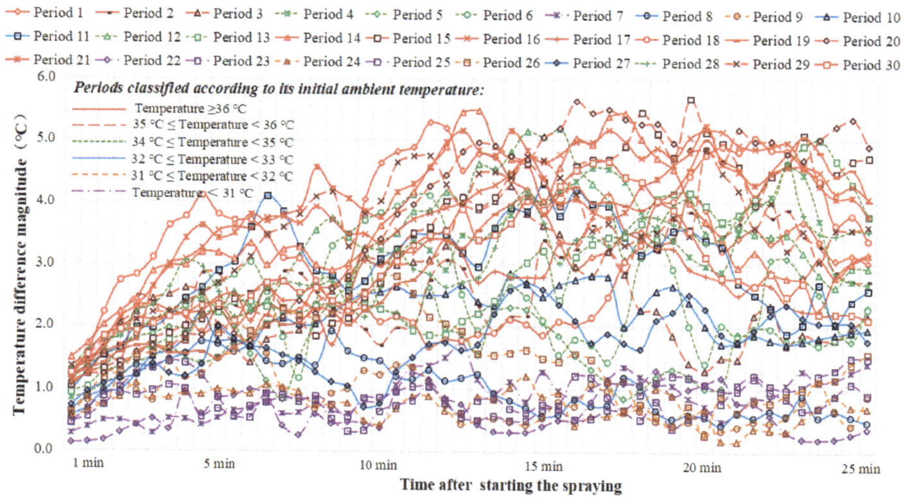

Figure 11. Variation in the magnitude of temperature difference during the operation of the spray.

3.2. Results and Analysis of the Questionnaire Survey

The number of participants willing to take part in the experiment was limited since the test was conducted during the COVID-19 pandemic. A total of 59 questionnaires were distributed, out of which 2 were invalidated due to incompleteness. Ultimately, 57 participants (29 males and 28 females) were involved in the experiment, with 86% aged between 19 and 29 (8 individuals aged under 20, 41 individuals aged 20–29, 5 individuals aged 30–39, and 3 individuals aged over 40) entering the mist area randomly. The average height was 167.5 cm (173.7 cm for men and 162.2 cm for women). None of the subjects were considered overweight or underweight. Throughout the experiment, participants completed questionnaires during 93% of the spraying periods, indicating that they entered the mist area under various ambient temperature and humidity conditions, as shown in Figure 12.

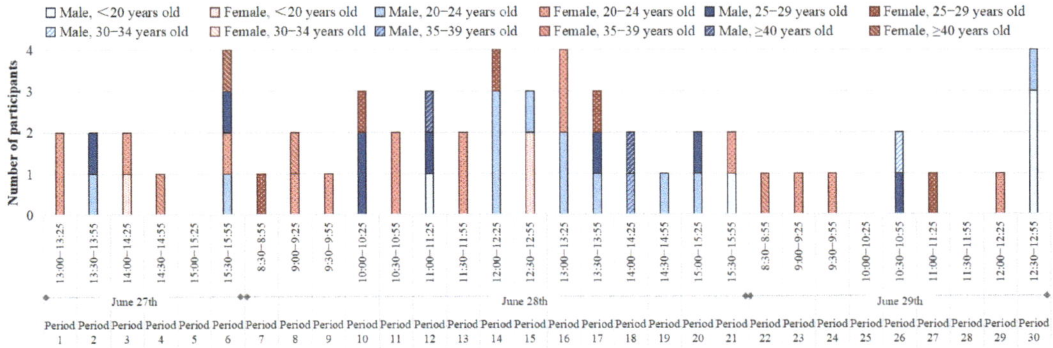

Figure 12. Distribution of participants of different genders and ages in each period.

Figure 13c–f presents statistical results of participants' thermal perception ratings at various times during the questionnaire survey, while Figure 13a,b display the air temperature and relative humidity of the environment where the participants were located at those moments.

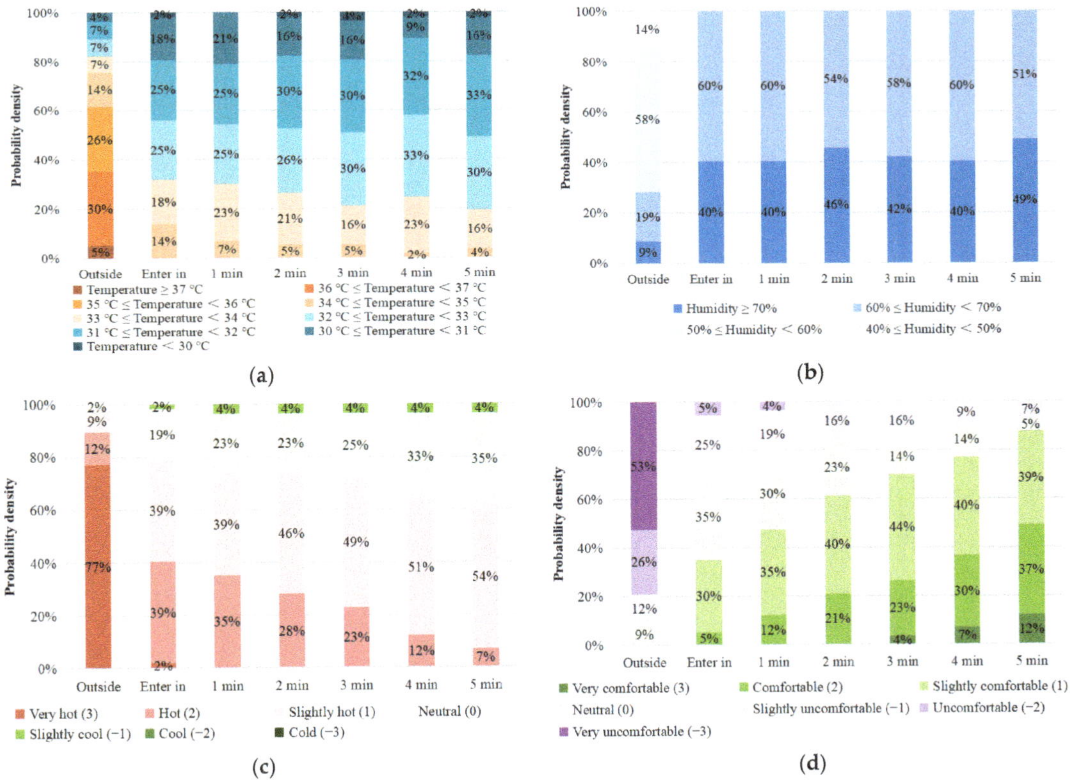

Figure 13. *Cont.*

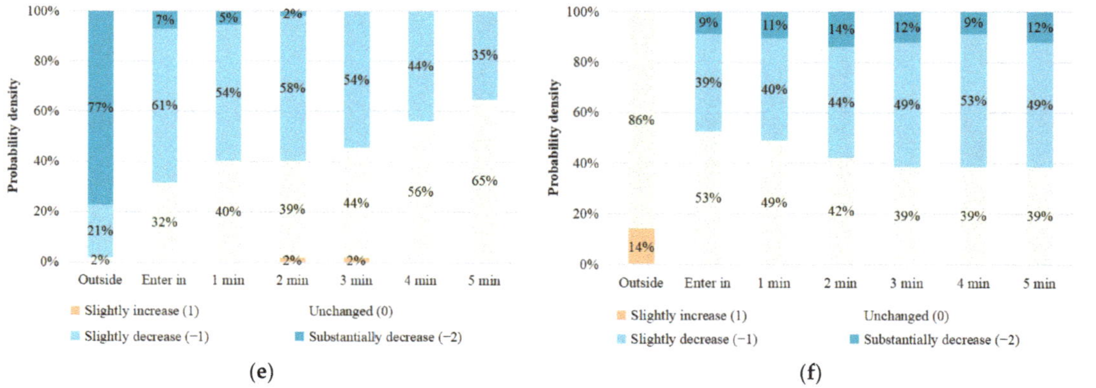

Figure 13. Questionnaire results and experimental results at different moments outside and inside the mist spray area: (**a**) Air temperature; (**b**) Relative humidity; (**c**) Thermal sensation; (**d**) Thermal comfort; (**e**) Temperature preference; (**f**) Humidity preference.

Prior to entering the misting area, 75% of the subjects were exposed to ambient temperatures above 34 °C, with 61% experiencing temperatures above 35 °C, 35% encountering temperatures surpassing 36 °C, and 5% even facing temperatures exceeding 37 °C. In terms of humidity levels, 58% of the participants were in an environment with humidity ranging from 50% to 60%. Meanwhile, 14% experienced humidity levels below 50%, while only 9% were exposed to humidity levels above 70%. Before entering the mist, 89% of the participants rated their thermal sensation as above 2 (Hot), with 77% reporting 3 (Very hot), resulting in an average rating of 2.65. In terms of thermal comfort, 79% of the subjects selected −2 (uncomfortable), with 53% choosing −3 (Very uncomfortable). The average rating result was −2.23. A total of 98% of the participants believed that the air temperature should be reduced. 86% of the participants were adaptive to the ambient humidity, while 14% were looking forward to a slight increase in humidity.

Upon entering the misting area, 50% of the subjects experienced temperatures ranging from 31 °C to 33 °C, while 32% were exposed to temperatures above 33 °C, and only 14% experienced temperatures exceeding 34 °C. Over the subsequent 5 min period, the number of subjects exposed to temperatures between 31 °C and 33 °C increased to nearly 65%, while the proportion of those exposed to temperatures above 34 °C decreased to less than 5%. Additionally, all participants were subjected to relative humidity of more than 60%, and within the following 5 min, the percentage of participants experiencing humidity levels above 70% rose to 49%.

The average temperature and humidity dynamics experienced by all the subjects during the test were statistically analyzed, as shown in Figure 14. Before and after entering the misting area, the temperature and humidity differences were as high as 2.7 °C and 10.22%, respectively. In the following five minutes, the average temperature and humidity of the spraying environment experienced by the subjects changed slightly, with maximum change values of 0.29 °C and 0.91%.

The thermal perceptions of the subjects underwent significant changes upon entering the misting area. Initially, only 2% of the participants reported feeling Very hot (rating 3), while 39% selected Slightly hot (rating 1), resulting in an average thermal sensation rating of 1.19. However, within the subsequent 5 min, the percentage of participants rating their thermal sensation as 0 and 1 increased to 35% and 54%, respectively, leading to a decrease in the average thermal sensation rating to 0.65. Notably, no participants reported feeling Very hot (rating 3) after 1 min in the mist. In terms of thermal comfort, 30% of the participants reported feeling Slightly uncomfortable (rating below −1), 35% felt Neutral (rating 0), and 35% felt Slightly comfortable (rating above 1) upon entering the misting area, resulting in an average rating of 0.05. Over the next 5 min, the percentage of participants rating 0 and

−1 decreased to 5% and 7%, respectively, with no participants reporting Uncomfortable (rating −2) after 2 min in the mist. As the participants continued to stay in the mist, the percentage of participants reporting feeling Comfortable (rating 2) gradually increased, reaching 37% towards the end, and the average rating for thermal comfort eventually reached 1.42.

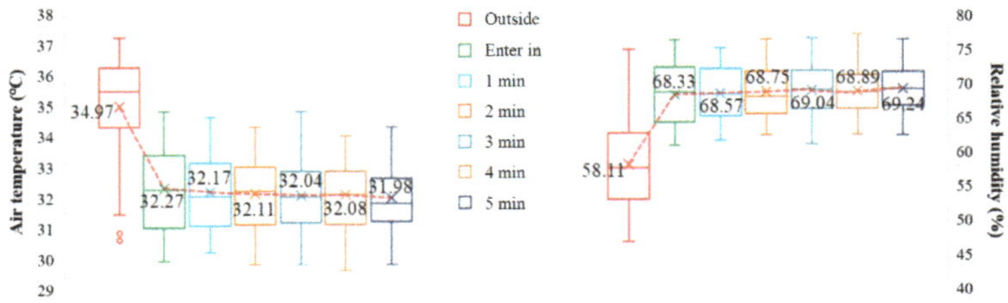

Figure 14. The average temperature and humidity dynamics to which all the subjects were exposed during the test.

The average thermal perception dynamics of all the subjects during their entire evaluation test are displayed in Figure 15. It is evident that there was a larger range of change in thermal comfort (−2.23 to 1.42) compared to thermal sensation (2.65 to 0.65), indicating that mist spraying provided a greater improvement in thermal comfort than in thermal sensation.

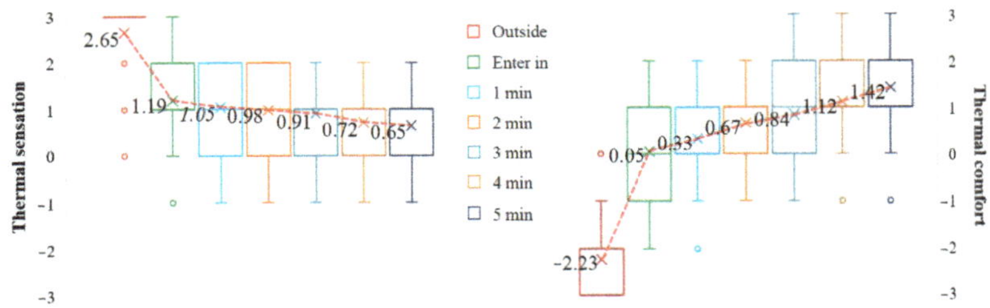

Figure 15. The average thermal perception dynamics of all the subjects during the test.

Upon entering the mist area, 68% of the participants believed that the air temperature should be reduced, while 32% preferred it to remain constant. However, the desire to lower the air temperature steadily decreased over the next 5 min, despite the temperature change not being obvious. Although 61% of participants felt Slightly hot or Hot, only 35% believed that the air temperature should be slightly decreased. Conversely, the percentage of participants who believed that the temperature could remain constant increased to 65% after staying in the mist for 5 min, indicating that they were gradually adapting to the temperature in the mist environment. Regarding humidity preference, most participants were comfortable with the ambient humidity outside the misting area. When entering the spraying, the percentage of participants who felt the humidity should remain unchanged dropped to 53%. Over the next 5 min, the percentage of participants who desired a slight decrease in humidity increased to 49%. Participants were able to feel the humid environment. As they spent more time in the mist, their demand for reduced humidity increased. However, participants tend to have a high tolerance for high humidity. By the 5 min mark, 61% of participants were requesting lower humidity, and only 7% felt Slightly uncomfortable.

The relationship between thermal perception evaluation results and ambient temperatures was analyzed before participants entered the mist area. As shown in Figure 16, a correlation is observed between them prior to participants entering the mist area. When the ambient temperature exceeded 31.3 °C, most participants reported feeling Slightly hot. When it exceeded 32.6 °C, most participants reported feeling Hot. The temperature reflecting thermal sensation neutrality was found to be 30.3 °C. Additionally, when the ambient temperature exceeded 32.4 °C and 34.4 °C, most participants reported feeling Slightly uncomfortable and uncomfortable, respectively. The temperature reflecting comfortable neutrality was 30.6 °C.

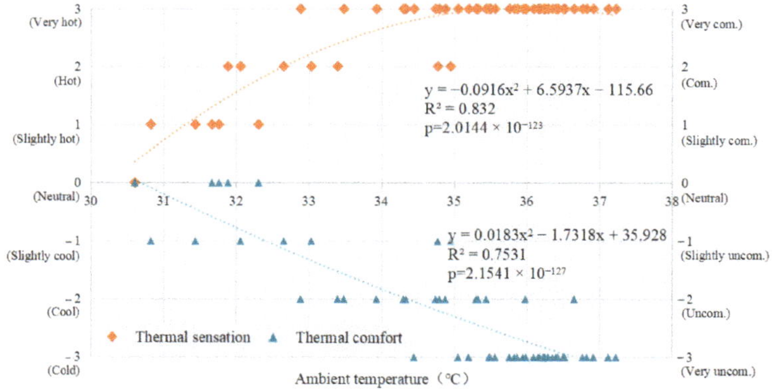

Figure 16. The relationship between evaluation result and ambient temperature outside the mist.

The total change in thermal sensation and thermal comfort during the entire evaluation process, both outside and inside the spray area, was −2 and +3.65, respectively. When the subjects first entered the spray area, the mean thermal sensation changed from 2.65 (between Very hot and Hot) to 1.19 (approaching Slightly hot), covering 73% of the total change range of thermal sensation. Upon entering the spray area, their mean thermal comfort changed from −2.23 (approaching Uncomfortable) to 0.05 (Neutral), which accounted for more than 62% of the total change range. The remaining small part of the change in thermal perception occurred as the participants gradually adapted to the mist area. These dramatic, instantaneous changes are related to the immediate temperature difference between inside and outside the mist area, as shown in Figure 17. The analysis demonstrates that as the instantaneous temperature difference increases, the magnitude of variation in thermal perception becomes more pronounced, and the increase in thermal comfort outweighs the decrease in thermal sensation. For every 1 °C increase in temperature difference, the thermal sensation decreases by 0.53, while thermal comfort increases by 0.64. The thermal sensation rating decreases for 95% of participants (54 subjects), while all individuals report feeling more comfortable. Despite the air temperature inside the mist remaining relatively high after cooling in some cases, participants still felt more comfortable.

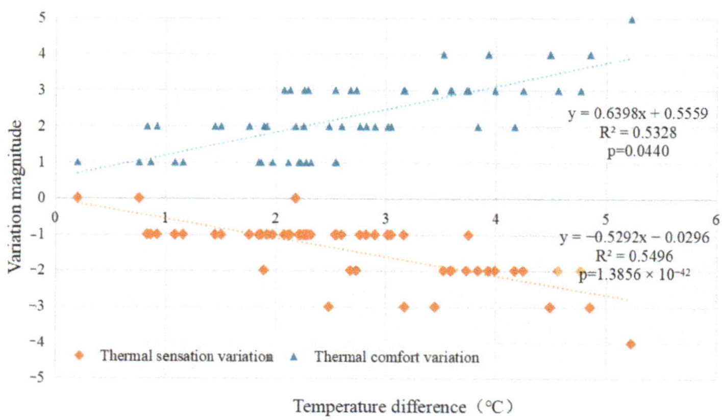

Figure 17. The relationship between the magnitude of variation in thermal perception and the instantaneous temperature difference.

After spending time in the mist area, the subjects' thermal sensation and thermal comfort continued to change, accounting for the remaining 27% and 38%, respectively, compared to the small changes in the mean temperature and humidity of the environment. As these subjects gradually acclimated to the spraying environment, the relationship between the objective micro-environment and thermal perception seemed to weaken. Furthermore, the impact of temperature on thermal perception was examined based on the duration that participants spent in the mist. Figure 18 illustrates the analysis findings for participants who stayed in the mist for 1 min, while Figure 19 shows the results for those who stayed for 5 min. The correlation gradually decreases as participants spend more time in the spraying area, indicating that the impact of temperature on thermal perception diminishes over time. These findings suggest that individuals may adapt to the misty and cool environment over time, leading to a reduced sensitivity to temperature changes as well as a diminished reliance on environmental temperature for thermal perception.

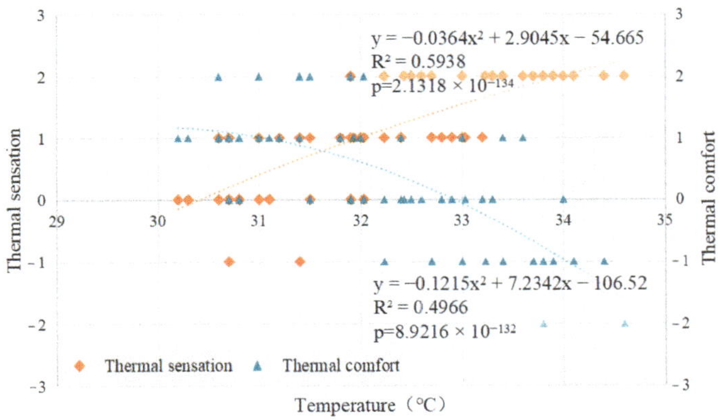

Figure 18. The relationship between thermal perception evaluation result and the temperature when the subjects stayed in the mist for 1 min.

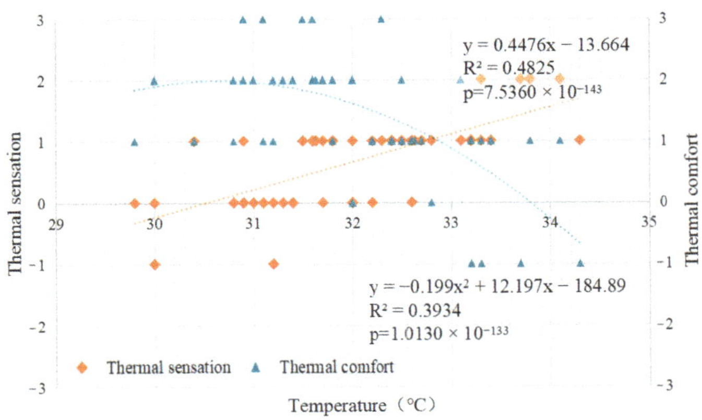

Figure 19. The relationship between thermal perception evaluation result and the temperature when the subjects stayed in the mist for 5 min.

4. Discussions

4.1. Cooling Efficiency of the Mist Spray System

In this study, the running period of the spraying resulted in a changing period of low temperature and high humidity in the mist, with the maximum and average temperature difference being 5.68 °C and 2.42 °C, respectively. This cooling effect aligns with previous literature [16,19,20], indicating that the mist spray system used in the experiment, which simulated an outdoor semi-open shaded space, has relatively high cooling efficiency. This is consistent with Meng's suggestions that combining appropriate shading treatments with spraying could have a better cooling effect [12]. Some studies have suggested that increased humidity from mist spraying may exceed the human body's tolerance limit, impacting subjective perception [41], but in our study, participants were able to perceive the humid environment, and this sensation did not diminish over time. However, the presence of moisture in the air and a relative humidity of up to 75% in the mist area did not compromise the effectiveness of the mist spray. It can be assumed that moderately high humidity does not necessarily prevent the perception of comfort. Individuals living in hot-humid climates generally have a higher tolerance for humidity and provide positive feedback when evaluating their thermal comfort, which aligns with findings from other studies in the literature [13,22,24].

From 12:00 to 14:00, when solar radiation is strong and the background temperature is high, the temperature difference is the largest in this study. The experimental results show that ambient temperature affects the magnitude of spray cooling and the rate of temperature difference. This conclusion is similar to Meng's view, who claimed that original temperature and humidity affect spray cooling [12]. Since the amount of spray is assumed to be constant and completely evaporates, the amount of cooling and humidification should not change as the ambient temperature changes. However, in the experiment, some of the mist might be attached to the subjects' hair, skin, or surrounding objects, such as the surface of the radiation box and the ground surface, without evaporating. In other words, the amount of spray is not equal to the amount of evaporation. Due to the changing participants and environment, the actual amount of mist evaporation was changed, which may be the reason why the amount of cooling and humidification varied with the ambient air temperature in this study.

4.2. Criteria for Activation of the Spraying

In this study, it was observed that the cooling range of the spray before 10:00 was limited, suggesting that the spray after 10:00 could achieve a better cooling effect, potentially leading to greater overall effectiveness and resource conservation. This finding aligns with Huang's recommendation to apply spray cooling technology during 10:00–16:00 in the summer [19]. However, it is important to note that the conclusion of the latter is based on specific temperature and humidity criteria (temperature $\geq$ 30 °C and humidity $\leq$ 70%) for time screening. Since temperature and humidity are relatively easy to obtain as environmental factors, they are often used as indicators to determine whether to start spraying. The automatic control condition used by Ishii in the experiment applied at the station in Japan was that the temperature was greater than 28 °C and the relative humidity was less than 70% [38]. Yamada suggested that mist spray is suitable when the temperature is above 30 °C and the humidity is below 70%, based on a numerical simulation analysis of three working conditions [34]. Huang reached the same conclusion from the analysis of two working conditions [31]. Several studies have used the neutral temperature as the criterion for activating the mist spray. According to Wong's regression model of the observed thermal sensation and the temperature recorded for the semi-outdoor experiment conducted in Singapore, it is suggested that mist spray should be turned on when the neutral temperature is above 29.2 °C [23]. The experimental results from this study suggest a neutral temperature of 30.3 °C, and further analysis is needed to determine the optimal timing for activating the spray system, taking into account both the environmental conditions and the desired cooling outcomes.

It is widely believed that temperatures above 30 °C are favorable for spray application. Figure 16 in this study shows that at an ambient temperature of 30 °C, the average thermal sensation rating is −0.29, and the average thermal comfort rating is 0.44. This indicates that local residents, accustomed to hot weather, can tolerate such outdoor thermal conditions. Although the mist can lower the temperature by approximately 0.40 °C (inferred from Figure 10), its necessity appears to be relatively low. By combining the results of Figures 10, 16 and 17, it could be concluded that when the ambient temperature reaches around 32.5 °C, the average level of thermal sensation is 1.88 (approaching Hot), and the average level of thermal comfort is −1.03 (around Slightly uncomfortable). Utilizing the mist spray at this point can lower the temperature by 1.58 °C. After participants enter the mist area, the average level of thermal sensation decreases to 1.02 (Slightly hot), while the average level of thermal comfort increases to 0.54 (between Neutral and Slightly comfortable). At an ambient temperature of 34 °C, the average level of thermal sensation is 2.64, while the average level of thermal comfort is −1.80. When participants enter the mist, which provides a cooling effect of 2.30 °C, the average level of thermal sensation decreases to 1.39, and the average level of thermal comfort increases to 0.23. Although most of the participants still feel slightly hot, they do not feel uncomfortable. This indicates that the mist system is effective. Spraying could help local residents maintain a physiological state close to slightly hot and neutral comfort.

4.3. Impact on Short-Term Thermal Perception

When participants spend a short time (5 min) in the mist, the change in thermal sensation and thermal comfort compared to being outside the spray is 2 and 3.65, respectively. This is consistent with the findings of Desert [21] in hot and humid extreme areas. The majority of this change occurs upon entering the mist, accounting for over 73% (thermal sensation) and 62% (thermal comfort) of the total change range. The sudden change in thermal perception was influenced by the temperature difference in the environment before and after the subjects were exposed. The longer the time spent in the mist, the more the subjects will gradually adapt to the environment, making their preference for temperature in misty areas more apparent. When comparing humidity preference with thermal comfort evaluation, we can see that participants tend to have a high tolerance for high humidity.

On the other hand, the correlation between the thermal perception assessment and the actual temperature in the spraying area is not very strong. Desert believed that the change in thermal comfort after cooling was not only caused by the change in the physical environment [21] but also by psychological factors [22], such as thermal expectation [42]. In other words, the subjects' anticipation of the cooling spot may enhance their perception of thermal delight [43], thus amplifying the impact of their subjective perception. This may be one of the reasons why the subjects in this study can feel comfortable in the mist even when the thermal sensation has not reached a neutral level.

4.4. Potential Application in the Actual Scenes

The practical application of this study is to implement spray cooling technology in outdoor environments during hot and humid summers with minimal wind. The selection of the experimental space and the setup of the spray equipment were based on common urban public settings and relevant studies. Therefore, the conclusions of this study can also be extended to comparable partially sheltered public settings (e.g., traffic waiting areas, pergolas, and pavilions), where conditions allow for the installation of typical mist spray hydro-power equipment. The mist spray system is capable of utilizing and filtering urban water sources, and the equipment can be installed underground, on walls, or in other structures to accommodate the host control equipment and high-pressure water pipes, with overhead structures for the installation of a specific number of nozzles. This study focused on the cooling efficiency of a mist-spray system in a simulated outdoor semi-open shaded space, considering its environmental influencing factors and short-term impact on the thermal perception of the crowd.

Based on this study's findings, it is recommended to activate the mist spray in outdoor spaces when the ambient temperature exceeds 32.5 °C in the summer. To achieve this, it is advisable to place the ambient temperature probe near public settings. Initiating the spray too early, such as before 10 a.m., or for extended periods is not conducive to a water-saving and energy-efficient spray mode. Given that the primary purpose of the spray is to enhance human thermal comfort, individuals experience a noticeable improvement in thermal perception upon entering the spraying area. Therefore, it is worth considering integrating a human body sensing device with the spray control system. The spray should not be activated when no one is around, and short bursts of spraying should only be triggered when human presence is detected approaching in high ambient temperatures. The initial minutes of spraying rapidly cool the environment, so the spraying duration should not be too short, lasting at least 2 min. By networking and controlling the mist spray system, temperature sensor, and body recognition device, an operation mode may be designed to efficiently create cool spots while conserving water and reducing energy consumption. The findings aim to offer control guidance for the application of mist spraying in outdoor environment design to enhance the micro-environment for citizens and improve thermal comfort levels.

4.5. Limitations and Future Research Directions

Due to the additional power consumption and challenges in controlling wind speed while operating mist-spray equipment, this study did not incorporate a fan into the test setup, in line with some published studies conducted in calm or breeze conditions [44]. This decision aimed to minimize variables (such as wind speed, wind direction, wind temperature, and so on) and investigate the quantitative relationship between spraying efficiency and initial temperature and humidity under typical weather conditions with minimal wind. Future studies on spray cooling will consider wind as a new variable. Some literature suggests that combining a fan with mist spraying can further reduce local temperatures [23,35]. Therefore, wind will be considered a new variable for future studies on spray cooling.

In order to fully explore the potential of spray cooling, future research should extend this study to multiple days in hot-humid areas. Increasing the number of subjects will enhance the credibility of the research conclusions, and participants' age, height, and weight can be more diverse. Conducting more detailed participant classification studies is expected to aid in the development of targeted and efficient spray control strategies for real-life public scenarios involving diverse audiences.

Furthermore, a crucial future research direction involves the comprehensive exploration of the practical application of spray system facilities in urban settings. Different urban scenarios may have varying requirements for technical parameters such as the number of nozzles and water consumption. This may include addressing issues such as integrating and filtering urban water sources in various urban settings, as well as strategically and flexibly planning the spray equipment and devices, seamlessly integrating them with municipal and landscape facilities, without compromising the cityscape or disrupting public activities. Additionally, it involves enhancing the intelligence of the spraying equipment, ambient temperature sensors, human body recognition devices, and other auxiliary devices, and future integration of them with the urban Internet of Things for more efficient control and operation of the mist spray system.

5. Conclusions

This paper studied the effects of periodic mist spraying on the outdoor thermal environment and the short-term thermal perception of local residents. The main conclusions are enumerated as follows:

(1) The intermittent on-off mist spray system caused periodic changes in the local environment. The spraying could reduce the air temperature by an average of 2.42 °C, with a maximum cooling value of 5.68 °C. The relative humidity was increased by an average of 9.02%, with a maximum humidification value of 22.21%. When comparing different time periods, it was observed that the temperature reduction was less than 1 °C before 10:00. The spraying system demonstrated its optimal cooling performance after 10:00 during the daytime in the summer.

(2) The quantitative analysis of each experimental period showed that higher ambient temperatures resulted in a more significant cooling and humidification effect. With every 1 °C increase in ambient temperature beyond 31 °C, the average temperature reduction caused by spraying increased by approximately 0.5 °C. Qualitative analysis also revealed that elevated ambient temperatures might lead to a faster increase in the rate of temperature difference between inside and outside the mist area after the beginning of each experimental period.

(3) With the exception of a few subjects who performed thermal perception assessments at low ambient temperatures, 95% of participants reported a decrease in thermal sensation, and all individuals reported feeling more comfortable. There was a larger range of change in thermal comfort (−2.23 to 1.42) compared to thermal sensation (2.65 to 0.65).

(4) Upon entering the spray area, the mean thermal sensation and thermal comfort change covered 73% and 62% of the total change ranges, respectively. These dramatic, instantaneous changes are related to the immediate temperature difference between inside and outside the mist. For every 1 °C increase in temperature difference upon entering the spray area, the thermal sensation decreased by 0.53, while thermal comfort increased by 0.64.

(5) During the 5 min stay in the mist, the subjects' preference for reducing the air temperature weakened, although the change in temperature and humidity was not obvious. However, their preference for reducing humidity strengthened. Their tolerance for high humidity causes them to still find the mist more comfortable. The impact of temperature on thermal perception diminishes as participants spend more time in the spraying area.

(6) The neutral temperature reflecting thermal sensation outside the mist is 30.3 °C. When the ambient temperature exceeds 32.5 °C, the average thermal sensation level approaches Hot, with the average thermal comfort level approaching Slightly uncomfortable. This information can serve as a reference for establishing criteria for activating spraying in hot-humid areas. Using the mist spray system at this point can reduce the temperature by 1.58 °C. The mist could help local residents maintain a physiological state close to slightly hot and neutral comfort.

In conclusion, the mist spray is expected to improve the outdoor thermal environment more efficiently and provide people with a more comfortable experience. Further research will be conducted to facilitate the application, control, and adjustment of the mist spray system in actual outdoor scenes.

Author Contributions: Conceptualization, P.W.; methodology, P.W., S.L. and N.L.; formal analysis, X.W. and J.T.; investigation, P.W.; resources, P.W., S.L. and N.L.; data curation, P.W.; writing—original draft preparation, P.W.; writing—review and editing, P.W., X.W. and J.T.; visualization, P.W.; supervision, P.W.; project administration, P.W.; funding acquisition, P.W., X.W. and J.T. All authors have read and agreed to the published version of the manuscript.

Funding: This work was supported by the Basic and Applied Basic Research Foundation of Guangdong Province (No. 2020A1515111115); the National Natural Science Foundation of China (No. 52178275); and the Humanities and Social Science Research Program of the Ministry of Education (No. 22YJAZH112).

Data Availability Statement: The data presented in this study are openly available in Mendeley Data at https://doi.org/10.17632/dc4y7v47ff.1.

Acknowledgments: Thanks to the property management department of the University Innovation Zone for providing the site, water, and electricity support for this experiment.

Conflicts of Interest: The authors declare no conflicts of interest.

References

1. Kolokotsa, D.; Lilli, K.; Gobakis, K.; Mavrigiannaki, A.; Haddad, S.; Garshasbi, S.; Mohajer, H.; Paolini, R.; Vasilakopoulou, K.; Bartesaghi, C.; et al. Analyzing the Impact of Urban Planning and Building Typologies in Urban Heat Island Mitigation. *Buildings* **2022**, *12*, 537. [CrossRef]
2. Deilami, K.; Kamruzzaman, M.; Liu, Y. Urban heat island effect: A systematic review of spatio-temporal factors, data, methods, and mitigation measures. *Int. J. Appl. Earth Obs. Geoinf.* **2018**, *67*, 30–42. [CrossRef]
3. An der Heiden, M.; Muthers, S.; Niemann, H.; Buchholz, U.; Grabenhenrich, L.; Matzarakis, A. Heat-Related Mortality An Analysis of the Impact of Heatwaves in Germany Between 1992 and 2017. *Dtsch. Arztebl. Int.* **2020**, *117*, 603. [CrossRef]
4. Lu, X.; Li, Q.; Tang, P.; Li, Q.; Zhao, L. Effect of heat mitigation strategies on the thermal environment of hardened pavement and pedestrian thermal comfort in a hot and humid area: A case study of Guangzhou, China. *J. Clean. Prod.* **2023**, *427*, 139089. [CrossRef]
5. Djekic, J.; Djukic, A.; Vukmirovic, M.; Djekic, P.; Brankovic, M. Thermal comfort of pedestrian spaces and the influence of pavement materials on warming up during summer. *Energy Build.* **2018**, *159*, 474–485. [CrossRef]
6. Park, J.; Kim, J.H.; Sohn, W.; Lee, D. Urban cooling factors: Do small greenspaces outperform building shade in mitigating urban heat island intensity? *Urban For. Urban Green.* **2021**, *64*, 127256. [CrossRef]
7. Haeri, T.; Hassan, N.; Ghaffarianhoseini, A. Evaluation of microclimate mitigation strategies in a heterogenous street canyon in Kuala Lumpur from outdoor thermal comfort perspective using Envi-met. *Urban Clim.* **2023**, *52*, 101719. [CrossRef]
8. Montazeri, H.; Toparlar, Y.; Blocken, B.; Hensen, J. Simulating the cooling effects of water spray systems in urban landscapes: A computational fluid dynamics study in Rotterdam, The Netherlands. *Landsc. Urban Plan.* **2017**, *159*, 85–100. [CrossRef]
9. Ulpiani, G. Water mist spray for outdoor cooling: A systematic review of technologies, methods and impacts. *Appl. Energy* **2019**, *254*, 113647. [CrossRef]
10. Wang, J.S.; Meng, Q.L.; Yang, C.; Ren, P. Spray optimization to enhance the cooling performance of transparent roofs in hot-humid areas. *Energy Build.* **2023**, *286*, 112929. [CrossRef]
11. Santamouris, M.; Ding, L.; Fiorito, F.; Oldfield, P.; Osmond, P.; Paolini, R.; Prasad, D.; Synnefa, A. Passive and active cooling for the outdoor built environment—Analysis and assessment of the cooling potential of mitigation technologies using performance data from 220 large scale projects. *Sol. Energy* **2017**, *154*, 14–33. [CrossRef]
12. Meng, X.; Meng, L.; Gao, Y.; Li, H. A comprehensive review on the spray cooling system employed to improve the summer thermal environment: Application efficiency, impact factors, and performance improvement. *Build. Environ.* **2022**, *217*, 109065. [CrossRef]
13. Ulpiani, G.; Di Giuseppe, E.; Di Perna, C.; D'Orazio, M.; Zinzi, M. Thermal comfort improvement in urban spaces with water spray systems: Field measurements and survey. *Build. Environ.* **2019**, *156*, 46–61. [CrossRef]
14. Ulpiani, G.; di Perna, C.; Zinzi, M. Water nebulization to counteract urban overheating: Development and experimental test of a smart logic to maximize energy efficiency and outdoor environmental quality. *Appl. Energy* **2019**, *239*, 1091–1113. [CrossRef]
15. Di Giuseppe, E.; Ulpiani, G.; Cancellieri, C.; Di Perna, C.; D'Orazio, M. Numerical modelling and experimental validation of the microclimatic impacts of water mist cooling in urban areas. *Energy Build.* **2021**, *231*, 110638. [CrossRef]
16. Sureshkumar, R.; Kale, S.R.; Dhar, P.L. Heat and mass transfer processes between a water spray and ambient air—II. Simulations. *Appl. Therm. Eng.* **2008**, *28*, 361–371. [CrossRef]
17. Yamada, H.; Okumiya, M.; Tsujimoto, M.; Harada, M. Study on cooling effect with water mist sprayer Measurement on Global Loop at the 2005 World Exposition. *Archit. Inst. Jpn. Conf. Summ.* **2006**, *D-1*, 677–678.
18. Oh, W.; Ooka, R.; Nakano, J.; Kikumoto, H.; Ogawa, O. Environmental index for evaluating thermal sensations in a mist spraying environment. *Build. Environ.* **2019**, *161*, 106219. [CrossRef]
19. Huang, C.; Ye, D.F.; Zhao, H.; Liang, T.; Lin, Z.; Yin, H.; Yang, Y. The research and application of spray cooling technology in Shanghai Expo. *Appl. Therm. Eng.* **2011**, *31*, 3726–3735. [CrossRef]
20. Zhang, M.; Xu, C.B.; Meng, L.; Meng, X. Outdoor comfort level improvement in the traffic waiting areas by using a mist spray system: An experiment and questionnaire study. *Sustain. Cities Soc.* **2021**, *71*, 102973. [CrossRef]
21. Desert, A.; Naboni, E.; Garcia, D. The spatial comfort and thermal delight of outdoor misting installations in hot and humid extreme environments. *Energy Build.* **2020**, *224*, 110202. [CrossRef]
22. Farnham, C.; Emura, K.; Mizuno, T. Evaluation of cooling effects: Outdoor water mist fan. *Build. Res. Inf.* **2015**, *43*, 334–345. [CrossRef]
23. Wong, N.H.; Chong, A.Z.M. Performance evaluation of misting fans in hot and humid climate. *Build. Environ.* **2010**, *45*, 2666–2678. [CrossRef]
24. Vanos, J.K.; Wright, M.K.; Kaiser, A.; Middel, A.; Ambrose, H.; Hondula, D.M. Evaporative misters for urban cooling and comfort: Effectiveness and motivations for use. *Int. J. Biometeorol.* **2022**, *66*, 357–369. [CrossRef]
25. Dhariwal, J.; Manandhar, P.; Bande, L.; Marpu, P.; Armstrong, P.; Reinhart, C.F. Evaluating the effectiveness of outdoor evaporative cooling in a hot, arid climate. *Build. Environ.* **2019**, *150*, 281–288. [CrossRef]
26. Dear, R.D.; Brager, G.J.A.T. Developing an adaptive model of thermal comfort and preference. *ASHRAE Trans.* **1998**, *104*, 73–81.
27. Binarti, F.; Koerniawan, M.D.; Triyadi, S.; Utami, S.; Matzarakis, A. A review of outdoor thermal comfort indices and neutral ranges for hot-humid regions. *Urban Clim.* **2020**, *31*, 100531. [CrossRef]

28. Hwang, R.-L.; Lin, T.-P. Thermal Comfort Requirements for Occupants of Semi-Outdoor and Outdoor Environments in Hot-Humid Regions. *Archit. Sci. Rev.* **2007**, *50*, 357–364. [CrossRef]
29. Xiong, J.; Zhou, X.; Lian, Z.W.; You, J.X.; Lin, Y.B. Thermal perception and skin temperature in different transient thermal environments in summer. *Energy Build.* **2016**, *128*, 155–163. [CrossRef]
30. Ji, W.J.; Cao, B.; Luo, M.H.; Zhu, Y.X. Influence of short-term thermal experience on thermal comfort evaluations: A climate chamber experiment. *Build. Environ.* **2017**, *114*, 246–256. [CrossRef]
31. Huang, C.; Cai, J.; Lin, Z.F.; Zhang, Q.R.; Cui, Y.Z. Solving model of temperature and humidity profiles in spray cooling zone. *Build. Environ.* **2017**, *123*, 189–199. [CrossRef]
32. Montazeri, H.; Blocken, B.; Hensen, J.L.M. CFD analysis of the impact of physical parameters on evaporative cooling by a mist spray system. *Appl. Therm. Eng.* **2015**, *75*, 608–622. [CrossRef]
33. Liu, L.S.; Wang, X.W.; Ge, M.H.; Zhao, Y.L. Experimental study on heat transfer and power consumption of low-pressure spray cooling. *Appl. Therm. Eng.* **2021**, *184*, 116253. [CrossRef]
34. Yamada, H.; Yoon, G.; Okumiya, M.; Okuyama, H. Study of Cooling System with Water Mist Sprayers: Fundamental Examination of Particle Size Distribution and Cooling Effects. *Build. Simul.* **2008**, *1*, 214–222. [CrossRef]
35. Oh, W.; Ooka, R.; Nakano, J.; Kikumoto, H.; Ogawa, O. Evaluation of mist -spraying environment on thermal sensations, thermal environment, and skin temperature under different operation modes. *Build. Environ.* **2020**, *168*, 106484. [CrossRef]
36. Ye, D.F.; Wu, L.; Tao, L.; Chen, H. Research and application of high-pressure spray-cooling technology to World Expo Axis. *Heat. Vent. Air Cond.* **2010**, *40*, 86–90.
37. Ulpiani, G.; di Perna, C.; Zinzi, M. Mist cooling in urban spaces: Understanding the key factors behind the mitigation potential. *Appl. Therm. Eng.* **2020**, *178*, 115644. [CrossRef]
38. Ishii, T.; Tsujimoto, M.; Yoon, G.; Okumiya, M. Cooling System with Water Mist Sprayers for Mitigation of Heat-island. *INTC Urban Clim.* **2009**, *29*, 2–3.
39. Xie, W.; Zhang, Y.; Zhang, W.; Li, T. Experimental study on cooling effect by fine water mist in confined space. *J. Eng. Thermophys.* **2019**, *40*, 2873–2878.
40. Zheng, K.; Yuan, C.; Wong, N.H.; Cen, C. Dry mist systems and its impact on thermal comfort for the tropics. *Sustain. Cities Soc.* **2019**, *51*, 101727. [CrossRef]
41. Oh, W.; Ooka, R.; Nakano, J.; Kikumoto, H.; Ogawa, O.; Choi, W. Development of physiological human model considering mist wettedness for mist-spraying environments. *Build. Environ.* **2020**, *180*, 106706. [CrossRef]
42. Auliciems, A. Towards a psycho-physiological model of thermal perception. *Int. J. Biometeorol.* **1981**, *25*, 109–122. [CrossRef] [PubMed]
43. Shooshtarian, S. Theoretical dimension of outdoor thermal comfort research. *Sustain. Cities Soc.* **2019**, *47*, 101495. [CrossRef]
44. Oh, W.; Kato, S. The effect of airspeed and wind direction on human's thermal conditions and air distribution around the body. *Build. Environ.* **2018**, *141*, 103–116. [CrossRef]

MDPI AG
Grosspeteranlage 5
4052 Basel
Switzerland
Tel.: +41 61 683 77 34

Buildings Editorial Office
E-mail: buildings@mdpi.com
www.mdpi.com/journal/buildings

www.ingramcontent.com/pod-product-compliance
Lightning Source LLC
Chambersburg PA
CBHW061824210926
51850CB00006B/800
9783725870844